Electrical Engineering
for
Non-Electrical Engineers

Second Edition

Electrical Engineering
for
Non-ElectricalEngineers

Second Edition

By
S. Bobby Rauf

Routledge
Taylor & Francis Group

LONDON AND NEW YORK

Published 2020 by River Publishers

River Publishers

Alsbjergvej 10, 9260 Gistrup, Denmark

www.riverpublishers.com

Distributed exclusively by Routledge

4 Park Square, Milton Park, Abingdon, Oxon OX14 4RN

605 Third Avenue, New York, NY 10017, USA

First issued in paperback 2023

Library of Congress Cataloging-in-Publication Data

Names: Rauf, S. Bobby, 1956- author.

Title: Electrical engineering for non-electrical engineers / by S. Bobby Rauf.

Description: Second edition. | Lilburn, GA : The Fairmont Press, Inc., [2016]
| Includes index. | Description based on print version record and CIP data
provided by publisher; resource not viewed.

Identifiers: LCCN 2016024346 (print) | LCCN 2016022465 (ebook) | ISBN
088173778X () | ISBN 0881737771 (alk. paper) | ISBN 9788770223362
(electronic) | ISBN 9781498799607 (Taylor & Francis distribution : alk.
paper)

Subjects: LCSH: Electrical engineering.

Classification: LCC TK146 (print) | LCC TK146 .R28 2016 (ebook) | DDC
621.3--dc23

LC record available at https://lccn.loc.gov/2016024346

Electrical Engineering for Non-Electrical Engineers Second Edition / by S. Bobby Rauf.
First published by Fairmont Press in 2017.

Routledge is an imprint of the Taylor & Francis Group, an informa business

Publisher's Note
The publisher has gone to great lengths to ensure the quality of this reprint but
points out that some imperfections in the original copies may be apparent.

ISBN 13: 978-87-7022-938-8 (hbk)

ISBN 13: 978-1-4987-9960-7 (hbk)

ISBN 13: 978-8-7702-2336-2 (online)

ISBN 13: 978-1-0031-5203-3 (ebook master)

While every effort is made to provide dependable information, the publisher, authors, and
editors cannot be held responsible for any errors or omissions.

The views expressed herein do not necessarily reflect those of the publisher.

Dedication

This book is dedicated to my wife Viktoriya Rauf,
my daughter Nida Rauf and my son Haseeb A. Rauf
for their support and understanding
during the countless hours I spent on this book.

Contents

Chapter 1

Electrical Engineering Basics 1
    Voltage 1
    Current 3
    Resistance. 5
    Resistors in Series 9
    Resistors in Parallel. 9
    Electrical Short and Open Circuit 11
    Capacitor and Capacitance 12
    Capacitors in Series. 18
    Capacitors in Parallel 21
    Capacitive Reactance. 23
    Inductor and Inductance 24
    Series Inductor Combination 33
    Parallel Combination of Inductors 34
    Inductive Reactance 37
    Impedance 38
    Magnetic Circuits vs. Electrical Circuits 39
    Basic Electrical Instruments 41

Chapter 2

DC Circuit Analysis and Basic Electronic Devices 49
    Ohms Law 50
    Kirchhoff's Voltage Law (KVL) 51
    Kirchhoff's Current Law (KCL) 54
    Voltage Division 58
    Current Division 60
    Multi-loop Circuit Analysis. 61
    Diodes. 69
    Half Wave Rectifier. 74
    Full Wave Rectifier 75
    Transistors 76
    Electronic device applications on printed circuit boards 78

Chapter 3

Alternating Current (AC) Fundamentals **87**

Alternating Current (AC) Versus Direct Current (DC). 87

Electrodeposition—DC and the Average Value of AC 90

Alternating Current and Impedance 94

Polar or Phasor Form 94

Rectangular Form. 95

Sinusoidal or Trigonometric Form 96

Exponential Form. 98

Impedance Analysis. 98

Transformers 102

Common Single Phase AC Transformer Applications 111

Autotransformers 114

Voltage Regulation, Voltage Regulators and
     Buck Boost Transformers 114

Three Phase AC. 117

Three Phase AC Systems vs. Single Phase AC Systems 118

Current Transformers. 119

Three Phase AC Transformers 121

Chapter 4

DC and AC Power **137**

Single Phase AC Power. 143

Three Phase AC Power 149

Efficiency. 152

Power Conversion from Steam to Electrical Form 153

Power Conversion from Water to Electrical Form 155

Power Conversion in Hydraulic Pump Systems—
Wire to Water Transformation of Power and Energy. 157

Chapter 5

Power Factor **167**

Power Factor 168

Lagging Power Factor 170

Lagging Power Factor from Impedance 170

Leading Power Factor and Impedance 175

Power Factor Correction 182

Chapter 6

Demand, Load Factor, Service Factor and
Electrical Power Bill Computation . . . . **.195**
    Demand 195
    Peak Demand 196
    Load Factor 198
    Service Factor 200
    Computation of Large Industrial or
        Commercial Electrical Power Bills 201

Chapter 7

Electric Motors and Generators . . . . **.209**
    DC Generator 209
    AC Alternator 214
    AC Induction Motor 217
    Motor Torque and Power 220
    Single Phase and Three Phase Motor Line 221
    Current Computation 221
    Three Phase AC Induction Motor Current 222
    Synchronous Motors 226
    Motor Starting Methods for Induction Motors 228
    Motor Classifications 231
    Typical Motor Nameplate Information for 232
    Large Three Phase Induction Motor 232

Chapter 8

Power Distribution Equipment, Instrumentation
and Electronic Safety Devices . . . . **. .235**
    Voltage Categories in Power Distribution Systems 235
    MCCs or Motor Control Centers 239
    Motor Starter—A-B Bulletin 2100 247
    Variable Frequency Drives—Up to 250HP 250
    Smart Motor Controllers (SMCs)—Up to 500A 251
    Comparison between SMCs, VFDs and Soft Starts 253
    PLC and I/O Chassis 254
    Main Switch Yard and Medium Voltage Switchgear 256
    Medium Voltage Switchgear—Loop Switch 257
    Circuit Breakers 258

Medium-voltage circuit breakers 261
High-voltage circuit breakers. 262

Chapter 9

National Electric Code, NFPA® 70 E and Electrical Standards 265

Ampacity of Conductors—Table 310-15. 272
Figure 9.4: Example of NFPA 70 E Class 4 PPE. 284
Electrical Safety Certifications 286
NEMA versus IP Enclosure Ratings 288
Common Electrical/Electronic Safety Devices. 291

Chapter 10

Electrical and Controls Drawings .307

Electrical Drawings. 307
One-line Schematic Diagram. 307
Current Carrying Conductor Size. 312
Conduit and Conduit Size 314
Overload Protection Setting 318
Starter Size 318
Overcurrent Protection Fuse Size. 318
Fusible disconnect switch size 319
Wiring Diagram. 324
Input Control Diagram 328
Output Control Diagram 330
Relay Ladder Logic. 332

Chapter 11

Electrical Power Rate Schedules and Electrical Energy
Cost Savings Opportunities 341

Three Pronged Approach to 341
Electrical Energy Cost Reduction* 341
Electric Utility Rate Schedules* .342
Rates Under OPT—I Schedule .346
Hourly Pricing Option/Schedule 348
Areas of Opportunity for Electrical Energy Cost Savings 348
Energy Performance Contracting and ESCO Opportunities* 351
Electrical Energy Related Measures
Typically Included in ESCO or EPC Endeavors 354

Chapter 12

Illumination and Lighting System Design . . . . **.363**
Lighting Terms, Concepts and Standards. 364
Luminous Efficacy or Luminous Efficiency 368
IESNA Recommended Illumination Levels and 371
Lighting Efficacy Comparison .371
Lambert's Law 376
Illuminance Inverse Square Law 377
Illuminance Cosine-cubed Law. 378
Lighting Design Considerations 381
Daylight Based Design Considerations. 381

Chapter 13

Batteries . . . . **.387**
Battery Classifications 389
Primary Batteries 389
Secondary Batteries. 390
Li-Ion Battery. 391
Lead-Acid or SLI Battery 394
Comparison of Electrical Energy and Charge Storage Systems . . . 402
Cost Comparison of Major Secondary Batteries 407
Discharging Characteristics of Batteries 408
Charging of Secondary Batteries 420

Appendices

Appendix A . . . . **.427**
Chapter 1—Solutions 427
Chapter 2—Solutions 436
Chapter 3—Solutions 446
Chapter 4—Solutions 457
Chapter 5—Solutions 465
Chapter 6—Solutions 472
Chapter 7—Solutions 476
Chapter 8—Answers/Solutions. .483
Chapter 9—Solutions 484
Chapter 10—Solutions 491
Chapter 11—Solutions 498
Chapter 12—Solutions 501

Chapter 13—Solutions 505

Appendix B—Common Units and Unit Conversion Factors. **508**

Appendix C— Greek Symbols Commonly Used in
Electrical Engineering. **512**

Index **513**

Preface

Many Non-engineering professionals as well as engineers who are not electrical engineers tend to have a phobia related to electrical engineering. One reason for this apprehensiveness about electrical engineering is due to the fact that electrical engineering is premised concepts, methods and mathematical techniques that are somewhat more abstract than those employed in other disciplines, such as civil, mechanical, environmental and industrial engineering. Yet, because of the prevalence and ubiquitous nature of the electrical equipment, appliances, and the role electricity plays in our daily lives, the non-electrical professionals find themselves interfacing with systems and dealing with matters that broach into the electrical realm. Therein lies the purpose and objective of this book.

This second edition includes an important new chapter on batteries. As renewable energy solutions gain a foothold and expand, we see challenges and opportunities in safe and effective storage of energy. That is where battery technology comes in. Hence the importance of this new chapter, which introduces the reader to different types of batteries, primary vs. secondary, traditional lead-acid batteries vs. the cutting edge Li-Ion batteries. In this chapter, the strengths, weakness, opportunities and threats faced by the new and traditional battery solutions are examined. In addition, this second edition updates some of the electrical power generation information; taking into account some of the remarkable shifts in the sources of energy for production of electrical power. Established and recognized electrical engineering drawing standards have been included, as well, in this edition. The NEC and NFPA 70 E Arc Flash information included in this book has been revised to comport with the current revisions of NEC (2014) and NFPA 70 E (2015).

This edition maintains the mission to serve as a resource for exploring and understanding basic electrical engineering concepts, principles, analytical strategies and mathematical strategies.

If your objective as a reader is limited to the acquisition of basic knowledge in electrical engineering, then the material in this book should suffice. If, however, the reader wishes to progress their electrical engineering knowledge to intermediate or advanced level, this book could serve as a useful platform.

As the aphorism goes, "a picture is worth a thousand words;" this book maximizes the utilization of diagram, graphs, pictures and flow charts to facilitate quick and effective comprehension of the concepts of electrical engineering.

In this book, the study of electrical engineering concepts, principles and analysis techniques is made relatively easy for the reader by inclusion of most of the reference data, in form of excerpts from different parts of the book, within the discussion of each case study, exercise and self-assessment problem solutions. This is in an effort to facilitate quick study and comprehension of the material without repetitive search for reference data in other parts of the book.

Certain electrical engineering concepts and terms are explained more than once as these concepts appear in different chapters of this text; often with a slightly different perspective. This approach is a deliberate attempt to make the study of some of the more abstract electrical engineering topics more fluid; allowing the reader continuity, and precluding the need for pausing and referring to chapters where those specific topics were first introduced.

Due to the level of explanation and detail included for most electrical engineering concepts, principles, computational techniques and analyses methods, this book is a tool for those energy engineers, engineers and non-engineers, who are not current on the subject of electrical engineering.

The solutions for end of the chapter self-assessment problems are explained in just as much detail as the case studies and sample problem in the pertaining chapters. This approach has been adopted so that this book can serve as a electrical engineering skill building resource for engineers of all disciplines. Since all chapters and topics begin with the introduction of important fundamental concepts and principles, this book can serve as a "brush-up," refresher or review tool for even electrical engineers whose current area of engineering specialty does not afford them the opportunity to keep their electrical engineering knowledge current.

In an effort to clarify some of the electrical engineering concepts effectively for energy engineers whose engineering education focus does not include electrical engineering, analogies are drawn from non-electrical engineering realms, on certain complex topics, to facilitate comprehension of the relatively abstract electrical engineering concepts and principles.

Each chapter in this book concludes with a list of questions or problems, for self-assessment, skill building and knowledge affirmation purposes. The reader is encouraged to attempt these problems and questions. The answers and solutions, for the questions and problems, are included under Appendix A of this text.

Most engineers understand the role units play in definition and verification of the engineering concepts, principles, equations, and analytical techniques. Therefore, most electrical engineering concepts, principles and computational procedures covered in this book are punctuated with proper units. In addition, for the reader's convenience, units for commonly used electrical engineering entities, and some conversion factors are listed under Appendix C.

Most electrical engineering concepts, principles, tables, graphs, and computational procedures covered in this book are premised on SI/Metric Units. However, US/Imperial Units are utilized where appropriate and conventional. When the problems or numerical analysis are based on only one of the two unit systems, the given data and the final results can be transformed into the desired unit system through the use of unit conversion factors in Appendix B.

Some of the Greek symbols, used in the realm of electrical engineering, are listed in Appendix C, for reference.

What Readers Can Gain from this Book

- Better understanding of electrical engineering terms, concepts, principles, laws, analysis methods, solution strategies and computational techniques.

- Greater confidence in interactions with electrical engineering design engineers, electricians, controls engineers and electrical engineering experts.

- Certain skills and preparation necessary for succeeding in electrical engineering portion of various certification and licensure exams, i.e. CEM, Certified Energy Manager, FE, Fundamentals of Engineering (also known as EIT, or Engineer in Training), PE, Professional Engineering and many other trade certification tests.

- A better understanding of the electrical engineering component of energy projects.

- A compact and simplified electrical engineering desk reference.

- Better understanding of electrical energy cost and tips on improvement of electrical energy intensity in industrial and commercial environment.

- Introduction to the generation of electricity from renewable sources.

- Better understanding of illumination principles and concepts, and an appreciation of efficient lighting/illumination design.

- Better understanding of myriad battery options available in the market; their strengths, weaknesses, opportunities that lie ahead and potential threats.

An epistemic advice to the reader: if you don't understand some of the abstract concepts the first time, don't give up. Read it again! Such is the nature, intrigue and challenge of engineering, physics, science and other subjects that require thinking, reflection and rumination.

Acknowledgements and Credits

Rockwell International ®
Many thanks to **John Keogh, Dave Dascenzo and Rachel R Schickowski** of Rockwell International® for provision of illustrative MCC pictures that, I am certain, will be appreciated by the readers.

National Fire Protection Agency, NFPA®
My gratitude to NFPA® for their permission to use NEC® information in this text.
NFPA 70®, *National Electrical Code and NEC®* are registered trademarks of the National Fire Protection Association, Quincy, MA.

Fluke Corporation
Many thanks to **Beverly Summers**, of Fluke Corporation for granting permission to use material associated with their multimeter and clamp ammeter.

Kim Craven, PE
My endless gratitude to **Kim Craven, PE**, for his guidance on electrical engineering topics included in this text.

Control Technologies, LLC
Many thanks to **Wes Lampkin** of Control Technologies, LLC, for his contribution of the PLC relay ladder logic examples.

Chapter 1
Electrical
Engineering Basics

Introduction

In this first chapter of the *Electrical Engineering for Non-Electrical Engineers* text, we will explore the basics of electrical engineering terms concepts, principles, and analytical techniques. Many readers who embark on investing time and effort in studying this text are likely to do so for the key purpose of gaining an introduction into the field of electricity. Many others, on the other hand, might be interested in refurbishing prior knowledge of electrical engineering terms, concepts, principles and basic analytical techniques. Regardless of whether you belong to one of these two groups—or are simply in pursuit of electrical engineering at the intermediate or associates degree level— in this chapter we will lay the foundations in the electrical engineering realm by covering basic electrical engineering terms, concepts and principles, without the understanding of which, discussion and study of terms that bear important practical significance, such as power factor, real power, reactive power, apparent power, load factor, etc. would not be feasible.

Most of the material in this chapter pertains to DC, or Direct Current, electricity. However, some entities we will discuss in this chapter, such as, capacitive reactance, inductive reactance and impedance are fundamentally premised in the AC, alternating current, realm.

Electrical engineering is rooted in the field of physics. Physics, and electrical engineering, as most other fields in science, depend on empirical proof of principles and theories. Empirical analysis and verification requires measurement tools or instrumentation. So, after gaining a better understanding of the basic electrical concepts, we will conclude this chapter with an introduction to two of the most common and basic electrical instruments, i.e. multi-meter and clamp-on ammeter.

VOLTAGE

Voltage is defined as an electromotive force that moves or pushes electrically charged particles like electrons, holes, negatively charged ions or positively charged ions. The term "electromotive" force stems from the early recognition of electricity as something that consisted, strictly, of the movement of "electrons." However, with today's more recent breakthroughs in the renewable and non-traditional electrical power generating methods and systems like microbial fuel cells, and hydrocarbon fuel cells, electrical power is being harnessed, more and more, in form of charged particles that may not be electrons.

Two, relatively putative, analogies for voltage in the mechanical and civil engineering disciplines are pressure and elevation. In the mechanical realm— or more specifically in the fluid and hydraulic systems—high pressure or pressure differential pushes fluid from one point to another and performs mechanical work. Similarly, voltage—in form of voltage difference between two points, as with the positive and negative terminals of an automobile battery—moves electrons or charged particles through loads such as motors, coils, resistive elements, lamps, etc. As electrons or charged particles are pushed through loads like motors, coils, resistive elements, light filaments, etc., electrical energy is converted into mechanical energy, heat energy or light energy. In equipment like rechargeable batteries, during the charging process, applied voltage can push ions from one electrode (or terminal) to another and thereby "charge" the battery. Charging of a battery, essentially, amounts to the restoration of battery terminals' or plates' chemical composition to "full strength." So, in essence, the charging of a battery could be viewed as the "charging" of an electrochemical "engine." Once charged, a chemical or electrochemical engine, when presented with an electrical load, initiates and sustains the flow of electrical current, and performs mechanical work through electrical machines.

Common symbols for voltage are: E, V, V_{DC}, V_{AC}, V_P, V_M, V_{Eff}, V_{RMS}. Symbols "E" and "V" are synonymous, and both represent voltage. The symbol E stands for electromotive force, while V, simply, denotes voltage. In the absence of further specification through a subscript, these symbols can be somewhat ambiguous, in that, they could be construed to represent either AC or DC voltage.

On the other hand, V_{DC}, specifically denotes DC voltage. See Chapter 3 for a detailed comparison between AC and DC, voltage and current. The AC and DC discussion, related to voltage and current, in Chapter 3, is supported by graphs that depict the differences in a graphical fashion.

The symbol V_{AC} represents AC voltage. When dealing with AC voltage, one needs to be specific about whether one is referring to "**peak**" voltage, V_P, or **RMS** voltage, V_{RMS}. Note that peak voltage, V_P, is synonymous with maximum voltage V_M. In addition, V_{Eff}, the effective (AC) voltage, is the same as, RMS voltage, V_{RMS}. The term RMS stands for **R**oot **M**ean **S**quare Value of

AC Voltage. The RMS or effective value of AC voltage is the work producing portion of the AC voltage. This implies that AC voltage, current and power, all, have the "work producing" components and the" non-work producing" components. The work producing components, in essence, transform into—or contribute toward the production of—various forms of energy, mechanical work and break horsepower—or, to be more accurate, break horsepower-hour. The concepts of power, break horsepower and other power related topics are discussed, in detail, in Chapter 4.

Even though detail discussion on the mathematical composition of RMS voltage is outside the scope of this text, the formula for RMS voltage is as follows:

$$V_{RMS} = V_{EFF} = \sqrt{\frac{1}{T}\int_0^T V^2(t).\,d(t)}$$

Eq. 1.1

AC voltage V_{RMS}, V_{Eff}, V_P, and V_M are inter-related through the following equations:

$$V_p = V_m = \sqrt{2}V_{RMS} = \sqrt{2}V_{EFF}$$

Eq. 1.2

$$V_{RMS} = \frac{V_p}{\sqrt{2}} = \frac{V_m}{\sqrt{2}}$$

Eq. 1.3

Voltage is measured in volts, or V's; named after the Italian physicist Alessandro Volta (1745–1827), who invented the first chemical battery.

CURRENT

Current consists of movement of electrons, ions, or simply charged particles. Movement of electrons can be oscillatory, vibratory or linear. When electrons vibrate or oscillate, the resulting current is AC current. AC current is established and sustained by AC voltage. When DC voltage is applied in an electrical circuit, electrons, ions or charged particles move in one direction. Such linear, unidirectional, movement of charged particles or electrons is DC current. DC electrical current is analogous to fluid flow in mechanical or hydraulic systems. Just as pressure, or pressure differential, cause fluid to flow from point A to point B, DC voltage drives electrically charged particles to move from one point to another. Important characteristics of electric current are illustrated in greater detail, in Chapter 3, under the topic of electrodeposition.

Detailed comparison between AC and DC currents is also illustrated in graphical form in Chapter 3.

Common symbols for current are: I, I_{DC}, I_{AC}, I_P, I_M, I_{Eff} and I_{RMS}. Similar to the voltage symbols, the symbols of current assume a more specific meaning through associated subscripts.

I_{AC} represents AC current. When dealing with AC current, one needs to be specific about whether one is referring to "peak" current, I_P, or RMS current, I_{RMS}. Note that peak current, I_P, is synonymous with maximum current I_M. In addition, I_{Eff}, the effective (AC) current, is the same as, RMS current, I_{RMS}. The term RMS current stands for **R**oot **M**ean **S**quare Value of AC current. The RMS or effective value of AC current is the work producing portion of the AC current. AC current, like AC voltage and AC power, has a "work producing" component and a "non-work producing" component. The work producing component of AC current contributes toward the production of mechanical work and break horsepower.

The formula for RMS current is as follows:

$$I_{RMS} = I_{EFF} = \sqrt{\frac{1}{T}\int_0^T I^2(t).\, d(t)}$$

Eq. 1.4

AC Current I_{RMS}, I_{Eff}, I_P, and I_M are inter-related through the following equations:

$$I_p = I_m = \sqrt{2}I_{RMS} = \sqrt{2}I_{EFF}$$

Eq. 1.5

$$I_{RMS} = \frac{I_p}{\sqrt{2}} = \frac{I_m}{\sqrt{2}}$$

Eq. 1.6

The unit for current is Ampere named after André-Marie Ampére (1775–1836), a French mathematician and physicist. André-Marie Ampére is revered as the father of electrodynamics. One amp of current is said to flow when electrical charge is flowing at the rate of *one Coulomb per second*. This leads to the following mathematical definition:

$$1 \text{ Amp of Current} = \frac{1 \text{ - Coulomb of Charge}}{\text{Second}}$$

Eq. 1.7

This definition of current is analogous to flow rate of fluid quantified in terms of mass flow rate $\dot{m}$ or volumetric flow rate $\dot{Q}$.

Example 1.1

In an AC system, a voltage source **V(t) = 156Sin(377t + 0°)** volts sets up a current of **I(t) = 15Sin(377t + 45°) amps**. The peak voltage and the peak current, in accordance with convention, are **156 V and 15 A**, respectively. Calculate the RMS values of voltage and current.

Solution:

According the Eq. 1.3:

$$V_{RMS} = \frac{V_p}{\sqrt{2}} = \frac{V_m}{\sqrt{2}}$$

$$\therefore V_{RMS} = \frac{V_p}{\sqrt{2}} = \frac{156}{\sqrt{2}} = 110 \; V_{rms}$$

Note: This is the voltage indicated by a true RMS voltmeter when measuring the AC voltage at a typical household or workplace wall receptacle. See more discussion toward the end of this chapter.

According to Eq. 1.6:

$$I_{RMS} = \frac{I_p}{\sqrt{2}} = \frac{I_m}{\sqrt{2}}$$

$$\therefore I_{RMS} = \frac{15}{\sqrt{2}} = 10.6 \; A_{RMS}$$

Note: This is the current indicated by a true RMS clamp-on ammeter when measuring AC current. See more discussion toward the end of this chapter.

RESISTANCE

Property of a material that opposes or resists the flow of current is known as electrical resistance, or simply, resistance. Electrical resistance is analogous to friction in mechanical systems; for instance, friction between the surfaces of two objects that slide against each other. In a fluid flow scenario, electrical resistance is analogous to friction between the fluid and the walls of the pipe. In electrical systems, resistance in conductors (wires) is an undesirable characteristic and results in wasted heat or heat losses. This is not unlike frictional head losses in fluid systems—frictional head losses governed by Darcy's equation and Hazen-William's equations. The symbol for electrical resistance is "**R.**" Resistance is measured in ohms, or simply, **Ωs**. The ohm symbol **Ω** is often prefixed with letters, such as, **k** for kilo or **M** for Mega. Where, **1kΩ** would represent 1,000 Ohms and **1MΩ** would represent 1,000,000 ohms.

$$R = \frac{V}{I} \text{ or, } V = I.R \qquad\qquad \textbf{Eq. 1.8}$$

An ohm can also be defined on the basis of the Ohm's Law. The Ohm's Law, and its application in AC and DC systems is discussed in greater depth in Chapter 2. At this point, upon examination of Ohm's Law, in form of Eq. 1.8, we can define **1 ohm** as the amount of resistance that would permit the flow of only **1-amp** of current when a voltage of **1-volt** is applied across a specific length of that conductor. In other words:

$$1Ω = \frac{1\text{-volt}}{1\text{-amp}}$$

The Ohm's Law equation stated above also stipulates that as the increase in electrical "demand" manifests itself in form of an increase in current, **I**, and as resistance, **R,** of the conductor increases, the voltage—or voltage drop—across the conductor increases; resulting in lower voltage at the point of delivery of power to the consumers. This is similar to the ***pressure drop, or loss***, experienced in a long compressed air pipe or header due to frictional head loss.

From physical characteristic and physical composition point of view, we could define resistance as being directly proportional to the **length** of the conductor and inversely proportional to the **area of cross-section** of the conductor. This proportional relationship can be transformed into a mathematical relationship or equation as follows:

$$R = \rho.\frac{L}{A} \qquad\qquad \textbf{Eq. 1.9}$$

Where
      ρ = Resistivity of the Conductor,
      L = Length of the Conductor and
      A = Area of Cross-section of the Conductor.

$\varrho_{copper} = 17.2 \text{ n } \Omega\text{m};$

$\varrho_{aluminum} = 28.2 \text{ n } \Omega\text{m}$

Where n = nano = 10^{-9}

Example 1.2

A cubic block of electrically conductive material measures 0.02 on each side. The resistivity of this material is 0.01 $\Omega \cdot$ m. What is the resistance between opposite sides of this block?

Solution:

$$R = \frac{\rho L}{A} = \frac{(0.01 \ \Omega\text{-m})(0.02 \ \text{m})}{(0.02 \ \text{m})^2} = 0.5 \ \Omega$$

Example 1.3

A phase conductor of a power distribution line spans, approximately, 500 ft and has a diameter of 1.5 inch. The conductor is composed of copper. Calculate the electrical resistance of this conductor, per phase.

Solution:

Solution Strategy: Since the resistivity value of copper, as stated above, is in metric or SI unit system, the length and diameter specifications stated in this problem must be streamlined in metric units before application of Eq. 1.9 for determination of resistance in ohms (Ωs).

L = 500 ft = 152.4 m

Diameter = 1.5 inch = 0.0381m; $\therefore$ **R** = Radius = D/2 = 0.019 m

A = Area of cross-section = $\pi.R^2$ = (3.14)(0.019)^2 = 0.00113m^2

ϱ_{copper} = 17.2 n Ωm = 17.2x10^{-9} Ωm

$$R = \rho.\frac{L}{A}$$

$$\therefore R = \rho.\frac{L}{A} = 17.2\text{x}10^{-9} \left(\frac{152.4}{0.00113}\right) = 0.00232\Omega$$

As described earlier, in electrical systems, resistance in conductors (wires) is an undesirable characteristic and results in wasted heat or heat losses. This energy lost, as heat, as the current flows through a conductor can be quantified through Eq. 1.10 below.

$$\text{Heat Loss} = P = VI = \frac{V^2}{R} = I^2 R, \text{ in Watts} \qquad \textbf{Eq. 1.10}$$

Although the subject of electrical power and energy is discussed, in detail, in Chapter 4, at this point, in an effort to illustrate the substantial impact of resistance on heat losses and design challenges associated with power transmission, let's continue examination of long transmission lines used to transport electrical power from power generating plants to power consumers. On a hot summer afternoon, those power transmission lines that we notice crisscrossing the country side are not only carrying higher currents due to the higher air conditioning loads, but experience an increase in resistance due to increases in resistivity, *p*, of the conductor in accordance with Eq. 1.11. In other words, the resistivity of aluminum, stated above as $\rho_{aluminum}$ = **28.2 n Ωm,** serves as a ***constant only at standard temperature of 20°C or 68°F.***

$$\rho = \rho_0 \left(1 + \alpha \left(T - T_0\right)\right) = \rho_0 \left(1 + \alpha \Delta T\right) \qquad \textbf{Eq. 1.11}$$

In Eq. 1.11:
- ρ = Resistivity at current temperature "**T**"
- ρ_0 = Resistivity at standard temperature "**T$_0$**"
- **T$_0$** = Standard temperature in °C
- **T** = Current temperature °C
- α = Thermal coefficient of resistance, in 1/°C

In addition, it's common knowledge that as temperature rises, most metals or conductors expand. So, with exposure to elevated summer season solar radiation, during those hot afternoons, the transmission lines elongate to a certain extent, resulting in higher "**L**." As those transmission lines elongate, the diameter of the conductors drops, resulting in lower area of cross-section **A**. So, the perceptible "sag" in the transmission lines during the hot summer afternoons is not simply an optical illusion or a myth.

As we account for the increase in **L**, reduction in **A** and the rise in **ρ** collectively, in accordance with Eq. 1.9, we see that all of these factors result in escalation of resistance. Furthermore, if we consider resistive heat loss equation, **Eq. 1.10**, the *exponential* effect of the rise in load current **I** and the increase in **R** precipitate in a "cascading," unfavorable, physical and electrical impact on the transmission line conductors.

RESISTORS IN SERIES

When electrical circuits, AC or DC, consist of multiple resistors, circuit analyses require simplification of such network of resistors into one, ***equivalent***, resistor or resistance, $\mathbf{R_{EQ}}$. Often, this equivalent, resistance is referred to as a "***total***" resistance. When "**n**" number of resistors are connected in a "daisy chained" or concatenated fashion as shown in Figure 1.1, they are said to be connected in series.

Figure 1.1: n –Resistors in series

When resistors are connected in series, they can be combined in a "linear addition" format, as stipulated in Eq. 1.12 below, for "**n**" number of resistors.

$$R_{EQ} = R_T = R_1 + R_2 + R_3 + ... + R_n \qquad \textbf{Eq. 1.12}$$

When multiple resistors are combined into an equivalent resistor, with resistance value $\mathbf{R_{EQ}}$, the simplified version of the original series circuit would appear as pictured in Figure 1.2.

RESISTORS IN PARALLEL

When electrical circuits, AC or DC, consist of multiple resistors, connected in a parallel fashion, as shown in Figure 1.3, circuit analysis would require simplification of the parallel network of resistors into one, equivalent resistor $\mathbf{R_{EQ}}$. Figure 1.3 shows "**n**" number of resistors connected such that the "heads" of all resistors are "bonded" are connected together,

Figure 1.2: Equivalent resistance for **n** –resistors in series

with and electrical connection to the anode (positive terminal) of the DC power supply. In addition, the "tails" of all resistors are connected together and to the cathode of the DC power source.

When multiple parallel resistors are combined into an equivalent resistor, with resistance value R_{EQ}, the simplified version of the original series circuit would appear as pictured in Figure 1.4. In parallel resistor networks, the calculation of R_{EQ} involves addition of the inverses of all resistors in the parallel network, and taking the inverse of the sum as stipulated in Eq. 1.13.

Figure 1.3: n –Resistors in parallel

$$R_{EQ} = \frac{1}{\dfrac{1}{R_1} + \dfrac{1}{R_2} + \dfrac{1}{R_3} + ... + \dfrac{1}{R_n}}$$ **Eq. 1.13**

In case of a simple two parallel resistor circuit, the equivalent resistance R_{EQ} could be calculated using the simplified equation, Eq. 1.14.

$$R_{EQ} = \frac{R_1 R_2}{R_1 + R_2}$$ **Eq. 1.14**

Figure 1.4: Equivalent resistance for **n** –resistors in parallel

As a special case, assume that the parallel resistor network shown in Figure 1.3 consists of "**n**" *equal* parallel resistors. Because, the resistors are assumed to be equal, calculation of R_{EQ}, or $R_{EQ\text{-}n}$, boils down to Eq. 1.15.

$$R_{EQ} = R_{EQ\text{-}n} = \frac{R}{n}$$

Eq. 1.15

ELECTRICAL SHORT AND OPEN CIRCUIT

When a conductor, wire or bus-bar, is used to pass current between two or more points in an electrical circuit—with smallest resistance feasible—such a connection is referred to as a "**short.**" A short segment of wire, assembled with an "*alligator clip*" on each end, is used by electrician and electrical engineers to establish a temporary short circuit between points in an electrical circuit. Such a specially fabricated pieces of wire are sometimes referred to as a "*jumper leads*," in electrical jargon.

A short circuit between two points implies zero or negligible resistance. The opposite of an electrical short is an "**open circuit.**" Examples of open circuits would be a switch that is open, a breaker that is turned off, or simply a wire that has been cut or "*clipped*." An open circuit between two points implies an infinite (∞) resistance. The concepts of open and short circuits are illustrated in Figure 1.5. Figures 1.5 (a) and (b) represent electrical circuit segments between points A and B. When the switch between points (1) and (2) is open, as depicted in Figure (a), we have an open circuit, and no current flows between points **A** and **B**. However, closure of the same switch, as shown in Figure (b), constitutes a short circuit. The closed switch scenario depicted in Figure (b) also represents "**continuity.**" The term "continuity" is used commonly by electrical engineers and electricians during trouble shooting of equipment. When trouble shooting electrical or electronic equipment failures, engineers and technicians often perform continuity tests on fuses to determine if they have cleared or opened. The instrument used for performing continuity checks is a multi-meter. A continuous piece of wire, shown in Figure 1.5 (c), represents continuity between terminals **A** and **B**.

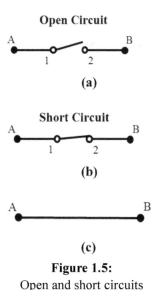

Figure 1.5:
Open and short circuits

Example 1.4

Determine the equivalent resistance for the DC circuit shown below if R_1 = 5Ω, $R_2 = R_3$ = 10Ω, and $R_4 = R_5$ = 20.

$R_1 = 5\Omega$

$R_2 = 10\Omega$ $R_3 = 10\Omega$

12 V

$R_4 = 20\Omega$ $R_5 = 20\Omega$

Solution:

The resistances need to be combined in a multistep process in order to determine R_{eq} for the overall circuit.

Combination of R_2 and $R_4 = R_{2,4} = R_2 + R_4 = 10\Omega + 20\Omega = 30\Omega$

Combination of R_3 and $R_5 = R_{3,5} = R_3 + R_5 = 10\Omega + 20\Omega = 30\Omega$

Combination of $R_{2,4}$ and $R_{3,5}$:

$$R_{2-5} = \frac{(30\Omega).(30\Omega)}{(30\Omega + 30\Omega)} = \frac{900}{60} = 15\Omega$$

$$\therefore R_{eq} = 15\Omega + 5\Omega = 20\Omega$$

CAPACITOR AND CAPACITANCE

A capacitor is a charge storage device, capable of storing charge in DC and AC applications. A charged capacitor stores electrical charge on two electrodes; one of the two electrodes is negative and the other one is positive. The negative electrode is called a *cathode* and the positive electrode is referred to as an *anode*. This separation of charge, and the quantity of charge separated, determine the electrical potential—or voltage—developed across the electrodes of the capacitor. The electrical potential difference between the capacitor plates, or electrodes, signifies the storage of electrical energy in the capacitor. Therefore, as an energy storage device—with a potential difference—capacitors are analogous to air receivers and pneumatic cylinders that store pressure energy—in form of higher pressure relative to the atmospheric pressure—which can be used to perform mechanical work.

Construction of a simple capacitor is depicted in Figure 1.6. As shown in Figure 1.6, a simple capacitor can be constructed with two parallel square plates, of equal size, separated by a dielectric substance like glass, mica, etc. The separation between the two plates (electrodes), "**r**," in conjunction with the area of the plates determines the "**capacitance**" of the capacitor. Capacitance, "**C**," of a capacitor is defined as the charge storage capacity of the capacitor.

Figure 1.6: A simple parallel plate capacitor

Capacitance can be defined, mathematically, through Eq. 1.16, below.

$$C = \varepsilon \frac{A}{r} \qquad\qquad \textbf{Eq. 1.16}$$

Where,
  C = Capacitance is quantified or specified in farads;
  A = The area of cross-section—or simply area—of the capacitor electrode plates;
  ε = Permittivity of the dielectric medium between the plates; and
  ε = $\varepsilon_r \cdot \varepsilon_0$

Where
  ε_r = Relative permittivity of a specific dielectric medium, and
  ε_0 = permittivity in vacuum or in air = 8.854×10^{-12} farads per meter (F·m−1).

One farad is rather large amount of capacitance for most common capacitor applications. Therefore, many capacitors—especially, at the circuit board level—are specified or labeled in terms of smaller units, such as, **mF** (milli-Farad), μ**F** (micro-Farad), or **nF** (nano-Farad). The capacitor shown in Figure 1.7 is rated 470μ**F** and designed to operate at a maximum of 35V.

Figure 1.7: A cylindrical 470 Micro-Farad Capacitor

The mathematical relationship stated as Eq. 1.16 stipulates that capacitance is directly proportional to the area **A** of the capacitor plates and inversely proportional to the separation **r** between the plates. In other words, if larger capacitance or ***charge storage capacity*** is desired, one must increase the area of the plates and/or decrease the separation between the capacitor plates. In addition to serving as a *"**constant of proportionality**"* for the equation, 7 $\in$ injects the property or characteristic of the dielectric medium into the computation of capacitance through the dielectric medium's characteristic $\in_r$ value. Although, for the sake of simplicity, the discussion on capacitance in this text is limited to flat plate capacitors, many capacitors have cylindrical construction such as the one shown in Figure 1.7.

Electrical energy stored in a capacitor can be determined through application of Eq. 1.17

$$\text{Energy (joules)} = \frac{1}{2}CV^2 \qquad\qquad \textbf{Eq. 1.17}$$

Charge storage characteristic of the capacitor should not be confused with the charge storage and power source function of a battery. One difference between capacitors and batteries is that when capacitors are charging or discharging, charge flows through a "dielectric" medium; while most batteries consist of electrolytes (i.e. sulfuric acid) that ionize readily and the ions sustain the flow of current. Capacitors allow the charge to move between electrode plates through a dielectric medium.

The dynamics of how a capacitor stores and dissipates charge are somewhat different between the DC and AC realms. Unlike resistive circuits, current and voltage associated with capacitors vary in a non-linear fashion. A common, series, RC circuit is shown in Figure 1.8, consisting of a capacitor, resistor and a switch that can be used to control the charging and discharging of the capacitor. The graphs in Figure 1.9 through 1.11 below compare voltage and current responses in circuits that are purely resistive versus a series RC circuit, the type illustrated in Figure 1.8. This "non-linear" charging and discharging of capacitors is referred to as ***transient behavior*** of RC circuits.

Figure 1.8: A series RC circuit

The straight line graph in Figure 1.9 illustrates and validates the Ohms law. This graph shows that voltage and current are directly proportional, with the "constant" resistor serving as the constant of proportionality. In a purely resistive circuit, the current would respond, instantaneously and linearly, to the application of voltage across a resistor.

Figure 1.9: Linear, voltage vs. current, response in a purely resistive circuit.

Voltage vs. Time, transient response in series RC circuit: Contrary to linear and instantaneous response in a purely resistive circuit, the current and voltage response in a common RC circuit is non-linear and non-instantaneous.

$$v_c(t) \; = \; v_c(0)\, e^{-\frac{t}{RC}} + V(1 - e^{-\frac{t}{RC}}) \qquad \text{Eq. 1.18}$$

The voltage response—or voltage variation—of a capacitor can be predicted through Eq. 1.18.

Where,

    R = Resistance in series with the capacitance
    C = Capacitance

$v_c(0)$ = Voltage across the capacitor, at time t = 0
$v_c(t)$ = Voltage across the capacitor, at a given time **t**
  V = Voltage of the power source
RC = τ = Time constant of an RC circuit

If the voltage variation or response of a capacitor were graphed, with respect to time, it would be resemble the v_c versus t graph shown in Figure 1.10.

Capacitor Voltage vs. Time

$$v_c(t) \;=\; v_c(0)\, e^{-\frac{t}{RC}} + V(1 - e^{-\frac{t}{RC}})$$

Figure 1.10: Non-linear, transient, voltage response in a circuit consisting of capacitance and resistance

Current vs. Time, transient, response in series RC circuit: The current response—or current variation—in a capacitor-resistor circuit can be predicted through Eq. 1.19.

$$i_c(t) = \left(\frac{V - v_c(0)}{R}\right) e^{-\frac{t}{RC}}$$ **Eq. 1.19**

Where,
   R = Resistance in series with the capacitance
   C = Capacitance, in farads
   $v_c(0)$ = Voltage across the capacitor, at time t = 0
   $i_c(t)$ = Current through the capacitor-resistor circuit, at a given time **t**
   V = Voltage of the power source

If the current response of a capacitor were graphed, with respect to time, it would resemble the i_c versus t graph shown in Figure 1.11. Note that the capacitor current versus time curve in Figure 1.11 validates Eq. 1.19. For instance, the graph shows that at $t = \infty$, or when steady state is achieved, current $i_c(t)$ through the RC circuit diminishes to zero; and if you substitute $t = \infty$ in Eq. 1.19, $i_c(t)$ becomes zero. In other words, at steady state, the capacitor transforms into an open circuit and the current ceases to flow.

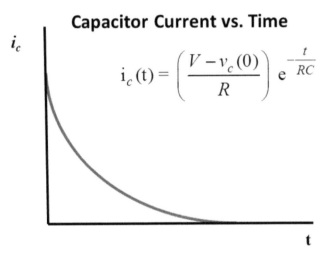

Capacitor Current vs. Time

$$i_c(t) = \left(\frac{V - v_c(0)}{R} \right) e^{-\frac{t}{RC}}$$

Figure 1.11: Non-linear, transient, current response in a circuit consisting of capacitance and resistance

The presence of resistance **R** in capacitive circuits results in what is referred to as a time constant "**τ**." The relationship between **τ**, **R and C** is stipulated by Eq. 1.20. The physical significance of time constant **τ** is that it represents the time it takes to charge a capacitor to **63.2% of the full value**, or **63.2% of the full voltage** of the source. Time constant **τ** also represents the time it takes to discharge a given capacitor to **36.8% of the full voltage**, or the voltage of the source.

τ = RC   **Eq. 1.20**

Example 1.5
Consider the RC circuit shown in the diagram below. The source voltage is 12V. The capacitor is in a discharged state before the switch is closed. The switch is closed at t = 0. What would the capacitor voltage be at t = 2**τ**?

$$R = 1k\Omega \qquad C = 5\mu F$$

Solution:

This particular case represents a capacitor charging scenario. Given the value of **R, C, $v_c(0)$** and the source voltage V, Equation 1.18 allows us to calculate the voltage after and elapsed time "**t**," during the capacitor charging phase.

$$v_c(t) \;=\; v_c(0)\, e^{-\frac{t}{RC}} + V(1 - e^{-\frac{t}{RC}})$$

In this case,

$\quad$ **R** = 1 kΩ = 1,000 Ω

$\quad$ C = 5μF = 5x10^{-6} F

$\quad$ $v_c(0)$ = 0V = Voltage across the capacitor at t = 0

$\quad$ $v_c(t)$ = Voltage across the capacitor, at a given time **t** = ?

$\quad$ *V* = Voltage of the power source = 12V

$\quad$ *RC* = τ = RC circuit time constant

$\quad$ *t* = 2τ = **2RC**

Substitution of the given values, stated above, Eq. 1.18 can be expanded and simplified as follows:

$$v_c(t) \;=\; v_c(0)\,.\,e^{-\frac{2\tau}{\tau}} + 12.(1 - e^{-\frac{2\tau}{\tau}})$$

$$= 0 + (12).(1 - e^{-2}) = 12(0.865)$$

$$= 10.38V$$

CAPACITORS IN SERIES

When a number of capacitors are connected in a "daisy-chained," or concatenated fashion, as shown in Figure 1.12, they are said to be connected in series. In Figure 1.12, "**n**" number of capacitors, C_1 through C_n, are shown connected in series.

Figure 1.12: Capacitors in series

When electrical circuits, AC or DC, consist of multiple capacitors, circuit analyses require simplification—or combination—of such network of capacitors into one, ***equivalent***, capacitor or capacitance, C_{EQ}. This equivalent, capacitance can also be referred to as a "***total***" capacitance.

Unlike series combination of resistors, when capacitors are connected in series, they can be combined in an "**addition of inverses**" format, as stipulated in Eq. 1.21 below, for "**n**" number of capacitors.

$$C_{EQ} = \cfrac{1}{\dfrac{1}{C_1} + \dfrac{1}{C_2} + \dfrac{1}{C_3} + ... + \dfrac{1}{C_n}}$$ **Eq. 1.21**

When multiple capacitors are combined into an equivalent capacitor, with capacitance value C_{EQ}, the simplified version of the original series circuit can be drawn as shown in Figure 1.13.

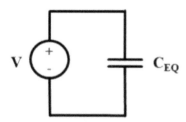

Figure 1.13: Equivalent capacitance for capacitors in series.

As a special case, suppose that the series capacitor network shown in Figure 1.12 consists of "**n**" series capacitors, with ***equal*** capacitance. Because the capacitors are assumed to be equal, calculation of C_{EQ}, or C_{EQ-n}, and application of Eq. 1.21 can be simplified to Eq. 1.22.

$$C_{EQ} = C_{EQ-n} = \frac{C}{n}$$ **Eq. 1.22**

If a series capacitive circuit consists of only three capacitors, as shown in Figure 1.14, Eq. 1.21 can be reduced to Eq. 1.23. Further simplification of Eq. 1.23 would result in a, simplified, three capacitor series equivalent capacitance equation, Eq. 1.24.

Figure 1.14: Equivalent capacitance for capacitors in series.

$$C_{EQ} = \frac{1}{\dfrac{1}{C_1} + \dfrac{1}{C_2} + \dfrac{1}{C_3}}$$ **Eq. 1.23**

$$C_{EQ} = \frac{C_1 C_2 C_3}{C_1 C_2 + C_2 C_3 + C_1 C_3}$$ **Eq. 1.24**

Example 1.6
Determine the equivalent capacitance for the DC circuit shown Figure 1.14 if $C_1 = 5\mu F$, and $C_2 = C_3 = 10\mu F$.

Solution:
Application of Eq. 1.24 to the three series capacitor circuit shown in Figure 1.14 yields:

$$C_{EQ} = \frac{(5x10^{-6})(10x10^{-6})(10x10^{-6})}{(5x10^{-6})(10x10^{-6}) + (5x10^{-6})(10x10^{-6}) + (10x10^{-6})(10x10^{-6})}$$

$C_{EQ} = 2.5\mu F$

CAPACITORS IN PARALLEL

When electrical circuits consist of capacitors connected in parallel, as shown in Figure 1.15, circuit analyses—as with a network of series connected capacitors—would require simplification of the parallel network of capacitors into one, equivalent capacitor C_{EQ}. Figure 1.15 shows "**n**" number of capacitors connected such the "heads" of all capacitors are (electrically) "bonded" or are connected together, with an electrical connection to the anode (or positive terminal) of the DC power supply, and the "tails" of all capacitors are connected together to the cathode (or negative terminal) of the power source.

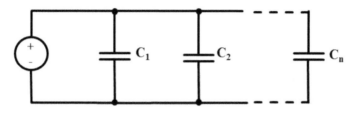

Figure 1.15: Equivalent capacitance for capacitors in series.

Simplification of a network of capacitors is parallel is similar to the approach utilized in the combination of *resistors in series*. When multiple *parallel* capacitors are combined into an equivalent capacitor, C_{EQ}, the simplified or condensed equivalent of the original parallel circuit could also be represented by the equivalent circuit diagram in Figure 1.13.

For a parallel capacitor circuit consisting of "**n**" number of parallel capacitors, equivalent capacitance can be calculated by applying equation Eq. 1.18

$$C_{EQ} = C_1 + C_2 + C_3 + ... + C_n \hspace{4cm} \text{Eq. 1.25}$$

Example 1.7

Determine the equivalent capacitance for the DC circuit shown below if $C_1 = C_2 = 5\mu F$, and $C_3 = 10F$.

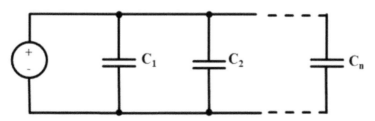

Solution:

Applying Eq. 1.25 to the three parallel capacitor circuit shown in diagram above yields:

$$C_{EQ} = C_1 + C_2 + C_3$$

$$C_{EQ} = (5x10^{-6})(10x10^{-6}) + (5x10^{-6})(10x10^{-6}) + (10x10^{-6})(10x10^{-6})$$

$$= 20x10^{-6} \text{ F, or } 20\mu F$$

Example 1.8

Determine the equivalent capacitance in series and parallel combination circuit shown below. The capacitance values are: $C_1 = C_2 = 5\mu F$, $C_3 = C_4 = 10\mu F$.

Solution:

Similar to the parallel and series combination approach described in the resistor section, our aim would be to simplify the circuit by first combining the linearly additive segment (or segments), followed by reciprocal combination of the remaining capacitors.

The capacitors in this circuit that lend themselves to linear combination are C_3 and C_4. Therefore, the combined capacitance, C_{34}, would be:

$$C_{34} = C_3 + C_4 = 10\mu F + 10\mu F = 20\mu F$$

Then, by applying Eq. 1.24 to this special hybrid capacitor combination case, with the 3rd capacitance being C_{34} instead of C_3:

$$C_{EQ} = \frac{C_1 C_2 C_{34}}{C_1 C_2 + C_2 C_{34} + C_1 C_{34}}$$

$$C_{EQ} = \frac{(5x10^{-6})(5x10^{-6})(20x10^{-6})}{(5x10^{-6})(5x10^{-6}) + (5x10^{-6})(20x10^{-6}) + (5x10^{-6})(20x10^{-6})}$$

$C_{EQ} = 2.22\mu F$

CAPACITIVE REACTANCE

When a capacitor is incorporated into an AC circuit, its impact in that circuit is quantified through an entity referred to as the capacitive reactance. The symbol for capacitive reactance is X_c. Capacitive reactance can be defined, mathematically, as:

$$X_c = \frac{1}{\omega C} = \frac{1}{2\pi f C}$$

Eq. 1.26

Where,

f = Frequency of the AC power source, i.e., 60 Hz in the US and 50 Hz in some other parts of the world.

ω = Rotational speed, in radians per second

C = Capacitance in farads

Capacitive reactance is measured in ohms, or Ω's. It is important to note that X_c is often misrepresented as Z_c. To the contrary, as explained in the impedance section, Z_c is the *impedance contribution* by the capacitor, and is represented as:

$$Z_c = -jX_c,$$
$$\therefore Z_c \neq X_c$$

Example 1.9

Assume that the circuit in Example 1.8 is powered by a 60 Hz, AC source instead of the DC source. Determine the total capacitive reactance, X_c, seen by the AC source.

Solution:

If the DC source is replaced by an AC source, the circuit would appear as follows:

As computed in Example 1.8, the combined or net capacitance contributed to the circuit by the parallel and series network of capacitors is C_{EQ} = 2.22μF. Then, by applying Eq. 1.26:

$$X_c = \frac{1}{\omega C} = \frac{1}{2\pi fC} = \frac{1}{2(3.14)(60)(2.22x10^{-6})} = 1194\Omega$$

INDUCTOR AND INDUCTANCE

Similar to capacitor, an inductor can be viewed as an energy storage device, which can be applied in DC or AC applications. Unlike a capacitor—where energy is stored in form of separation of charges, resulting in a potential difference—in an inductor, the energy is stored in the magnetic field that is produced through the flow of electric current. This phenomenon was first discovered by Michael Faraday in 1831. The magnetic field—which can be referred to as magnetic flux—is not established instantaneously upon flow of current through an inductor. Instead, much like a "time constant,τ" based charge build up in a capacitor, the current change and magnetic field build up in an inductor ramps up, or down, at a non-linear rate. This non-linear rate is a function of the inductor's time constant,τ. Inductor's time constant concept will be explained later in this chapter.

Since the energy stored in the magnetic field of an inductor is due to the flow of current—and, ultimately, due to the movement of electrons—we could view an inductor as being analogous to a rotating "flywheel," in the mechanical realm—where the energy is stored in a rotating mass. As with a flywheel—where any attempt to stop the rotation of the flywheel is opposed by momentum and kinetic energy in the rotating mass of the flywheel—*any attempt to change the flow of current in an inductor is opposed by the "inductance"* of the inductor. It is due to the inductance of an inductor that if you, for instance, have 10 amps flowing in an electrical circuit, and you try to "break" the circuit by opening a switch, current flow is maintained briefly through an *electric arc*, where permitted, across the opening switch. Where the electric arc—which is, in essence, plasma or ionized air—serves as a temporary "channel" for the flow of current. Conversely, when a switch—in a de-energized inductive circuit— is closed to connect a power source to an inductive device, no current flows through the circuit, initially, as *inductance* of the inductor in the electrical circuit opposes the *change* in the flow of current from "*zero*" to some measurable "*non-zero*" level. Therefore, inductance is defined as the capacity or tendency of an inductor to resist the *change in flow of current.*

In addition, it is worth noting that just like the kinetic energy stored in a rotating flywheel can be "tapped" to perform mechanical work, the energy stored in the *magnetic field* of a "charged" inductor can be released to push a ferromagnetic cylindrical core, in one direction or another, to open or close a valve, or to open or close an electrical switch. The former application is an example of a **solenoid operated valve**, while the later represents the operation of a **contactor or a relay**.

The principle of inductance and physical aspects of inductors (or coils) are illustrated in Figure 1.16 (a) and (b). Basically, if you take a straight piece of wire, as shown in Figure 1.16 (a), and wind it around a cylindrical core, the final product would be a coil or an inductor. As shown in the Figure 1.16 (a), current flowing through a straight wire produces a "weak" magnetic field. While, current flowing through a "coiled" conductor produces a stronger and denser magnetic field, capable for conducting "work," such as, pushing of a "plunger" against the restraint of a spring' to open or close a valve or to open or close an electrical circuit in a relay or a contactor.

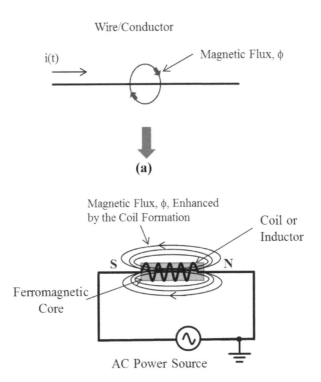

Figure 1.16: (a) Straight current carrying conductor (b) "Coiled" current carrying conductor.

Figure 1.16 (b) represents a conceptual view of a solenoid or a coil, illustrating the fundamental principle of electromagnetism. On the other hand, physical construction of a simple solenoid or inductor is depicted in Figure 1.17, in a diametrical cross-sectional view. When the coil of the solenoid is energized, the plunger—or core—responds to the magnetic flux by moving to the left. As the plunger moves to the left—against the spring—with a force that is proportional to the magnetic flux, the "pin" attached to the tip of the plunger pushes mechanical devices such as relay contacts or valves, etc., to change their state from open to closed, or vice and versa.

Figure 1.17: Construction of a solenoid.

Inductance is denoted by "**L**," and it can be defined, mathematically, through Eq. 1.27, below.

$$L = \frac{\mu . N^2 . A}{1}$$
                                                                                      Eq. 1.27

Where,
    μ = Permeability of the medium, in **H/m**
    N = Number of turns of coil (unit-less)
    A = Cross-sectional area of the core (in **m^2**)
    l = Mean length through the core (in **m**)
    Unit for Inductance: **H (henry)**

The mathematical relationship stated as Eq. 1.27 stipulates that inductance is directly proportional to the area of cross-section "**A**" of the core. This equation also states that the inductance is directly proportional to "**N^2**," the "square" of the number of turns in the coil, and is inversely proportional to the mean length "**l**" of the core. In other words, if larger inductance or ***energy storage capacity*** is desired, one must increase the area of cross-section of the core, increase the number of turns, or reduce the mean length of the core and the coil. Of course, the values of these variables can be increased or decreased, simultaneously, to achieve the desired results. In addition to serving as a "*constant of proportionality*" for the equation, permeability "*μ*" injects the physical characteristics of the core into the computation of inductance. So, if a material with higher relative permeability, μ_r, is chosen, the inductance of a coil would be greater. The permeability *μ* of a specific medium can be defined, mathematically as follows:

$$\mu = \mu_r \cdot \mu_0$$

Where,

μ_r = Relative permeability of the core material. Relative permeability of steel is 100 and that of an insulating materials like wood and Teflon is 1.0; which is the same as free space, vacuum or air.

μ_0 = Permeability of free space or vacuum = $4\pi \times 10-7$ = **1.257 x 10^{-6} H/m**

Electrical energy stored in an inductor can be determined through Eq. 1.28, below:

$$E_{stored} = \frac{1}{2} LI^2 \qquad\qquad \textbf{Eq. 1.28}$$

Where, the energy is measured in joules (or N-m), **L** in H (henry) and **I** in amps.

Most inductive electrical systems—that is electrical circuits with inductance—and inductive devices like motors and transformers, contain inductance **L**, as well as resistance, **R**. In order to understand the current and voltage response in inductive circuits—or to understand current and voltage variation in inductive systems—we will examine a simple series RL circuit diagram in Figure 1.18.

Figure 1.18: Series RL circuit

In a typical inductive-resistive circuit, as the one shown in Figure 1.18, the voltage and current relationships, as a function of time, are governed by Equations 1.29, 1.30 and 1.31.

$$v_L(t) = L\frac{di_L}{dt} \qquad\qquad\qquad \textbf{Eq. 1.29}$$

$$v_L(t) = -i(0)Re^{-\frac{R}{L}t} + Ve^{-\frac{R}{L}t} \qquad\qquad \textbf{Eq. 1.30}$$

$$i_L(t) = i_R(t) = i(0)e^{-\frac{R}{L}t} + \frac{V}{R}(1 - e^{-\frac{R}{L}t}) \qquad \textbf{Eq. 1.31}$$

Equations 1.30 and 1.31 hold greater practical significance, in that, they can be used to predict the changes in voltage and current—with respect to time—for given values of source voltage, **V**, series resistance **R** and series inductance **L**. The relationship between resistance **R** and inductance **L**, or the *relative* size of **R** and **L**, determine the charging and discharging rate of the inductor via the time constant "**τ** ." The time constant **τ** can be defined, mathematically, in terms of **R** and **L**, in form of **Eq. 1.32**.

$$\tau = \frac{L}{R} \qquad\qquad\qquad\qquad \textbf{Eq. 1.32}$$

It is ostensible from examination of Eq. 1.30 and 1.31 that in circuits that consist of inductance and resistance, unlike purely resistive circuits—but similar to RC circuits—current and voltage associated with inductors vary in a non-linear fashion. This "non-linear" charging and discharging of an inductor in an RL circuit—similar to RC circuits—is referred to as *transient behavior*. The graphs in Figure 1.19 and 1.20 compare voltage and current responses—on the basis of Eq. 1.30 and 1.31, respectively.

Voltage vs. Time, transient, response in Series RL Circuit: Examination of Eq. 1.30 and Figures 1.18 and 1.19 reveals the following facts:

a) At time, t = 0, which is when the switch in the circuit is closed, *i*(0) = 0; so the first segment of Eq. 1.30 becomes zero and drops out. Also, at t = 0, the exponent of "e" becomes zero, which makes $e^{-Rt/L} = 1$. This results in $v_L(t) = V$, (i.e., the source voltage) at t = 0. In other words:

Figure 1.19: Non-linear *voltage* response in a circuit consisting of inductance and resistance.

$v_L(t) = V$ at the instant the switch is closed in the series RL circuit.

b) On the other end of the time spectrum, where $t = \infty$, or when steady state condition has been achieved:

$$v_L(\infty) = -i(0)Re^{-\infty} + Ve^{-\infty}$$

$$= -(0) \cdot \left(\frac{1}{\infty}\right) + V\left(\frac{1}{\infty}\right) = 0$$

The analyses above support the following basic tenets of series RL inductive circuits:

i. The voltage across the inductor, at the instant the switch in a series RL circuit is closed, is the same as the source voltage, implying that no current flows through the inductor.

ii. The voltage across the inductor, after a large amount of time has elapsed, diminishes to zero. And, with voltage drop across the inductor zero, the inductor acts as a short in a series RL circuit under steady state conditions—with steady state current, V/R, flowing through it.

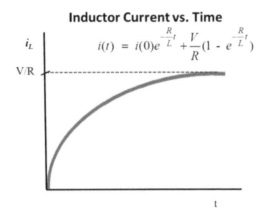

Figure 1.20: Non-linear *current* response in a circuit consisting of inductance and resistance

Current vs. Time, transient, response in series RL circuit: Examination of Eq. 1.31 and Figures 1.18 and 1.20 reveals the following facts associated with current versus time response in a series RL circuit:

a) At time, t = 0, which is when the switch in the circuit is closed, **i(0) = 0**; so the first segment of Eq. 1.31 becomes zero and drops out. Also, at t = 0, the exponent of "**e**" becomes **zero**, which makes $e^{-Rt/L} = 1$. This results in $i_L(t) = 0 + V/R(1-e^0) = 0$, this means that at **t = 0**, or at the instant the switch is closed, the inductance of the inductor resists the initiation of current flow, hence, no current flows through the inductor at **t = 0.**

b) On the other end of the time spectrum, when t = ∞, or when steady state condition has been achieved:

$$i_L(\infty) = i_R(\infty) = i(0)e^{-\frac{R}{L}t} + \frac{V}{R}(1 - e^{-\frac{R}{L}t})$$

$$i_L(\infty) = i_R(\infty) = (0)(e^{-\infty}) + \frac{V}{R}(1 - e^{-\infty})$$

$$i_L(\infty) = i_R(\infty) = \frac{V}{R}(1 - 0) = \frac{V}{R}$$

The analyses stated above support the following basic characteristics of inductive circuits:

i.   No current flows through the inductor and resistor combination at the instant the switch is closed, or at t = 0, which is congruent with one of the fundamental characteristics of an inductor described earlier. In other words, at the outset, the inductance of the inductor, successfully, resists the rise of the current to a non-zero value.

ii. Current through the inductor and resistor combination develops to the maximum level after a long span of time. The maximum level of current in the inductor and resistor combination is equal to **V/R**. This also implies that the inductor acts as a "***short***" when steady state condition is achieved or after 10τ amount of time.

Example 1.10
Consider the series RL circuit shown in the diagram below. The source voltage is 12V and R = 10Ω. The switch is closed at t = 0. What would be magnitude of current flowing through this circuit at t = τ?

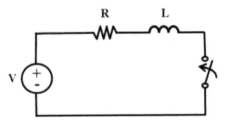

Solution:
In most series RL cases, the current value at a certain time "t" can be predicted through **Eq. 1.31.**

$$i_L(t) \;=\; i_R(t) \;=\; i(0)e^{-\frac{R}{L}t} + \frac{V}{R}(1 - e^{-\frac{R}{L}t})$$

Note: In this case, the value of **L** is not given, but the elapsed time is given as a function of time constant as, "**1τ.**"

Since **τ = L/R, Eq. 1.31** can be rewritten, in **τ** form as:

$$i_L(t) = i_R(t) = i(0)e^{-\frac{t}{\tau}} + \frac{V}{R}(1 - e^{-\frac{t}{\tau}})$$

Then, by substituting t = τ, and given the fact that i(0) = 0, the $i_L(t)$ equations simplifies into the following form:

$$i_L(t) = i_R(t) = i(0)e^{-\frac{\tau}{\tau}} + \frac{V}{R}(1 - e^{-\frac{\tau}{\tau}})$$

$$= \frac{V}{R}(1 - e^{-1})$$

$$= \frac{V}{R}(0.632) = \frac{12}{10}(0.632) = (1.2)(0.632) = 0.759 \text{ A}$$

This analysis of current response validates a characteristic fact about inductors: current develops to **63.2%** of its full potential in "one time constant" or **1τ** worth of time.

Example 1.11

Consider the series RL circuit given in Example 1.9, in discharge mode, with voltage source removed. Inductor L = 10mH. The switch has been closed for long period of time, such that the current has developed to the maximum or steady state level 1.2 A. How much time would need to elapse for the current to drop to 0.8A after the switch is opened.

R = 10Ω   L = 10mH

Solution:
Apply series RL current equation, **Eq. 1.31**.

$$i_L(t) = i_R(t) = i(0)e^{-\frac{R}{L}t} + \frac{V}{R}(1 - e^{-\frac{R}{L}t})$$

Given:

$t = ?$

$L = 10 \times 10^{-3}H$

$R = 10\Omega$

$V = 0$

$i(0) = 1.2A$

$i_L(t) = 0.8A$

$$i_L(t) = (0.8) = (1.2)e^{-\frac{10}{0.01}t} + (0)(1 - e^{-\frac{10}{0.01}t})$$

$$0.8 = (1.2)e^{-\frac{10}{0.01}t}$$

$$0.667 = e^{-\frac{10}{0.01}t}$$

$$\ln(0.667) = \ln(e^{-\frac{10}{0.01}t})$$

$$-0.4055 = -1000t$$

$$t = 0.00041s \text{ or } 0.41 \text{ ms}$$

SERIES INDUCTOR COMBINATION

When analyzing dc circuits with inductors connected in series, derivation of equivalent inductance, **L_{eq}**, can accomplished by, simply, adding the inductance values linearly as represented by Eq. 1.33. Figure 1.21 depicts "**n**" inductors, **L_1** through **L_n**, connected in series.

$$L_{EQ} = L_1 + L_2 + L_3 \text{ .... } + L_n \qquad \text{Eq. 1.33}$$

Figure 1.21: Series combination of "**n**" inductors

Derivation of the combined equivalent inductance L_{EQ}, permits us to represent Figure 1.21 in form of a condensed version depicted in Figure 1.22.

Figure 1.22: L_{EQ}, equivalent inductance representing series combination of "**n**" inductors.

Example 1.12

Determine the equivalent inductance for three inductors connected in a series combination circuit shown below. The inductance values of the three inductors are: $L_1 = 5mH$, $L_2 = 5mH$, and $L_3 = 10mH$.

Solution:

Apply Eq. 1.33:

$$L_{EQ} = L_1 + L_2 + L_3$$
$$= 5mH + 5mH + 10 \ mH$$
$$= 20mH$$

PARALLEL COMBINATION OF INDUCTORS

The circuit depicted in Figure 1.23 shows "n" number of inductors connected in parallel. The formula for determining the equivalent inductance L_{eq} for this circuit is represented by Eq. 1.34.

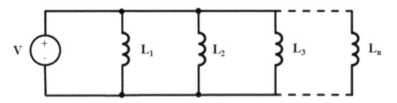

Figure 1.23: Parallel combination of "n" inductors

$$L_{EQ} = \frac{1}{\dfrac{1}{L_1} + \dfrac{1}{L_2} + \dfrac{1}{L_3} + ... + \dfrac{1}{L_n}}$$

Eq. 1.34

If a parallel resistor circuit or network consists of only three inductors, the circuit would appear as shown in Figure 1.24 and the L_{eq} equation for this circuit would reduce to Eq. 1.36.

Figure 1.24: Combination of three parallel inductors

$$L_{EQ} = \frac{1}{\dfrac{1}{L_1} + \dfrac{1}{L_2} + \dfrac{1}{L_3}}$$

Eq. 1.35

$$L_{EQ} = \frac{L_1 L_2 L_3}{L_1 L_2 + L_2 L_3 + L_1 L_3}$$

Eq. 1.36

Example 1.13
Determine the equivalent inductance L_{EQ} for three parallel inductor DC circuit shown Figure 1.24 if L_1 = 1H, and L_2 = 5H and L_3 = 10H.

Solution:
Apply Eq. 1.36 to compute L_{EQ} for the three parallel inductor circuit shown in Figure 1.24:

$$L_{EQ} = \frac{L_1 L_2 L_3}{L_1 L_2 + L_2 L_3 + L_1 L_3}$$

$$= \frac{(1H)(5H)(10H)}{(1H)(5H) + (5H)(10H) + (1H)(10H)}$$

$$= 0.77H$$

Example 1.14

Calculate the net or total inductance as seen from the 24V source vantage point in the circuit shown below.

Solution:

We need to focus on the parallel combination of L_2, L_3, and L_4 first. Apply Eq. 1.36 to calculate the equivalent inductance L_{234} for the three parallel inductors:

$$L_{234} = \frac{L_2 L_3 L_4}{L_2 L_3 + L_3 L_4 + L_1 L_4}$$

$$= \frac{(10mH)(20mH)(30mH)}{(10mH)(20mH) + (20mH)(30mH) + (10mH)(30mH)}$$

$$= 5.45mH$$

This reduces the circuit as shown below:

Inductors L_1 and L_{234}, in this reduced circuit, lend themselves to a linear combination. Therefore, the equivalent inductance L_{EQ} for the entire parallel and series inductor hybrid circuit would be:

$$L_{EQ} = L_1 + L_{234} = 5mH + 5.45mH = 10.45mH$$

INDUCTIVE REACTANCE

When an inductor is incorporated into an AC circuit, its impact in that circuit is quantified through an entity referred to as the ***inductive reactance***. The symbol for inductive reactance is X_L. Inductive reactance can be defined, mathematically, as:

$$X_L = \omega L = 2\pi f L \qquad\qquad \textbf{Eq. 1.37}$$

Where,

f = Frequency of the AC power source, i.e., 60 Hz in the US and 50 Hz in some other parts of the world.

ω = Rotational speed, in radians per second

L = Inductance in henry, or **H**.

Inductive reactance is measured in ohms, or Ω's. It is important to note that, X_L is often misconstrued as Z_L. To the contrary, as explained in the impedance section, Z_L is the ***impedance contribution by the inductor***, and is represented as:

$$Z_L = jX_L,$$
$$\therefore Z_L \neq X_L$$

Example 1.15

Assume that the circuit in Example 1.14 is powered by a 60 Hz AC source. Calculate the inductive reactance, X_L, as seen by the AC voltage source.

Solution:

If the DC source is replaced by an AC source, the circuit would appear as follows:

L_{EQ}, as seen by the AC voltage source, is shown in the simplified equivalent circuit below:

As computed in Example 1.13, the combined or net inductance contributed to the circuit by the parallel and series network of inductors is L_{EQ} = 10.45mH. Then, by applying Eq. 1.37, the inductive reactance, $\mathbf{X_{L\text{-}EQ}}$ as seen by the AC voltage source V_{AC}, would be:

$$X_{L\text{-}EQ} = \omega.L = (2\pi f).L_{EQ} = 2(3.14)(60Hz)(10.45mH) = 3.94\ \Omega$$

IMPEDANCE

The narrative definition of impedance would be that it is the current resisting and impeding characteristic of load or conductor, in an AC circuit. As implied in the definition of this term, impedance is an alternating current entity. While in DC circuits the factor that opposes the flow of DC current is resistance, the entity that influences the flow of AC current, in AC circuits, is impedance. Impedance, like AC current, voltage and power, is a vector entity. By definition, a vector can be completely defined by two key characteristics, namely, the **magnitude** and the **direction**. Therefore, impedance and other entities in AC circuits, such as current, voltage and power, can be defined completely through specification of their magnitude and direction.

Symbol for impedance is **Z**. When typewritten, as with most AC entities, the symbol **Z** for impedance is represented in bold font. When handwritten, most AC entities are denoted by the respective symbol with a half arrow. So, impedance would be denoted by " $\vec{Z}$." The unit for impedance is **ohm, or Ω**; similar to the unit for resistance, **R**, capacitive reactance $\mathbf{X_c}$ and Inductive reactance $\mathbf{X_L}$. Consistency of units between **R**, $\mathbf{X_L}$, $\mathbf{X_C}$ and **Z** is one justification for the following mathematical definition for **Z**:

$$\mathbf{Z = R + jX_l - jX_c} \qquad\qquad \textbf{Eq. 1.38}$$

Where,

$$\mathbf{jX_l = Z_l} \qquad\qquad \textbf{Eq. 1.39}$$
$$= \text{Impedance contribution by the inductance in the circuit.}$$

And,

$$-jX_c = Z_c$$ Eq. 1.40

= Impedance contribution by the capacitance in the circuit.

Concept of impedance, other AC entities such as AC current, AC voltage, AC power, and associated computational methods are discussed in greater depth in Chapter 3.

MAGNETIC CIRCUITS VS. ELECTRICAL CIRCUITS

The two diagrams shown in Figure 1.25 below illustrate similarities and differences between a basic electrical circuit and a magnetic circuit. In depth discussion on the subject of magnetism is beyond the scope of this text. Nevertheless, a contrast between the basic magnetic and electric circuits below will allow the reader an opportunity to gain basic understanding of magnetism, more specifically, electromagnetism.

The circuit shown in Figure 1.25 (a) represents a basic DC electrical circuit. This circuit consists of a DC voltage source labeled "**V**." As explained earlier, voltage is synonymous to the term electromotive force. So, we can explain the phenomenon in the electrical circuit as the electromotive force "**V**" driving DC current "**I**" through the load or resistor "**R**." Note the direction of the current is from the left to right, in a clockwise loop, emerging from the positive electrode of the DC voltage source and terminating into the negative electrode of the voltage source. This clockwise current flow is assigned on the basis of an electrical convention that stipulates that the current flow consists of "*holes*," or positively charged particles, being repelled or driven out of the positive terminal. This convention also affirms the fact that electrical current is not Necessarily, always, due to the flow or movement of electrons. Electrical current, as explained in the section on the topic of current, can be due the flow or movement of negatively or positively charged particles. The positively or negatively charged particles, at the atomic or molecular level are referred to as ions.

The relationship between V, I and R in the electrical circuit is governed by the Ohms Law. The Ohms Law, in conjunction with other basic electrical laws—used to analyze electrical circuits—will be explained in more detail in Chapter 2. For now, note that the Ohms Law is stated, mathematically, in form of equation 1.41. In other words, according to Ohms Law, electromotive force is equal to the product of current and resistance.

Electromotive Force, V = I.R = (Current) x (Resistance)   Eq. 1.41

The circuit shown in Figure 1.25 (b) represents a basic electromagnetic circuit. This circuit consists of a toroid or donut shaped core—typically constructed out of iron. In this magnetic circuit, a conductor, or wire, is wrapped in four turns around the left side of the toroid core. When current is passed through wound conductor, magnetic field is established in the core as represented by the dashed circular line, with an arrow pointing in clockwise direction. This magnetic field is referred to as magnetic flux, ϕ. Magnetic flux is measured in weber. The unit weber is named for the German physicist **Wilhelm Eduard Weber** (1804–1891). In the magnetic realm, the flux serves as a counterpart to the current, **I**, from the electrical realm. Just like the electromotive force, **EMF**, or voltage, drives the current through the resistor, **R**, the magnetomotive force (**MMF**), F, drives the magnetic flux, ϕ, through the toroid magnetic core. Magnetomotive force is measured in ampere-turns. In electrical systems, load is represented by the resistor **R**. In the magnetic circuit, the flow of magnetic flux is opposed by reluctance $\mathcal{R}$. Just as the Ohm's Law, represented by Eq. 1.41, governs the relationship between electromotive force (voltage), current and resistance in the electrical realm, **Eq. 1.42** represents the relationship between the magnetomotive force, F, the magnetic flux, ϕ, and the reluctance $\mathcal{R}$, in the magnetic domain.

$$F = \phi . \mathcal{R} = \text{(Magnetic Flux)} \times \text{(Reluctance)} \qquad \textbf{Eq. 1.42}$$

Equation 1.42 is referred to as the ***Hopkinson's Law***.

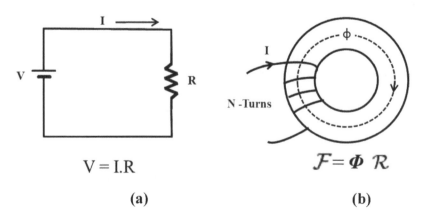

$$V = I.R$$

(a)

$$\mathcal{F} = \Phi \, \mathcal{R}$$

(b)

Figure 1.25: Magnetic and electrical circuit comparison

Magnetic reluctance can also be perceived as magnetic resistance; a resistance that opposes the flow of magnetic flux. Like resistance, reluctance is a scalar entity, but unlike electric resistance, it stores magnetic energy instead of dissipating it. Reluctance is measured in ampere-turns per weber, or turns per henry. Ferromagnetic substances such as iron have low reluctance while dielectric substances like air and vacuum offer high reluctance to magnetic flux. That is the reason why transformers, contactors, relays and other similar electromagnetic devices utilize iron—or iron alloy—cores.

Analogous to Eq. 1.1, which represents the relationship between resistance, resistivity, length and area of cross-section, the reluctance of a uniform magnetic circuit can be calculated as:

$$\mathcal{R} = \left(\frac{1}{\mu_r \mu_o} \right) \cdot \frac{l}{A} \qquad\qquad \textbf{Eq. 1.43}$$

Or,

$$\mathcal{R} = \left(\frac{1}{\mu} \right) \cdot \frac{l}{A} \qquad\qquad \textbf{Eq. 1.44}$$

Where,

l is the mean length of the circuit or core, in meters.

μ_o is the permeability of free space or vacuum = $\mathbf{4.\pi.10^{-7}}$ henry per meter.

μ_r is the relative magnetic permeability of the core material. This is a dimensionless number.

μ is the permeability of the core material in henry per meter.

A is the area of cross-section of the core or the magnetic circuit defined in $\mathbf{m^2}$.

BASIC ELECTRICAL INSTRUMENTS

Two of the most fundamental instruments used to make electrical measurements and perform electrical system trouble shooting are as follows:

I. Multi-meter, or a VOM, Volt-Ohm-Meter
II. Clamp-on Ammeter

Multi-meter

The modern multi-meter, sometimes just written as "multimeter," has evolved from its basic predecessor, the ***Ohm-meter.*** The original Ohm-meters ware designed to measure resistance of electrical components and to verify continuity and integrity of electrical or electronic circuits. Voltage measuring feature was later added to the basic resistance measuring function of the Ohm-meter in form of a more versatile instrument called the ***Volt-Ohm-Meter***, or **VOM**. Due to the miniaturization of electronic components, additional functions were added to the basic VOM resulting in the contemporary multi-meter that transitioned from analog to digital format. See the diagram of a digital multi-meter in Figure 1.26. Some of the following features and functions are common among most multi-meters available on the market today:

a) Voltage measurement, AC and DC
b) Resistance measurement
c) Current measurement
d) Temperature measurement
e) Capacitor testing
f) Diode testing
g) Transistor testing

Standard accuracy of a portable digital multi-meter is, approximately, 0.3%. Bench top and lab grade multi-meters are known to offer accuracy that is better than ±0.1%. Many multi-meters include the "***Peak-Hold***" feature, which allows one to capture peak reading when the measured parameter is not steady. Some of the more sophisticated multi-meters have the ability to interface with PCs for direct data transfer, plotting and storage.

When used for measuring voltage, multi-meters operate in "high imped-ance" mode. In other words, when multi-meters are used to measure voltage, the circuit being tested "sees' high impedance between the probes. This allows for the application of a multi-meter to be a ***non-invasive*** to the circuit being tested. This means that for most voltage measurements, a multi-meter can be applied to an electrical circuit without turning off the power to the system being tested. On the other hand, if a multi-meter is being used for measuring current, the circuit must be turned off, interrupted and severed at the point of measure-ment, and the meter must be injected in series to route the current through the multi-meter for measurement.

One must consider certain important safety measure associated with the application of multi-meters. Multi-meters, similar to other electrical instrumen-tation, are designed to operate within specified voltage range. Most, commonly used, multi-meters are rated for *low voltage* applications. Low voltage category ranges from 0 to 600 VAC or VDC. Application of a low voltage instrument on higher voltages can result in a potentially catastrophic failure of the instrument and a potentially lethal arc flash incident. In addition, from safety point of view,

Backlit Digital
LCD Display

Transistor
Testing Socket

Rotary Switch for
Selection of
Measurement Mode

Anode, or Positive,
Probe Port

Cathode, or Negative,
Probe Port

Cathode Probe

Anode Probe

Capacitor and
Diode Testing
Socket

Cathode Probe
Ports for
Current
Measurement

Figure 1.26: VOLTCRAFT® Digital Multi-meter

it is important to inspect the instrument, periodically, for signs of mechanical or electrical damage. The probes must always be inspected for signs of wear, fraying, and signs of "*mechanical* stress or strain." The outermost surface of the cathode and anode probes is, essentially, insulation. Therefore, if the insulation is damaged or atrophied, person handling the probes could be exposed to electrical shock hazard. In most cases, prudent and safe course of action would be to *decommission and discard* faulty probes, and replace them with, suitable, manufacturer recommended replacements.

The Clamp-on Ammeter

When current in an electrical or electronic circuit must be measured without interruption of an electrical or electronic circuit, a clamp-on ammeter, sometimes referred to as "clamp meter," can be used. Current measurement with a clamp-on ammeter, unlike a multi-meter, requires no probes. See Figure 1.27. When measuring current, the rotary selector switch on the face of the ammeter is switched to AC current measurement setting. The spring loaded "clamping" current transformer is opened by pressing the knob on one side of the ammeter. The current carrying conductor is surrounded by the open current transformer clamp and the current transformer clamp is allowed to close around the conductor. As the current transformer closes or loops around current carrying conductor, it develops magnetic flux that is proportional to the flow of current through the conductor. This magnetic flux is *transduced* into voltage that is subsequently scaled and displayed, digitally. The current indicated on the display of a typical AC ammeter is an RMS value. In single phase AC applications, the clamp-on ammeter should be clamped around the ungrounded or energized conductor. If the neutral and the energized conductors are clamped together, the currents through the two conductors cancel each other resulting in null or zero readout on the display. Similarly, while measuring line currents in multiphase AC systems, *only one phase* must be enclosed in the clamp-on ammeter current transformer. Some clamp-on ammeters are equipped with auxiliary features like voltage measurement; such is the case with the clamp-on ammeter shown in Figure 1.27. The ports for connecting the voltage measuring anode and cathode probe leads are located at the bottom of the ammeter.

The voltage rating related safety precaution, advised in the use of multi-meters, applies to clamp-on ammeters, as well. The clamp-on ammeter depicted in Figure 1.27 is a low voltage (0-600 VAC) device. Use of low voltage clamp-on ammeters on **Medium or High Voltage** electrical circuits is unsafe and can result in catastrophic failures, such as, *arc flash incidents*. Proper precaution must also be exercised concerning the maximum current rating of

Spring Loaded Clamping Current Transformer

Rotary Switch for Selection of Measurement Mode

LCD Display

Single Current Carrying Conductor

Push to Open Current Transformer

Cathode and Anode Probe Ports for Voltage Measurement

Figure 1.27: Fluke® Digital Clamp-on Ammeter

a clamp-on ammeter. Current and voltage ratings of a clamp-on ammeter must *not* be exceeded. Miss-application of test instrumentation can result in catastrophic faults or arc flash incidents.

As noted earlier, most clamp-on ammeters used routinely by electrical engineers and electricians are designed to detect and measure AC current. Although uncommon, DC clamp-on ammeters, operating on **Hall Effect Principle**, are available for "non-invasive" DC current measurement.

Self-assessment Problems & Questions—Chapter 1

1. In an AC system, a voltage source $V(t) = 120Sin(377t + 0°)$ volts, RMS, sets up a current of $I(t) = 5Sin(377t + 45°)$ amps, rms. Calculate the maximum values of voltage and current in this case.

2. A phase conductor of a transmission line is one mile long and has a diameter of 1.5 inch. The conductor is composed of aluminum. Calculate the electrical resistance of this conductor.

3. What is the resistance of the following circuit as seen from the battery?

4. Consider the RC circuit shown in the diagram below. The source voltage is 12V. The capacitor before the switch is closed is 2V. The switch is closed at t = 0. What would the capacitor voltage be at t = 5 sec?

5. Determine the equivalent capacitance for the DC circuit shown in the circuit diagram below if $C_1 = 5\mu F$ and $C_2 = 10F$.

6. Determine the equivalent capacitance for the DC circuit shown below if this circuit consists of twenty $100\mu F$ capacitors in series.

7. Determine the equivalent capacitance in series and parallel combination circuit shown below. The capacitance values are: $C_1 = 10\mu F$, $C_2 = 10\mu F$, $C_3 = 20\mu F$, $C_4 = 20\mu F$.

8. Assume that the circuit in problem 4 is powered by a 60 Hz AC source instead of the DC source. Determine the total capacitive reactance, X_c, seen by the AC source.

9. Consider the series RL circuit shown in the diagram below. The source voltage is 12V, $R = 10\Omega$ and $L = 10mH$. The switch is closed at $t = 0$. What would be magnitude of current flowing through this circuit at $t = 2ms$?

10. Consider the series RL circuit given in problem 9, in discharge mode, with voltage source removed. Parameters such as R = 10Ω and L = 10mH, are the same. The switch has been closed for long period of time, such that the current has developed to the maxi-

mum or steady state level 1.04 A. How much time would need to elapse for the current to drop to 0.5 A after the switch is opened.

11. Determine the equivalent inductance L_{EQ} for three parallel inductor DC circuit shown in the diagram below if L_1 = 2mH, and L_2 = 5mH and L_3= 20mH.

12. Calculate the net or total inductance as seen from the 24V source vantage point in the circuit shown below.

13. Assume that the circuit in Problem 12 is powered by a 60 Hz AC source. Calculate the inductive reactance, X_L, as seen by the AC voltage source.

Chapter 2

DC Circuit Analysis and Basic Electronic Devices

Introduction

In this chapter, we will review three fundamental laws of physics and electrical engineering, namely, (1) Ohms Law, (2) Kirchhoff's Voltage Law and (3) Kirchhoff's Current Law. These three laws are instrumental in performing basic circuit analysis. Laws serve a vital purpose in electrical engineering analysis and problem solving; in DC as well as AC domains. Most engineering problems entail determination of unknown values of key parameters, under certain known circumstances or conditions. Laws allow us the opportunity to formulate equations that can be solved for values of unknown parameters. Solving for *one unknown* variable requires a minimum of *one equation*. Solution for determination of the values of *two unknown variables* requires *two equations*, and so on. Therefore, laws afford us the opportunity to model an engineering problem or scenario in form of set of equations that can be solved to adequately define and identify unknowns. Similar to the three basic laws of static equilibrium, i.e. $\Sigma F_x = 0$, $F_y = 0$, and $\Sigma M = 0$—laws that a civil or mechanical engineer might employ to solve for unknown forces in a static equilibrium beam or truss scenario—the Ohms law, the Kirchhoff's voltage law and the Kirchhoff's current law are used to solve for important electrical parameters like current, voltage, resistance, power, energy, reactance, impedance, reactance, etc. The later three parameters are AC entities; discussed in greater detail in Chapter 3.

In addition to the basic laws, important electrical circuit principles and circuit simplification techniques, such as, voltage division and current division will be discussed and their application will be illustrated through sample problems in this chapter. Note that there are many other principles and methods that are at electrical engineer's disposal to solve or analyze complex circuits. Such as, the conversion of **Y** load (or resistor) configuration to Delta load configuration, or vice and versa; conversion of complex circuits to Thevenin equivalent, Norton equivalent, two port networks, etc. However, these and many other advanced circuit analysis methods are outside the scope of this text.

We will introduce the reader to two basic electronic devices, namely, a diode and a transistor, their basic characteristics and some of their applications. We will conclude this chapter with a pictorial "tour" of a typical electronic printed circuit board to allow the reader an opportunity to gain a measure of familiarity with basic electronic devices.

OHMS LAW

Ohm's law was introduced, briefly, in Chapter 1. Ohm's law stipulates that voltage, or voltage drop, in a DC (or AC) circuit is equal to the product of current flowing in the circuit and the resistance (or impedance) in the electrical circuit. In other words, voltage or voltage drop in an electrical circuit is directly proportional to the resistance (or impedance) of the circuit and the current flowing through it. Ohm's law can, therefore, be stated mathematically as follows:

$$V = IR, \ I = \frac{V}{R}, \ or \ R = \frac{V}{I} \quad \text{for DC Circuits} \qquad \textbf{Eq. 2.1}$$

$$\vec{V} = \vec{I}\vec{Z}, \ \vec{I} = \frac{\vec{V}}{\vec{Z}}, \ or \ \vec{Z} = \frac{\vec{V}}{\vec{I}} \quad \text{for AC Circuits} \qquad \textbf{Eq. 2.2}$$

The first mathematical representation of Ohm's law, in form of Eq. 2.1, pertains to DC circuits, and in this statement of Ohm's law, all three parameters—**V, I** and **R**—are scaler. On the other hand, mathematical representation of Ohm's law, in form of Eq. 2.2 applies to AC circuits, where voltage "**V**" and current "**I**" are vectors or complex entities and **R**, as always, is scalar. Note that the symbols with "half-arrows" above them denote complex AC, or vector, entities. These half arrows are shown here to introduce the reader to this method for representing vector or complex entities. Later in this text, vector or complex entities will be represented mostly in bold fonts. As one examines the statement of Ohm's law in form of Equations 2.1 and 2.2, it becomes obvious that the Ohm's law can be interpreted and applied in several ways. This fundamental aspect of Ohm's law, its versatility and wide application will become more evident through various circuit analysis problems in this chapter, and others in this text, beginning with Example 2.1.

Example 2.1

The DC circuit shown below consists of a hybrid, parallel-series, network or resistors: $R_1 = 10 \ \Omega$, $R_2 = 5 \ \Omega$, $R_3 = 1 \ \Omega$, and $R_4 = 10 \ \Omega$. Calculate the following parameters in this circuit: (a) R_{eq} or R_{total} for the entire circuit. (b) The amount of current "i" flowing through resistor R_1.

Solution:

a) R_{eq} or R_{total} for the entire circuit:

$R_{eq} = R_1 + R_4 + (R_2 \cdot R_3)/(R_2 + R_3)$

$= 10\ \Omega + 10\ \Omega + (5/6\ \Omega) = 20.833\ \Omega$

b) The amount of current "i" flowing through resistor R_1:

By Ohms Law: $V = I \cdot R$, or, $I = V/R$.

Therefore, $i = 12V/R_{eq}$

$= 12V/20.833\ \Omega = 0.576$ Amps.

KIRCHHOFF'S VOLTAGE LAW (KVL)

Kirchhoff's Voltage Law stipulates that the algebraic sum of voltage drops around any closed path, within a circuit, is equal to the sum of voltages presented by all of the voltage sources. The mathematical representation of KVL is as follows:

$$\Sigma V_{Drops} = \Sigma V_{Source} \qquad\qquad \textbf{Eq. 2.3}$$

Kirchhoff's voltage law can also be stated as:

Sum of ALL voltages in a circuit loop = 0, or:

$$\Sigma V = 0 \qquad\qquad \textbf{Eq. 2.3a}$$

Some electrical engineers find the later representation of Kirchhoff's voltage law somewhat easier to apply when performing circuit analyses because, with this version, once the voltages and respective polarities have been identified, you simply sum up all the voltage values with appropriate signs, as observed, while going around the loop in the chosen direction. This importance of this approach, and alternative approaches, is illustrated through the Example 2.2 below.

Example 2.2—KVL

Variation of current in the circuit shown below needs to be studied as a function of the three resistors and the voltage source, V_s. Using the Ohm's Law and KVL, develop an equation that can be used to compute the value of current I for various values of R_1, R_2, R_3 and V_s.

Solution:

Similar to other engineering disciplines, in most electrical engineering problems, multiple methods can be employed to derive the solution. The suitability of one method over another depends on the known parameters and the complexity of the circuit.

Approach I:

Reduce or simplify the given circuit to a "net" voltage source and equivalent resistance R_{eq}. Since R_1, R_2 and R_3 are in series:

$$R_{eq} = R_1 + R_2 + R_3$$

Based on the assumption that V_s is indeed the source driving this circuit, by electrical convention, the current in this circuit would flow "out" of the positive terminal, or anode, of the voltage source. Hence, the current would flow in the clockwise direction as shown in the diagram, below.

By the same token, application of the electrical current convention to the voltage device, or load voltage, V_L would mean that it would *try* to set up current in the counter-clockwise direction. However, because we assumed that voltage source V_s is driving the net flow of current through the circuit, its dominance over V_L is implied, and the net voltage in the circuit would be:

$$V_{Net} = V_S - V_L$$

This results in the simplification of the circuit as depicted below:

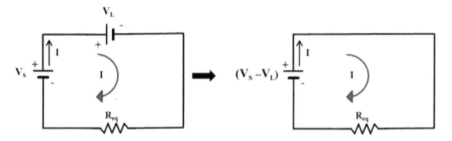

Then, application of Ohm's law yields:

$$I = \frac{(V_s - V_L)}{R_{eq}} \text{ or, } I = \frac{V_s - V_L}{R_1 + R_2 + R_3}$$

Approach II:

This approach is premised on the application of KVL to the given circuit after the circuit has been annotated with voltage designations, voltage polarities and current direction.

According to another electrical convention, voltage polarities are assigned such that the current enters the resistances (or loads, in general) on the *positive* side and exits from the *negative* side. The voltage sources, or existing "voltage loads" retain their stated polarities. The aforementioned steps result in the transformation of the original (given) circuit as shown on the following page.

Apply the Ohm's law to define the voltages, or voltage drops, across the three resistors.

$$V_{R1} = IR_1 \qquad V_{R2} = IR_2 \qquad V_{R3} = IR_3$$

With all voltages—voltage source, voltage load and voltage drops across the resistors—identified and their polarities noted, apply KVL by "walking" the annotated circuit beginning at the cathode or negative electrode of the voltage source, V_s. Note the voltages and respective polarities as you make a complete loop around the circuit in the clockwise direction of the current. This results in the following equation:

$$\Sigma V = 0$$
$$-V_s + V_{R1} + V_L + V_{R2} + V_{R3} = 0$$

Expansion of this KVL based equation through substitution of the resistor voltage drop formulas, derived earlier, yields:

$$-V_s + IR_1 + V_L + IR_2 + IR_3 = 0$$

Further rearrangement and simplification results in:

$$I(R_1 + R_2 + R_3) = (V_s - V_L)$$

$$I = \frac{V_s - V_L}{R_1 + R_2 + R_3}$$

KIRCHHOFF'S CURRENT LAW (KCL)

According to Kirchhoff's Current Law, total current flowing *into* a *node* is equal to the total current that flows *out* of the *node*. The mathematical representation of KCL is as follows:

$$\Sigma\, i_{in} = \Sigma\, i_{out}$$ **Eq. 2.4**

Proper identification of a ***node*** before application of KCL is pivotal in application of KCL. The concept of a node is not unique to electrical circuit. Intersections and junctions in fluid piping systems are sometimes referred to as nodes, by Mechanical Engineers and Technicians. The term finds its use in disciplines as diverse as human anatomical "lymph node" system and the algorithm nodes in computer systems. In the electrical realm, a node is sometimes construed as a point where two conductors merge or get connected. However, as illustrated through application of KCL in Example 2.3, a more meaningful definition of a node in electrical circuits is that it is a point where ***three or more conductors are electrically terminated or connected together***. Just as significance and effectiveness of KVL was illustrated through Example 2.2, we will demonstrate the utility of KCL, Kirchhoff's current law, and selection of a meaningful node in a "parallel" electrical circuit through Example 2.3 below.

Example 2.3

Determine the value of voltage source ***current*** in the parallel circuit below.

Solution:

Similar to Example 2.2, we will present two different approaches for determining the value of unknown source current. The first approach simply utilizes the Ohm's law and the parallel circuit simplification method. The second approach, on the other hand, utilizes KCL and "***nodal***" analysis technique.

Approach I:

Reduce or simplify the given circuit to a voltage source and equivalent resistance R_{eq}. Since R_1, R_2 and R_3 are in ***parallel***, application of Eq. 1.13 yields:

$$R_{eq} = \frac{1}{\dfrac{1}{R_1} + \dfrac{1}{R_2} + \dfrac{1}{R_3}}$$

$$R_{eq} = \frac{R_1 R_2 R_3}{R_1 R_2 + R_2 R_3 + R_1 R_3}$$

This simplifies the given parallel DC circuit as follows:

Next, Ohm's law is applied to determine the source current:

$$I = V_s \left(\frac{R_1 R_2 + R_2 R_3 + R_1 R_3}{R_1 R_2 R_3} \right)$$

Approach II:

This approach is premised on the application of KCL to the given circuit after the node had been identified, circuit has been annotated with voltage designation, voltage polarity, branch currents and current directions. See circuit diagram below:

Subscribing to the definition of a node as a point where three or more conductors merge, the shaded segment in the diagram above is designated as the node for this circuit. Next, before KCL can be applied to determine the source current, the individual currents, through each of the resistors, need to be defined—using the Ohm's law—in terms of the specific resistance values and the voltages around them:

$$I_1 = \frac{V_1}{R_1} \quad I_2 = \frac{V_2}{R_2} \quad I_3 = \frac{V_3}{R_3}$$

Then, application of KCL at the designated node yields the following equation:

$$I = I_1 + I_2 + I_3$$

Substitution of the values of branch currents, as defined earlier, yields:

$$I = \frac{V_1}{R_1} + \frac{V_2}{R_2} + \frac{V_3}{R_3}$$

At this juncture, it is important to note that when circuit elements are in parallel—as is the case with $\mathbf{R_1}$, $\mathbf{R_2}$ and $\mathbf{R_3}$—their voltages (or voltage drops around them) are equal. In fact, not only are the voltages around the parallel circuit elements equal to each other but they are the same as the source voltage, $\mathbf{V_s}$. In other words:

$$V_s = V_1 + V_2 + V_3$$

Therefore, the current equation can be rewritten as:

$$I = \frac{V_s}{R_1} + \frac{V_s}{R_2} + \frac{V_s}{R_3} = V_s \left(\frac{1}{R_1} + \frac{1}{R_2} + \frac{1}{R_3} \right)$$

And the source current would be:

$$I = V_s \left(\frac{R_1 R_2 + R_1 R_3 + R_2 R_3}{R_1 R_2 R_3} \right)$$

Which is the same as the answer derived through Approach 1.

VOLTAGE DIVISION

Voltage division is a shortcut for determination of voltage across a series resistor. According to the voltage division rule, the voltage across resistance **R**, in a DC circuit, with total resistance **R**$_{total}$, and a voltage source **V**, can be determined through the following formula:

$$V_R = \frac{R}{R_{total}} \cdot V$$

Eq. 2.5

For AC circuits, the voltage on impedance Z_i, in a loop with total impedance Z_{total}, with a voltage source **V**, would be:

$$V_i = \left(\frac{Z_i}{Z_{total}}\right).V$$

Eq. 2.6

Example 2.4

Determine the following for the DC circuit shown below:

a)   Equivalent resistance for the entire circuit, if $R_1 = 5\Omega$, $R_2 = R_3 = 10\Omega$, and $R_4 = R_5 = 20$

b)   Current flowing through resistor R_1

c)   Voltage across resistor R_5

Solution:

a)   Combination of R_2 and $R_4 = R_{2,4} = R_2 + R_4 = 10\Omega + 20\Omega = 30\Omega$
      Combination of R_3 and $R_5 = R_{3,5} = R_3 + R_5 = 10\Omega + 20\Omega = 30\Omega$
      Combination of $R_{2,4}$ and $R_{3,5} =$

$$R_{2-5} = \frac{(30\Omega).(30\Omega)}{(30\Omega + 30\Omega)} = \frac{900}{60} = 15\Omega$$

$R_{eq} = R_1 + R_{2-5} = 5\Omega + 15\Omega = 20\Omega$

b)   Current through R_1 would be the same as the current through the 12V supply:

$$I = \frac{V}{R_{eq}} = \frac{12V}{20\Omega} = 0.6A$$

c)   One method for determining $\mathbf{V_{R5}}$, voltage across $\mathbf{R_5}$, is to first calculate $\mathbf{V_{R2\text{-}5}}$, the voltage across the combined resistance of resistances $\mathbf{R_2}$, $\mathbf{R_3}$, $\mathbf{R_4}$, and $\mathbf{R_5}$. Then, by applying voltage division, calculate $\mathbf{V_{R5}}$ as follows:

According to Ohm's law,
$$\mathbf{V_{R2\text{-}5} = I.(R_{2\text{-}5}) = (0.6A).(15\Omega) = 9V}$$

Then, by applying the voltage division rule:

$$V_{R5} = (9V).(\frac{R_5}{R_5 + R_3})$$
$$= (9V).(\frac{20\Omega}{20\Omega + 10\Omega})$$
$$= (9V).(0.67) = 6V$$

Example 2.5
      What is the voltage across the 6Ω resistor?

Solution:

The right-hand side of the circuit, consisting of the two 10Ω resistors and the 5Ω resistor, is irrelevant insofar as the determination of voltage across the 6Ω resistor is concerned.

Parallel combination of the two 8 Ω resistors results in an equivalent resistance of 4 Ω as follows:

$$R_{8\Omega//8\Omega} = \frac{(8\Omega).(8\Omega)}{8\Omega + 8\Omega}$$

Using voltage division, the voltage across the 6 Ω resistor would be

$$V_{6\Omega} = (10V).(\frac{6\Omega}{4\Omega + 6\Omega})$$
$$= 6V$$

CURRENT DIVISION

The current through a resistor **R** in parallel, or in **shunt**, with another resistance **R**$_{parallel}$ and a current into the node of **I** is:

$$I_R = \left(\frac{R_{parallel}}{R_{total}}\right).I$$

Eq. 2.7

Where,

R$_{total}$ = The sum of the resistances in parallel (and **not** the parallel combination **R**$_{EQ}$).

R$_{parallel}$ = Resistance value of the resistor opposite the "subject" resistor.

I = Current through the source.

When current division is applied in **AC** circuits, the formula for current through an impedance **Z**, in parallel with another impedance $Z_{parallel}$, would be:

$$I_R = \left(\frac{Z_{parallel}}{Z_{total}}\right).I$$

Eq. 2.8

Where,

I = Current, in its complex AC form, flowing into the node formed by the parallel impedances.

Z_{total} = The sum of the impedances in parallel (and **not** the parallel combination Z_{EQ}).

$Z_{parallel}$ = The impedance of the load opposite to the subject impedance.

I = AC current through the source.

Example 2.6

Determine the current flowing through the 10Ω resistance in the circuit shown below.

Solution:

We must to determine the value of source current **I**, first. In order to determine the value of current **I** flowing through the source and the 5Ω resistor we must consolidate all resistors into an equivalent resistance R_{EQ} and then apply the Ohm's law.

$$R_{EQ} = 5\Omega + \left(\frac{(10\Omega).(20\Omega)}{(10\Omega)+(20\Omega)}\right) = 5\Omega + 6.67\Omega = 11.67\Omega$$

$$I = \frac{V}{R_{EQ}} = \frac{24V}{11.67\Omega} = 2.06 \text{ A}$$

Apply current division formula in form of **Eq. 2.7**

$$I_{10\Omega} = I_1 = \left(\frac{R_{parallel}}{R_{total}}\right) I = \left(\frac{20\Omega}{10\Omega + 20\Omega}\right) 2.06A = 1.36 \text{ A}$$

MULTI-LOOP CIRCUIT ANALYSIS

As expected, analyses of circuits that consist of more than one current loop, tend to be more complex, require formulation of multiple equations and utilization of additional conventions and principles. Similar to single loop circuit analysis covered earlier, in most cases, there are multiple strategies and approaches available for solving multi-loop circuits. We will illustrate one approach through Example 2.7.

Example 2.7

Using the KVL method, determine the current flows, I_1 and I_2 in the circuit below.

Solution:

Current I_1 and I_2 are two of the five unknown parameters in the circuit above. The other unknown parameters are, $V_{1\Omega}$, $V_{2\Omega}$ and $V_{4\Omega}$. One approach for determining the values of I_1 and I_2 would be to formulate two equations, using the Kirchhoff's Voltage Law, such that each equation includes the same two unknown variables, I_1 and I_2. Then, by applying the simultaneous equation technique to the two equation system, with two unknowns, we can determine the values of I_1 and I_2.

Application of KVL requires that voltage around each circuit element be defined in terms of known values and the unknown variables. In addition, the sign or polarity of each voltage must be assigned. See the diagram below.

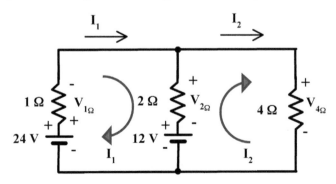

Rules for assuming the current direction and assigning voltage polarities to various loads (resistors) and sources (voltage sources) are as follows:

1) As shown in the figure above, by convention, the currents are assumed to be emanating from the positive pole (positive electrode or anode) of the voltage source and are assumed to be terminating into the negative pole (negative electrode or cathode) of the voltage source.

2) As shown in the figure above, the end or side of the resistor or load that the current enters from is labeled as positive.

3) The polarity for a voltage source is assumed as encountered in the direction of current flow.

Examination of the circuit above reveals that there are three loops in the given circuit. The left loop will be referred to as loop 1, the loop on the right segment of the circuit is loop 2. The third loop in this circuit is formed by the outer perimeter. We will focus on the first two loops to derive two equations for the determination of the two unknown currents. Assume that I_1 is greater than I_2. Conventionally, it is acceptable to make such assumptions as long as the assumptions are, strictly, adhered to in deriving all equations necessary for the solution.

"Walking" loop 1, beginning at the negative terminal of the 24 V dc source, yields the following equation:

$$-24V + V_{1\Omega} + V_{2\Omega} + 12V = 0 \qquad\qquad \textbf{Eq. 2.9}$$

"Walking" loop 2, beginning at the negative terminal of the 12 V dc source, yields the following equation:

$$-12V - V_{2\Omega} + V_{4\Omega} = 0 \qquad\qquad \textbf{Eq. 2.10}$$

Based on Ohm's Law:

$$V_{1\Omega} = (I_1).(1\Omega) = I_1 \qquad\qquad \textbf{Eq. 2.11}$$
$$V_{4\Omega} = (I_2).(4\Omega) = 4I_2 \qquad\qquad \textbf{Eq. 2.12}$$
$$V_{2\Omega} = (I_1 - I_2).(2\Omega) = 2.(I_1 - I_2) \qquad\qquad \textbf{Eq. 2.13}$$

Then, by substituting Eq. 2.11, Eq. 2.12 and Eq. 2.13 into Eq. 2.9 and Eq. 2.10, we get:

$$-24 + I_1 + 2.(I_1 - I_2) + 12 = 0$$
$$3I_1 - 2I_2 = 12 \qquad \qquad \textbf{Eq. 2.14}$$

$$-12 - 2.(I_1 - I_2) + 4I_2 = 0$$
$$-2I_1 + 6I_2 = 12 \qquad \qquad \textbf{Eq. 2.15}$$

Equations 2.14 and 2.15 represent the two simultaneous equations that were needed to solve for currents I_1 and I_2. These equations will be solved simultaneously to determine the values of I_1 and I_2.

$$3I_1 - 2I_2 = 12 \qquad \qquad \textbf{Eq. 2.14}$$
$$-2I_1 + 6I_2 = 12 \qquad \qquad \textbf{Eq. 2.15}$$

For simultaneous equation solution, multiply left hand side and the right hand side of Eq. 2.14 by 3 and add it to Eq. 2.15:

$$9I_1 - 6I_2 = 36$$
$$\underline{-2I_1 + 6I_2 = 12}$$
$$7I_1 = 48$$
$$\therefore I_1 = 6.86 \text{ Amps}$$

Then, by substituting this value of I_1 into Eq. 2.15 yields:
$$-2(6.86) + 6I_2 = 12$$

Or,

$$I_2 = 4.29 \text{ Amps}$$

Note: The values of unknown currents I_1 and I_2 can also be determined by applying Cramer's Rule to Eq. 2.14 and 2.15, in matrix format, and linear algebra.

Circuit analysis using Cramer's Rule and Linear Algebra

Cramer's Rule can be applied to solve for unknowns, in lieu of simultaneous equations or substitution methods, after set of equations have been formulated using the loop analysis method described above. Since the Cramer's Rule involves matrices and linear algebra, it is the method that electrical circuit analysis computer software is premised on. The application of Cramer's Rule is illustrated through Example 2.8 below.

Example 2.8

The values of all known parameters for the following multi-loop circuit are listed in the table below. Find the values of currents I_1, I_2, and I_3.

R_1	10Ω
R_2	2Ω
R_3	3Ω
R_4	4Ω
R_5	7Ω
R_6	3Ω
R_7	5Ω
V_1	20
V_2	5
V_3	12

Solution:

In this example, the fundamental strategy, conventions and principles needed to calculate the three unknown currents I_1, I_2 and I_3 would be the same as the ones described in Example 2.4. The obvious difference is that this circuit has a total of four loops—including the outermost loop—and has three unknown variables in form of currents I_1, I_2 and I_3. Therefore, we will need a minimum of three equations. Those three equations, as before, are derived by applying the Kirchhoff's Voltage Law to each of the three inner loops. The voltage drops across each of the load components (or resistors) are defined on the basis of the Ohm's law, $V = I.R$.

Before we embark on the formulation of current computation equations, let's ensure that the circuit is in its most simplified form. In that vane, by inspection, we notice that the two series resistors in the bottom loop, R_2 and R_7, can be added together or combined as follows:

$R_{2-7} = R_2 + R_7$

This simplification and assignment of voltage drop polarities results in the following circuit schematic:

Application of KVL in loop 1, with the assumptions that $I_1 > I_2$ and I_3, and $I_2 > I_3$ yields:

$$I_1R_1 - V_1 + (I_1 + I_2)R_5 + (I_1 - I_3)R_4 = 0$$
$$I_1R_1 - V_1 + I_1R_5 + I_2R_5 + I_1R_4 - I_3R_4 = 0$$
$$(R_1 + R_5 + R_4)I_1 + I_2R_5 - I_3R_4 = V_1 \qquad \text{Eq. 2.16}$$

Application of KVL in loop 2, maintaining the assumption that $I_1 > I_2$ and I_3, and $I_2 > I_3$ yields:

$$-V_2 + I_2R_3 + (I_1 + I_2)R_5 - V_3 + (I_2 + I_3)R_6 = 0$$
$$-V_2 + I_2R_3 + I_1R_5 + I_2R_5 - V_3 + I_2R_6 + I_3R_6 = 0$$
$$R_5I_1 + (R_3 + R_5 + R_6)I_2 + I_3R_6 = V_2 + V_3 \qquad \text{Eq. 2.17}$$

Application of KVL in loop 3, maintaining the assumption that $I_1 > I_2$ and I_3, and $I_2 > I_3$ yields:

$$I_3R_{2\text{-}7} - (I_1 - I_3)R_4 - V_3 + (I_2 + I_3)R_6 = 0$$
$$I_3R_{2\text{-}7} - I_1R_4 + I_3R_4 - V_3 + I_2R_6 + I_3R_6 = 0$$
$$-R_4I_1 + R_6I_2 + (R_4 + R_6 + R_{2\text{-}7})I_3 = V_3 \qquad \text{Eq. 2.18}$$

The three simultaneous equations thus derived are:

$$(R_1 + R_5 + R_4)I_1 + R_5I_2 - R_4I_3 = V_1 \qquad \text{Eq. 2.16}$$
$$R_5I_1 + (R_3 + R_5 + R6)I_2 + R_6I_3 = V_2 + V_3 \qquad \text{Eq. 2.17}$$
$$-R_4I_1 + R_6I_2 + (R_4 + R_6 + R_{2\text{-}7})I_3 = V_3 \qquad \text{Eq. 2.18}$$

Substitution of the given resistor and voltage source values into Eq. 2.16, Eq. 2.17, and Eq. 2.18 yields the following simultaneous equations:

$$21I_1 + 7I_2 - 4I_3 = 20$$
$$7I_1 + 13I_2 + 3I_3 = 17$$
$$-4I_1 + 3I_2 + 14I_3 = 12$$

Apply the Cramer's rule to solve for the three unknown currents I_1, I_2 and I_3. The augmented matrix thus developed would be:

$$\begin{vmatrix} 21 & 7 & -4 & \vdots & 20 \\ 7 & 13 & 3 & \vdots & 17 \\ -4 & 3 & 14 & \vdots & 12 \end{vmatrix}$$

The coefficient matrix, denoted as **A**, would be:

$$\begin{vmatrix} 21 & 7 & -4 \\ 7 & 13 & 3 \\ -4 & 3 & 14 \end{vmatrix}$$

The determinant of the coefficient matrix, denoted as $|A|$, would be:

$$|A| = 21\{(13\text{x}14) - (3\text{x}3)\} - 7\{(7\text{x}14) - (-4\text{x}3)\} - 4\{(7\text{x}3) - (-4\text{x}13)\} = \mathbf{2571}$$

The determinant of the substitutional matrix, A_1, for determining the value of I_1, is denoted as $|A_1|$, and

$$A_1 = \begin{vmatrix} 20 & 7 & -4 \\ 17 & 13 & 3 \\ 12 & 3 & 14 \end{vmatrix}$$

$$|A_1| = 20\{(13\text{x}14) - (3\text{x}3)\} - 7\{(17\text{x}14) - (12\text{x}3)\} - 4\{(17\text{x}3) - (12\text{x}13)\} = \mathbf{2466}$$

The determinant of the substitutional matrix, A_2, for determining the value of I_2, is denoted as $|A_2|$, and

$$A_2 = \begin{vmatrix} 21 & 20 & -4 \\ 7 & 17 & 3 \\ -4 & 12 & 14 \end{vmatrix}$$

$$|A_2| = 21\{(17\text{x}14) - (12\text{x}3)\} - 20\{(7\text{x}14) - (-4\text{x}3)\} - 4\{(7\text{x}12) - (-4\text{x}17)\} = \mathbf{1434}$$

The determinant of the substitutional matrix, A_3, for determining the value of I_3, is denoted as $|A_3|$, and

$$A_3 = \begin{vmatrix} 21 & 7 & 20 \\ 7 & 13 & 17 \\ -4 & 3 & 12 \end{vmatrix}$$

$$|A_3| = 21\{(13\times12)-(3\times17)\} - 7\{(7\times12)-(-4\times17)\} + 20\{(7\times3)-(-4\times13)\} = \mathbf{2601}$$

Applying the Cramer's rule, the unknown variables, currents I_1, I_2 and I_3, can be calculated by dividing the determinants of substitutional matrices A_1, A_2 and A_3, respectively, by the determinant of the coefficient matrix A.

Therefore,

$$I_1 = \frac{|A_1|}{|A|} = \frac{2466}{2571} = 0.959 \text{ A}$$

$$I_2 = \frac{|A_2|}{|A|} = \frac{1434}{2571} = 0.558 \text{ A}$$

$$I_3 = \frac{|A_3|}{|A|} = \frac{2601}{2571} = 1.012 \text{ A}$$

DIODES

A diode, unlike its "functional" predecessor—the vacuum tube rectifier—is constructed out of *semiconductor* materials such as silicone, germanium, gallium arsenide, etc. While a semiconductor is not a good conductor at room temperature, it doesn't fall distinctly in the category of insulators, such as, glass, ceramics, urethanes, plastics, PVC's, etc. A semiconductor can, however, be transformed into a "partially" or "selectively" conductive substance through a process called "**doping**." The term doping implies addition of "impurities" into a pure substance like silicone. If these impurities are added to create a region with a predominant concentration of negative charge carriers, or electrons, the end result would be the formation of a region called **n-type** semiconductor. At the same time, if impurities are added, adjacently, to create a region with predominant positive charge carriers, or holes, a **p-type** semiconductor is formed. The plane where the **p** and **n** doped materials interface with each other is called the **p-n junction**. A p-n junction is where the essential function of a diode takes place in response to the application of proper voltage.

A diode can be viewed as a device that functions as an electronic "check valve." As we know, the function of a check valve, in mechanical or hydraulic systems, is to permit the flow of fluids in one specific direction. In other words, an attempt by the fluid to move in the reverse direction is blocked by a check valve. A diode performs the same function in the flow of current. A diode permits the flow of current only from a higher voltage (or electrical potential) point in an electrical circuit to a lower voltage or ground potential point. This unidirectional behavior is called **rectification**, and this function of a diode finds a common application in the conversion of alternating current (AC) to direct current (DC). Common application of diodes in rectifiers and other equipment are discussed later in this chapter.

The symbol and drawing of common circuit board type diode are shown in Figure 2.1.

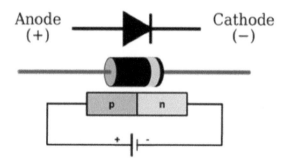

Figure 2.1: Symbol, diagram and schematic of a basic diode circuit.

The left side of the diode, labeled "Anode" is normally connected to the positive or higher voltage point in the circuit. The right side of the diode, labeled "cathode," on the other hand, is normally connected to the ground, negative, or lower potential point in the circuit. When a diode is connected in this manner, it is said to be *forward biased*. As apparent from the diagram of a typical diode in Figure 2.1, a band on one side of the diode denotes the cathode side of the diode. If, however, the voltage is reversed such that the anode of the diode is connected to the negative voltage potential and the cathode is connected to the positive potential, the diode is said to be *reverse biased.* The current response of a diode in forward and reverse bias modes of operation is depicted in Figure 2.2.

In Figure 2.2, V_D represents the forward bias voltage. Forward bias voltage of a diode is the voltage at which the diode begins to conduct current; or is the voltage where the diode is said to be "turned on." A diode can be perceived as a "self-actuating" electronic switch, as well. The forward bias voltage, V_D, is ap-

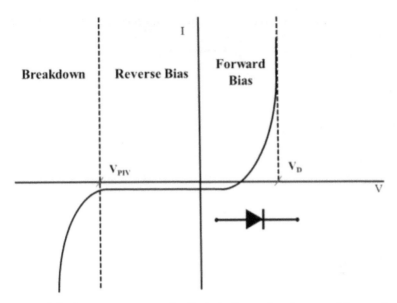

Figure 2.2: Current response of a diode in forward and reverse bias modes

proximately, **0.7V** for silicone type diodes and **0.3 volts** for germanium diodes.

As depicted in Figure 2.2, when a diode is reverse biased the amount of current (or reverse current) it conducts is negligible, until the magnitude of the reverse bias voltage approach approaches V_{BR}, breakdown voltage. The breakdown voltage is also called the V_{PIV}, Peak Inverse Voltage or V_{PRV}, Peak Reverse Voltage. The reverse voltage is considerably higher than the forward bias voltage; approximately *ten* times the magnitude of V_D. Although, technically, all diodes are capable of rectifying AC, the term "**rectifier**" is essentially reserved for diodes designed to operate at currents in excess of 1A, and therefore, utilized frequently to convert AC into DC. Conventionally, the term diode refers to applications involving currents less than or equal to 1A.

Special Purpose Diodes

Two of the most common, special purpose, diodes that find applications in mainstream electrical or electronic equipment, i.e. variable frequency drives, inverters or converters are SCRs and Zener diodes.

SCR: The acronym "SCR" stands for "silicone controlled rectifier," or "semiconductor controlled rectifier." These specialty diodes are also referred to as "thyristors." Schematic of an SCR is shown in Figure 2.3.

Figure 2.3: Symbol or diagram for an SCR or Thyristor

As apparent in Figure 2.3, an SCR differs from a regular diode, mainly, due to an added feature called the "gate." The gate serves as "trigger" or "firing" mechanism for an SCR. Specific voltage application at the gate triggers or "pulses" the SCR on, and allows the current to flow. The diode portion of the SCR continues to conduct after the gate voltage dissipates. The SCR, or diode portion of the SCR, stops conducting once the forward bias voltage drops below the threshold voltage, V_D.

SCRs are mainly used in devices associated with the control of high power and high voltage. Their innate characteristic and mode of operation makes them suitable for use in medium to high-voltage AC power control applications, such as lamp dimming, regulators and motor control. Thyristors are also commonly used for rectification of high power AC in high-voltage direct current power transmission applications. They are also used in the control of welding machines.

Zener Diode: Zener diodes are similar, in construction, to basic diodes. The key difference is that Zener diodes, unlike regular diodes, are capable of recovering from avalanche reverse bias break down mode when the reverse bias is removed. Therefore, as a mechanical analogy, one could compare zener diodes to "relief valves" that open under abnormally high pressure, and reinstate their normal blocking function once the pressure subsides to a nominal level. The Zener diode is named after Clarence Zener, who discovered its electrical property. The symbol or electronic representation of a Zener diode is depicted in Figure 2.4.

Figure 2.4: Symbol and diagram for Zener diode

A Zener diode allows current to flow in the forward direction in the same manner as an ideal diode, but will also permit it to flow in the reverse direction when the voltage is above the breakdown voltage. The breakdown voltage is also referred to as the "zener knee voltage," "zener voltage" or "avalanche point." Because of this basic characteristic of a Zener diode, it is commonly used to provide a reference voltage for voltage regulators, or to protect other semiconductor devices from momentary voltage pulses or excessive voltage "spikes." The current response of a Zener diode is shown in Figure 2.5.

Figure 2.5: Current response of a Zener diode

Common Applications of Diodes

Some common applications of diodes are as follows:
1) Half wave rectifier
2) Full wave rectifier
3) Clamping circuit
4) Base clipper
5) Peak clipper

Among the diode applications listed above, the two most common ones are (1) Half Wave Rectifier and (2) Full Wave Rectifier. These two applications are explored below.

HALF WAVE RECTIFIER

Half wave rectifier is a circuit consisting of two core components, a diode and resistor. See Figure 2.6(a). As shown in Figure 2.6(a), the sinusoidal AC waveform is applied to the input side of the diode-resistor circuit. The diode, as shown in Figure 2.6(b), is the core "active" component. A half wave rectifier circuit banks on the innate characteristic of a diode to allow current to flow only "one-way," in the forward biased direction. In other words, the diode permits the current to flow only during the positive half of the AC cycle. When the AC voltage "dives" into the negative realm—acting as a "check valve"—the diode shuts off the flow of current. This response of the diode is plotted graphically in Figure 2.6(c), in form of a series of positive wave crests; average of which, represents the DC voltage produced.

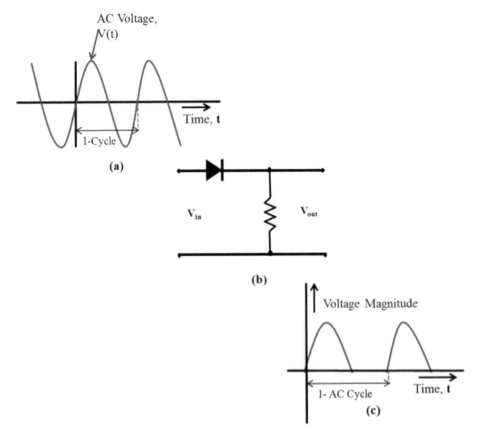

Figure 2.6: Half wave rectifier

Note, however, that half wave rectifiers only produce *one* positive DC crest per AC cycle. The DC output of a half wave rectifier can be computed through Eq. 2.19.

$$V_{dc} = \frac{V_p}{\pi}$$ **Eq. 2.19**

FULL WAVE RECTIFIER

Common full wave rectifier is a circuit consisting of four diodes and a resistor. See Figure 2.7. The sinusoidal AC waveform applied to the input side of the *four diode bridge* and resistor circuit is shown in Figure 2.7(a). The four diode bridge configuration shown in Figure 2.7(b) is the essential power conversion segment of the overall full wave bridge rectifier circuit. When AC voltage is applied between terminals **A** and **B** on the input side of the diode bridge, as V_{in}, the positive crest of the sinusoidal waveform drives current through terminal **A**, diode **CR₁**, resistor **R**, diode **CR₄** to terminal **B**. During this positive crest initiated flow of current, diodes **CR₁** and **CR₄** are forward biased. As the current assumes this course, it "drops" a positive "half" wave across the resistor, or terminals **C** and **D**. This positive DC crest is the first crest from the left, in Figure 2.7(c). During the positive half of the AC cycle, diodes **CR₂** and **CR₃** are reverse biased; therefore, they do not conduct, and *no* current flows through the **CR₂**, **R** and **CR₃** path.

During the negative half of the AC cycle, diodes **CR₁** and **CR₄** are reverse biased and no current flow through them. However, because terminal **A** is negative during the negative half of the AC cycle, diodes **CR₂** and **CR₃** are forward biased and the current flow is driven from terminal **B**, through diode **CR₃**, resistor **R**, diode **CR₂**. This path of current flow also results in a positive voltage drop, or positive voltage crest across terminals **C** and **D**. This positive DC crest is represented by the *second* crest in Figure 2.7(c). Therefore, *one* AC cycle on the V_{in} side results in *two* positive crests on the V_{out} side. The average value of the positive crests shown in Figure 2.7(c) represents the DC output of this full wave bridge rectifier circuit. Of course, the DC output in most DC power supplies is refined or corrected into a straight line form through application of resistors, capacitors and inductors. The formula for full wave rectified DC voltage is represented by Eq. 2.20.

$$V_{DC} = 2 \cdot \left(\frac{V_p}{\pi} \right)$$ **Eq. 2.20**

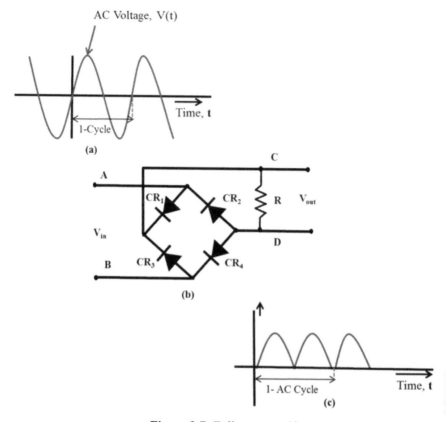

Figure 2.7: Full wave rectifier

TRANSISTORS

Similar to diodes, transistors are semiconductor devices. The approach to constructing transistors is similar to the approach used for fabrication of diodes. The "n and "p" doping approach employed with the construction of transistors is shown in Figure 2.8. As somewhat evident from Figure 2.8, a transistor—functionally, and from construction point of view—appears as set of two diodes connected "back to back." There are many types of transistors. In this text, we will focus on the type of transistors referred to as the "bipolar" or FET, field effect transistors. Within the bipolar junction transformer category, there are two subcategories: The "**npn**" and the "**pnp**" transistors. We will limit our discussion to the introduction of **npn** type transistors in this text.

Unlike the diode, an npn transistor consists of three "doped" zones: (a) An n-doped segment that known as the "collector," (b) A center p-doped segment

Figure 2.8: Bipolar (FET) junction, npn transistor

referred to as the "base," and (c) A second n-doped segment known as the "emitter." See bottom section of Figure 2.8 for the relative location and the construction of an npn transistor. The interface lines where the collector—base and the base to emitter regions meet are referred to as the n-p, collector to base junction, and the p-n, base to emitter junction, respectively. As shown in Figure 2.8, in a normal npn transistor application, the collector to base junction is reverse biased and the base to emitter junction is forward biased. As shown in Figure 2.8, common configuration and application of a transistor resembles a back to back connection of two diodes; the one on the left being reverse biased and the one on the right, forward biased.

The symbol used to represent a transistor in an electrical or electronic circuit is depicted in the top segment of Figure 2.8, with proper labeling of the three terminals of the transistors, in terms of collector, base and emitter.

Common application of FET, bipolar junction transistors

Application of a transistor in a typical "output module" of a PLC:
A transistor can be applied as an "electronic switch" to execute a PLC micro-processor command in the field. This is illustrated in form of a transistor based PLC output, operating a solenoid valve, as shown in Figure 2.9. Applying ample positive bias to the left of R_1 will turn "on" the npn transistor, causing a "short" to ground. This permits the current to flow through the solenoid coil. Flow of current through the solenoid coil generates magnetic flux through the core of the coil, thus applying magnetic force on the "plunger" or "pin" in the core of the solenoid coil. Depending on the design of the solenoid, the magnetic force can either open or close the solenoid valve.

Figure 2.9: Bipolar (FET) junction application in a PLC output module

ELECTRONIC DEVICE APPLICATIONS ON PRINTED CIRCUIT BOARDS
    This section is designed to provide a brief introduction to a common printed circuit (PCB) board and an assortment of common electronic devices

typically installed on a PCB to perform control, monitoring and data collection functions. Some of the electronic and electrical devices identified on this board are explained in greater detail in the preceding and subsequent chapters.

The devices pin pointed in Figure 2.10 are listed below, in clockwise order, beginning from the top right corner:

1. A toroid type ***inductor or coil*** applied for DC power refining and filtering purpose.

2. DC ***voltage regulator*** applied for voltage regulation purposes. See Chapter 3 for a comprehensive discussion on regulation.

3. Solid state logic gate ***integrated circuit*** micro-chip applied to perform logic, algorithm and computations for control purposes. This type of IC consists of logic gates such as OR gates, NOR gates, Exclusive OR gates, AND gates, NAND gates, Flip Flops, etc.

4. A typical, low voltage electrical ***fuse*** for protection of the board and up-stream power source against faults and shorts.

5. A ***transistor***, with collector, base and emitter pins visible to the right.

6. A ***diode***, labeled using the conventional "CR" prefix based nomenclature.

7. A typical *"**proprietary" integrated circuit (IC)*** device. While the functional specifications of such devices are made available to electronic control engineers for application purposes, the contents and design of such devices are, typically, kept confidential and are not published.

Figure 2.10: Electrical and electronic devices on a printed circuit board- I

Figure 2.11 depicts another segment of the same printed circuit control board. The devices noted in this segment of the control board, in clockwise order, beginning from the top right corner are:

1.   Power transistor, npn (or NPN); with collector current rating of 30A at 40V, capable of switching at a maximum frequency 200kHz. Maximum power rating of 150W DC. The black heat dissipating fins surrounding the power transistor are designed to radiate waste heat and to protect the transistor against overheating.

2.   A low wattage resistor with visible color coded bands identifying the resistance value and the tolerance specification of the resistor.

3.   Higher wattage ceramic resistor. Notice the remarkable difference in the physical size and construction of this higher wattage and higher operating temperature ceramic resistor and its lower wattage counterpart described above. The ceramic resistor in Figure 2.11 is encased in a ceramic enclosure to withstand higher temperature and heat.

4.   A capacitor.

5.   A row of LEDs designed to *annunciate* the *state* of various inputs *received* and outputs *transmitted* to electrical and electronic equipment in the field. Each of the output terminals represents a hardware based command from on-board controls out to the field.

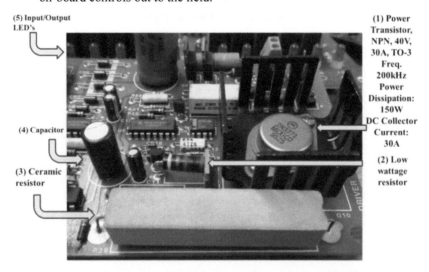

Figure 2.11: Electrical and electronic devices on a printed circuit board- II

An important feature that is instrumental in minimizing the time it takes to replace a control board, in case of board malfunction, is the "quick connect/ disconnect" type terminal strip—labeled as (1)—shown on the right side to the control board (PCB), as shown in Figure 2.12. Prior to the advent of this quick connect/disconnect type terminal strip, board replacement required meticulous examination and care in reading the electrical drawings, identification of each wire and terminal, and finally, exercise of proper craftsmanship in termination of wires to ensure correct and reliable electrical connections. The lack of such diligence resulted in miswiring, electrical faults, extensive troubleshooting period, prolonged commissioning time, etc.

When disconnecting a control board, the quick connect/disconnect terminal strip, shown in Figure 2.12, allows technicians and engineers to simply pull off the connector, with reasonable certainty that wires will remain securely intact for quick reinstallation. Since all wires normally remain terminated in the connector, once the new (replacement) board is physically secured in place, the connector is pushed/plugged onto the connecting edge of the new board. Note that the circuit boards and respective connectors are, typically, equipped with an interlocking feature to prevent incorrect orientation of the connector. As obvious, incorrect insertion of the connector onto the board can result in electrical faults and damage to the electrical devices.

The picture of the circuit board in Figure 2.12 also shows a set of two LED's, labeled as item (2). These LEDs serve as indicators of certain control conditions or signals.

Figure 2.12: Electrical and electronic devices on a printed circuit board—III

Three devices pointed out on the circuit board pictured in Figure 2.13 are as follows:

1.   A large size circuit board capacitor

2.   Higher wattage ceramic resistor. Notice the legend on the right side of the resistor. Typically, this legend includes the resistance value and power capacity specifications.

3.   Metal oxide varistor, rated 100J (100 Joules) and 1 kVA. Metal oxide varistor, also referred to as MOV, serves as stray energy arresting device on electronic circuits and electrical systems, in general. In this capacity, an MOV serves to absorb voltage spikes and rogue energy that might otherwise spread around various parts of a circuit board or electrical control system. Stray energy or voltage spikes, left unchecked, can damage integrated circuits, or IC semiconductor chips. ICs are relatively sensitive. They operate at low voltages, typically, at around 5 volts DC. The MOV in this particular case, as visible in the picture, is designed to absorb a maximum of 100 Joules of electrical energy. The "rate of absorption of this energy," or the power absorption capacity, is labeled on the MOV as 1 kVA; to be exact, the apparent power (S) rating of this MOV is 1,000 volt amperes.

Figure 2.13: Electrical and electronic devices on a printed circuit board- IV

Self-assessment Problems & Questions—Chapter 2

1. Determine the following for the DC circuit shown below if $R_1 = 5\Omega$, $R_2 = R_3 = 10\Omega$, and $R_4 = R_5 = 20\Omega$:
    a) Current flowing through resistor R_1
    b) Voltage across resistor R_5

2. What is the current through the 6 Ω resistor in the circuit shown below?

3. Find the current through the 0.5 Ω resistor in the circuit shown below.

4.   Determine the value of currents I_1, I_2 and I_3 in the circuit shown below if
      the voltage source V_3 fails in short circuit mode. The specifications of all
      components are listed in the table below:

R_1	10Ω
R_2	2Ω
R_3	3Ω
R_4	4Ω
R_5	7Ω
R_6	3Ω
R_7	5Ω
V_1	20
V_2	5
V_3	12

5.   Use current division to determine the value of current I_1 in the circuit be-
      low:

6.   Using Kirchhoff's voltage law, calculate the current circulating in the series resistor network below:

7.   Determine the value of voltage source *current* in the parallel circuit below using KCL, Kirchhoff's Current Law.
     Ancillary question: If one of the 5Ω resistors is removed (or replaced with an open circuit) and the other one is replaced with a short circuit, what would be the source current?

Chapter 3

Alternating Current (AC) Fundamentals

Introduction

This chapter begins with an introduction to AC, i.e. alternating current, and segues into a comparison between AC and DC. This comparison accentuates the complexity of AC as compared with DC due to the use of complex numbers and vectors for complete representation of AC entities and parameters. In this chapter, we get an opportunity to appreciate how the complexity of AC is amplified by three phase AC consideration in more common, industrial and commercial applications. In an effort to allow readers ample opportunity to explore and learn basic, single and three phase AC principles and mathematical computations, we introduce various single and three phase AC functions, equations and mathematical techniques; reinforced by examples and self-assessments problems. The depth to which the readers may endeavor to explore the topics in this chapter depends on their need, appetite and aptitude.

Due to the practical applications of the process of electrodeposition, in the coatings industry—and the innate relationship between electrodeposition and electricity—this process will be defined and illustrated through equations and practical numerical examples. This chapter provides the reader an introduction to AC electrical transformers of various types; single phase and three phase. Of course, similar to other topics in this text, the concepts, principles, equations and applications in this chapter are illustrated with, analogies, numerical examples and end of the chapter problems. All in all, this chapter provides an adequate introduction to alternating current and prepares the reader for intermediate level study of AC concepts.

ALTERNATING CURRENT (AC) VERSUS DIRECT CURRENT (DC)

The contrast between AC and DC isn't just rooted in physics but goes as far back as the 1800s when, for a period of time, both vied for the residential, commercial and industrial markets. Electricity was first discovered and harnessed, mostly, in the DC, or direct current form. However, due to the physical constraints associated with transmission, distribution and application of DC in industrial, commercial and residential applications, AC vanquished DC in the power distribution arena. Nevertheless, today, DC, holds its own in low voltage control and digital logic applications. In fact, albeit rare, DC is being reintroduced in the ultra-high voltage realm. The 900,000 Volt DC transmission line system project, currently underway in India, would serve as a more contemporary testimonial in that regard. This ultra-high DC voltage system is expected to redistribute surplus power from India's hydroelectric rich north east region to the energy starved, high growth, central region.

If we could attribute the acceptance of AC for residential, commercial and industrial applications to one pivotal entity, it would be the ***transformer*** and concept of ***voltage transformation***. In essence, when DC was first distributed and transmitted to end users, it was at a fixed voltage level because DC voltage transformation was not feasible. Due to resistance in the transmission lines—as stipulated by Ohm's Law: $\mathbf{V = IR}$—the DC power generators had to be located in the vicinity of the consumers to mitigate the undesirable voltage regulation issues. Longer transmission runs resulted in higher resistance, or "$\mathbf{R}$," which, in turn, resulted in larger voltage drop, "$\mathbf{V}$." In the mechanical fluids realm, this would be analogous to unacceptable pressure drop—due to frictional head losses—if fluid is distributed over extended distances without some sort of booster pumps.

Introduction of AC accompanied practical means for "stepping up" and "stepping down" of voltage through the use of transformers. So, AC could be generated at, for instance, 4,000 volts, stepped up to 100,000 volts—through the application of transformers—for transmission purposes, and miles away, it could be stepped down to usable levels, such as, 480 volt, 240 volts or 120 volts. That is, mostly, the way AC is generated, transmitted and distributed to consumers, today.

We learnt earlier that movement of electrons, or other charged particles, constitutes electrical current. When the charged particles move in one specific direction—such that there is a net displacement in their position, or when the charged particles travel a net distance—***direct current, or DC*** is said to exist. On the other hand, when the electrons or charged particles oscillate or vibrate about a point—similar to a pendulum—such motion of charged particles constitutes ***alternating current, or AC***.

If voltage and current were plotted on Cartesian Coordinates, the resulting graphs would be as depicted in Figures 3.1 and 3.2, respectively. In fact, if AC voltage, DC voltage and current were compared using an ***oscilloscope****, the

---

*****Oscilloscope*** is an electrical instrument used to graphically view and analyze voltage, current, frequency or power performance of electrical or electronic systems or circuits.

graph displayed on the screen would be similar to voltage and current graphs shown in Figures 3.1 and 3.2, respectively.

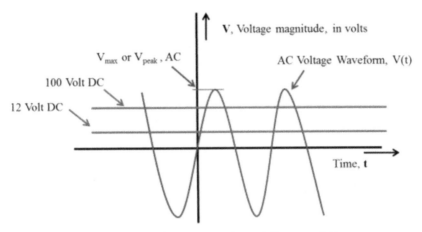

Figure 3.1: Voltage comparison, AC versus DC

As we examine Figure 3.1, we see that the two horizontal flat lines represent DC voltage, plotted on two dimensional X-Y coordinates. The lower flat line represents a typical 12 volt automotive battery. The upper flat line, labeled 100 Volt DC, represents the DC equivalent of 110 V_{RMS}, or 156 V_{peak}—i.e. the typical US electrical outlet voltage. The 100 Volt DC equivalent of 110 V_{RMS} applies to the DC power derived through full wave rectification of AC input power. The conversion of AC to DC and associated computations are illustrated through Example 3.1 and equations introduced in the electrodeposition section.

Further exploration of Figure 3.1 reveals that the x-axis (or abscissa) represents time, **t**, in seconds. The Y-axis (or ordinate) represents the voltage magnitude, **V**, in volts. The sine waveform, oscillating about the time axis, represents the AC voltage, varying as a function of time, **t**.

DC and AC *currents* can also be contrasted using graphical representations, in a manner similar to the DC and AC voltage comparison conducted above. The DC and AC *current* functions are depicted in Figure 3.2 in a form that is similar to their voltage counterparts. The horizontal flat line represents the DC current set up by virtue of the DC voltage of the DC power source; which can be a battery or a DC power supply. The Y-axis of this *current vs. time* plot represents the current magnitude, **I**, in Amps. The DC current, in this case, is assumed to be a constant 3A. The sine waveform, oscillating about the time axis, represents the AC current, I(t), varying as a function of time, t. The peak or maximum value of the AC current, in this case is assumed to be 10A. Note that in this illustration, the AC current appears to be surfacing into the positive current territory at time t_1. If one were to assume that this current was produced or driven by the voltage depicted in Figure 3.1, where the voltage broaches through the x-axis at time t = 0, one could say that the current is *lagging behind the voltage*. As we will explore later, in greater depth, in the power factor section, such a situation where current lags behind the voltage is said to cause a ***lagging power factor***.

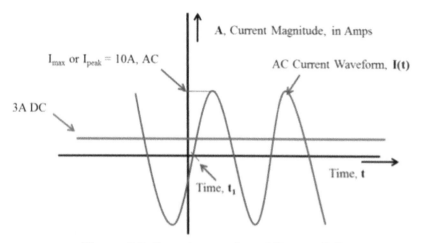

Figures 3.2: Current comparison, AC versus DC

ELECTRODEPOSITION—DC AND THE AVERAGE VALUE OF AC

Electrodeposition, as mentioned earlier, plays an important role in the coatings operations. Simply put, it involves the use of electrical charge for the deposition of paint pigments to metallic surfaces. When electrical charge is used to transport paint pigments to a metallic surface, the pigments tend to penetrate and adhere to the metal surface more strongly, uniformly and durably than if the surface were painted through the traditional spray or brush painting methods. Since electro deposition process involves net displacement of relocation of the pigments, DC electrical current and voltage have to be employed. AC voltage, due to the periodic oscillation or reversal of voltage in the positive and negative realms, would ***not*** deliver the net displacement of the pigment particles.

In electro deposition or electroplating scenarios, often the amount of electroplating required is specified in form of the amount of charge that is needed to be transferred in order to achieve a certain thickness of electroplating or electrodeposition. The basic equation that is useful in determining how ***long*** a known amount of DC current must be passed in order to achieve the required charge transfer is:

$$I_{DC} = \frac{q}{t}, \text{ or, } t = \frac{q}{I_{DC}} \qquad \textbf{Eq. 3.1}$$

Where, q = charge in coulombs of faradays, I_{DC} = DC current and t = time in seconds. Note that symbols "q" and "Q" may be used interchangeably to denote charge.

Conversion of AC voltage into DC voltage, accomplished through the full wave rectification, can be quantified using the following equation:

$$V_{DC} = 2 \cdot \left(\frac{V_{max}}{\pi}\right) \qquad \textbf{Eq. 3.2}$$

Where,

$$V_{max} = \sqrt{2} \cdot V_{rms} \qquad \textbf{Eq. 3.3}$$

In order to derive the DC current needed to compute the duration of electrodeposition, the effective resistance of the plating tank, with paint—or liquid pigment solution—is needed. This effective resistance can be measured. With the effective resistance known, the DC current can be calculated using the Ohm's law:

$$I_{DC} = \frac{V_{DC}}{R} \qquad \textbf{Eq. 3.4}$$

A common computation associated with application of AC and DC in the electrodeposition process is illustrated in Example 3.1.

Example 3.1
A plating tank with an effective resistance of 150 ohm is connected to the output of a full-wave rectifier. The AC supply voltage is $120V_{rms}$. Determine the amount of time it would take to perform 0.01 Faradays worth of electroplating?

Solution:
Background/Theory: The amount or coating or electroplating being accomplished in this example is specified in terms of the electrical charge that

must be moved or transferred; which in this case is specified in Faradays. **Faraday** is a unit for quantifying electrical charge, similar to **Coulomb**.

Or,

$$1 \text{ amp} = 1 \text{ Coulomb/sec}$$

$$1(A).(s) = 1 \text{ C}$$

And,

$$96,487 \text{ Coulomb} = 1 \text{ faraday}$$

Therefore, the amount of charge transfer, in Coulombs, in this electrodeposition case, would be:

$$q = \left(\frac{96,487 \text{ Coulombs}}{1 \text{ Faraday}} \right) . (0.01 \text{ Faraday}) = 964.87 \text{ Coulombs}$$

Since we are interested in the amount of time it takes to transfer a known amount of charge, rearrangement of Eq. 3.1 results in:

$$t = \frac{q}{I}$$

The next step entails determination of the DC current produced by the full wave rectification of 120 V_{max} AC. The net effect of full wave rectification of AC into DC is illustrated through the graphs depicted in Figures (a) and (b), below. The graph in Figure (a) shows the AC waveform on the **input** side of a full wave rectifier.

Figure (a)

The graph in Figure (b) shows the **DC output** side of a full wave rectifier; the net effect being the "flipping" or reversing of the negative troughs of the AC waveform such that the resulting output waveform has a definite *average value*. Note that the average of the sinusoidal AC waveform is *zero* because the AC voltage or current dwells the same amount of time in the positive realm—or above the x-axis—as it does in the negative realm, or below the x-axis. In other words, on the AC side, since the total **positive** area portended by the voltage or current graphs equals the corresponding **negative** areas on the **AC input** side of the rectifier, the average value is zero, and the DC content—voltage or current—on the AC input side is *zero*.

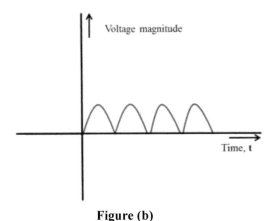

Figure (b)

Based on the data provided, we must first convert the given V_{rms} voltage into V_{max} using Eq. 3.3, and then compute V_{DC} by applying Eq. 3.2. As shown below, these two steps are followed by the application of Ohms Law, in form of Eq. 3.4, to compute the DC current, I_{DC}.

$$V_{max} = \sqrt{2} \cdot V_{rms} = (1.414).(120V) = 170 \text{ V}$$

Eq. 3.3

$$V_{DC} = 2 \cdot \left(\frac{V_{max}}{\pi}\right) = 2 \cdot \left(\frac{170}{3.14}\right) = 108 \text{ V}$$

Eq. 3.2

And,

$$I_{DC} = \frac{V_{DC}}{R} = \frac{V_{Ave}}{R} = \frac{108}{150} = 0.72A$$

Eq. 3.4

Then, application of Eq. 3.1 yields:

$$t = \frac{964.87 \text{ Coulombs}}{0.72 \text{ Coulombs/sec}} = 1340 \text{ seconds}$$

ALTERNATING CURRENT AND IMPEDANCE

As explained briefly in Chapter 1, impedance is the current opposing or current impeding characteristic of a load (or conductor) in an AC circuit. As implied in the definition of this term, impedance is an alternating current (AC) entity. While in DC circuits the factor that opposes the flow of current is resistance, the entity that influences the flow of AC current is impedance. Impedance, like AC current, voltage and power, is a vector entity and, therefore, it can be fully defined by two key characteristics, the ***magnitude*** and the ***direction***.

Mathematical and Geometric Representation of Complex AC Entities

Vector, complex number based entities, such as impedance, AC current and voltage can be expressed or quantified in the following forms:

a)   Polar or Phasor form
b)   Rectangular form
c)   Sinusoidal form
d)   Exponential form

POLAR OR PHASOR FORM

The terms polar and phasor are used synonymously. Phasor or polar representation of AC entities such as impedance, current, voltage and power requires definition of the magnitude of the respective entity and the direction in degrees (°). The polar/phasor representation of *AC **current*** would be represented as follows:

$$\mathbf{I} = I_m \angle \theta° \qquad\qquad\qquad \text{Eq. 3.5}$$

Where, $\mathbf{I}$ is the vector or complex value of AC current. The magnitude of the overall AC current, unless otherwise specified, is the maximum or peak value, and is denoted by $\mathbf{I_m}$. In the phasor representation of AC current above, $\mathbf{\theta°}$ is the angle of the AC current.

As an example of phasor/polar representation of AC current, consider a current $\mathbf{I} = 10\angle30°$ A. In this phasor representation of AC current, 10 represents the peak or maximum ***magnitude*** of AC current in amps, and 30° represents the ***angle*** of the current. This is illustrated in Figure 3.3. This perception or understanding of AC current, as a vector, is analogous to the role of a *force* vector in mechanical or civil engineering realms; such as, in the study of statics, where, the magnitude of a vector entity such as force may be defined in Newtons (N) or pounds force (lbf) and the direction of the force in degrees.

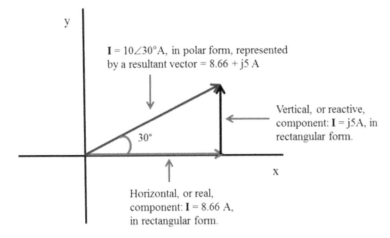

Figure 3.3: Vector representation of phasor value of AC current

Note that in most practical applications and computations, the root mean square value of current, or I_{rms}, is used instead of the maximum/peak magnitude of current I_{max}. For instance, the overcurrent protection devices, i.e., fuses, breakers—and circuit isolation apparatus like the disconnect switches—to name a few, are specified in rms terms. The same is true for AC voltages.

RECTANGULAR FORM

Rectangular representation of AC entities such as impedance, current, voltage and power entails numerical definition of those entities in form of horizontal and vertical, vector, components. An AC current of 10∠30° A rms, represented in the polar or phasor form, can be translated to the corresponding rectangular for as: 8.66 + j5 A rms; where the horizontal component of 8.66A represents the "real" component of the overall AC current and 5A represents the "imaginary" or "reactive" component of the overall AC current. This conversion of the AC current from its phasor to rectangular form can be accomplished through a scientific calculator or by performing trigonometric calculations. The trigonometric approach would involve the computation of horizontal and vertical components of the rectangular form, as follows:

$$I_{real} = 10Cos30° = 10(0.866) = 8.66 \text{ A}$$

And,

$$I_{reactive} = 10Sin30° = 10(0.5) = 5 \text{ A}$$

In the mechanical engineering or kinetics realm, the rectangular representation of AC current would be similar to the representation of a $10\angle30°$ mph velocity vector in form of its horizontal and vertical components as 8.66 mph and 5 mph, respectively.

SINUSOIDAL OR TRIGONOMETRIC FORM

The sinusoidal representation of AC parameters or entities involves the application of trigonometry. Sinusoidal representation of an AC entity, such as AC voltage, would be a follows:

$$V = V(t) = V_m Sin(\omega t + \theta) \qquad\qquad \textbf{Eq. 3.6}$$

Where, "**V**" and V(t) denote sinusoidal AC voltage, V_m is the maximum or peak voltage, ω represents the angular frequency, in rad/sec, and θ represents the angle of the AC voltage, in degrees. Figure 3.4, depicts this AC voltage function in the graphical form.

Figure 3.4 Graphical representation of sinusoidal voltage

Note that the AC current **I**, or **I(t),** would be represented by a graph, very similar to the voltage **V(t)** graph, with the exception of the fact that the **I(t)** wave form would be shifted to the left or right of the voltage wave form, depending upon whether the reactance of the AC load is predominantly inductive or capacitive. This is illustrated in Figure 3.5 by a sinusoidal voltage and current graph of a scenario with predominantly inductive load; where, the current is lagging the voltage.

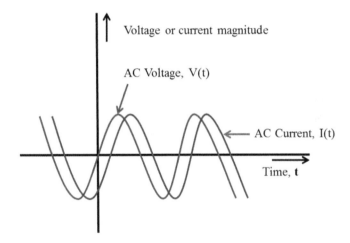

Figure 3.5 Graphical contrast between AC voltage and current wave forms

Continuing with our example of the $10\angle 30°$ A rms AC current, we can illustrate its sinusoidal AC form by converting the phasor AC rms current value into its sinusoidal form as:

$I(t) = I_{rms}$ Sin $(\omega t + \theta°)$ A-rms, in the general form, **Eq. 3.7**

And,

$I(t) = 10$ Sin $(377t + 30°)$ A-rms, in this specific case.

Since,

$$I_{max} = \left(\sqrt{2}\right)I_{rms} = (1.41).(10) = 14.1A$$

In peak or maximum value form, the sinusoidal representation would be as follows:

I(t) = 14.1 Sin (377t + 30°) A

Where, the coefficient 14.1 represents the ***maximum value*** of the current, in amps (A), and 30° represents the angle of the current. The 377 coefficient of time "**t**" represents, ω, the angular frequency of AC in radians per second. This value of 377 rad/sec is derived from the standard US alternating current source frequency of 60 Hz, or 60 cycles/sec. The AC frequency in Great Britain and some of the former British colonies is 50 Hz. The conversion from 60 Hz to 377 rad/sec can be explained mathematically as follows:

$$\omega = 2.\pi.f \qquad\qquad\qquad\qquad\qquad \textbf{Eq. 3.8}$$
$$= 2.(3.14).(60) = 377 \text{ rad/sec}$$

Exponential Form

The exponential representation of an AC parameter—whether it is current, voltage, impedance or power—is somewhat similar in form to the polar/phasor form. This is because the exponential form consists of the maximum value and the angle. For instance, the $14.1\angle 30°$amp AC current example we discussed earlier, would be represented in the exponential form as follows:

$$I = I_m e^{j\theta} \qquad\qquad\qquad\qquad\qquad \textbf{Eq. 3.9}$$
$$= 14.1 e^{j30} \text{ A}$$

Impedance Analysis

The concept of impedance was introduced in Chapter 1. In this section, we will cover mathematical and graphical representation and basic computational methods pertaining to impedance. Because impedance is a vector, it can be drawn or depicted as an arrow whose length represents the magnitude of the impedance while its orientation, expressed in degrees, represents the angle of the impedance. Since a vector can be analyzed or split into its horizontal and vertical components, impedance "**Z**" can be drawn in the vector diagram format as shown in Figure 3.6.

As introduced in Chapter 1, when an AC circuit is consists of resistance, capacitance and inductance, impedance **Z** can be stated mathematically in form of equations 3.10 and 3.11.

The unit for impedance is **ohm**, or Ω; similar to the unit for resistance, **R**, capacitive reactance, X_c, and inductive reactance, X_L. Consistency of units between **R**, X_L, X_C and **Z** is requisite for the following mathematical definitions for **Z** to hold true:

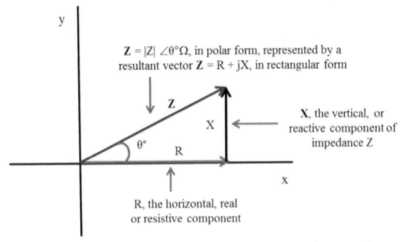

Figure 3.6: Graphical representation of impedance **Z** in vector form

$$\mathbf{Z} = \mathbf{R} + \mathbf{Z_l} + \mathbf{Z_c} \qquad\qquad\qquad \text{Eq. 3.10}$$

$$\mathbf{Z} = \mathbf{R} + \mathbf{jX_l} - \mathbf{jX_c} \qquad\qquad\qquad \text{Eq. 3.11}$$

It's obvious from examination of the above two equations that:

$$\mathbf{Z_l} = \mathbf{jX_l} \qquad\qquad\qquad \text{Eq. 3.12}$$

$$\mathbf{Z_c} = \mathbf{-jX_c} \qquad\qquad\qquad \text{Eq. 3.13}$$

Where, $\mathbf{Z_l}$ is the impedance contribution by the inductance in the AC circuit and $\mathbf{Z_c}$ is the impedance contribution by the capacitance in the AC circuit. The units for all impedances are **ohms,** or Ω.

Complete comprehension of AC circuit analysis and associated computations requires basic appreciation of how AC circuit components, such as resistances, capacitances and inductances, are converted into their respective impedance contributions before they are consolidated into an overall equivalent or combined impedance. In most cases, the derivation of other AC circuit parameters, such as current, voltage and power, is undertaken after determination of the total or equivalent impedance.

Example 3.2

Determine the equivalent, or total, impedance $\mathbf{Z_{Eq}}$ and the source rms current, **I,** in the AC series circuit below.

Solution:

According to Eq. 3.10 and Eq. 3.11:

$$\mathbf{Z = R + Z_l + Z_c}$$ 　　　　　　　　　　　　　　　　**Eq. 3.10**

Or, $\mathbf{Z = R + jX_l - jX_c}$ 　　　　　　　　　　　　　　**Eq. 3.11**

$$\therefore Z_{Eq} = R + \frac{1}{j\omega C} + j\omega.L$$

$$= R + \frac{1}{j(2).(\pi).(f).(C)} + j(2).(\pi).(f).(L)$$

$$= 4 + \frac{1}{j(2).(3.14).(60).(700 \times 10^{-6})} + j.(2).(3.14).(60).(3 \times 10^{-3})$$

$$= 4 + \frac{1}{j(0.2638)} + j.(1.1304)$$

$$= 4 - j(3.79) + j1.1304 = 4 - j2.66 = 4.8\angle{-33.63}\ \Omega$$

Current "**I**" calculation:

$$V(t) = V_{max}\ Sin(\omega t + \theta) = 156 Sin(377t + 30°) \qquad - \text{Given}$$

$$\therefore \left|V_P\right| = \left|V_{max}\right| = 156\ V$$

$$\text{And,}\ \left|V_{rms}\right| = \frac{156}{\sqrt{2}} = 110\ V$$

$$\therefore V(t), \text{ in rms form, would be} = V_{rms}\ Sin(\omega t + \theta) = 110 Sin(377t + 30°)$$

And, V_{rms} ,in polar or phasor form would be $= 110 \angle{30°}\ V_{rms}$

Then, according to Ohm's Law, $I = \dfrac{V}{Z_{Eq}} = \dfrac{110 \angle 30°}{4.8\angle\text{-}33.63}$

Or, $I = 23 \angle\{30°\text{-}(\text{-}33.63)\}$

$\therefore$ The source rms current, $I = 23 \angle 63.63°A$

Example 3.3

Determine the equivalent, or total, impedance $\mathbf{Z_{Eq}}$ and the **rms** value of the source current **I** in the AC parallel circuit below.

$V_{ac} = 156 \sin(377t + 30°)$ $R = 4\Omega$ $L = 3$ mH

Solution:

Solution strategy: Convert the given inductance value of $L = 3$ mH into its equivalent inductive reactance X_L, and subsequently into its impedance contribution Z_L. Then, combine the resistance R and Z_L values into the equivalent impedance Z_{Eq} by applying the formula for combination of parallel circuit elements.

$$X_L = \omega.L = (377).(3 \times 10^{-3}) = 1.13 \ \Omega$$
$$\therefore Z_L = j1.13 \ \Omega = 1.13\angle 90°\Omega$$

The parallel combination formula is:

$$\frac{1}{Z_{Eq}} = \frac{1}{R} + \frac{1}{Z_L}$$

Or, $Z_{Eq} = \dfrac{(R).(Z_L)}{R + Z_L} = \dfrac{(4).(1.13\angle 90°)}{4 + j1.13} = \dfrac{(4.52\angle 90°)}{4 + j1.13}$

$$Z_{Eq} = \frac{4.52\angle 90°}{4.16\angle 15.77} = 1.09\angle 74.22°\Omega$$

Current **"I"** calculation:

V(t), in rms form, $= 110 \angle 30° V_{rms}$ as determined in the previous example.

Then, according to Ohm's Law, $I = \dfrac{V}{Z_{Eq}} = \dfrac{110 \angle 30°}{1.09 \angle 74.23}$

∴ The source rms current, $I = 101 \angle -44° A$

TRANSFORMERS

Earlier in this chapter, we introduced the concept of transformers and the historically pivotal role transformers played in the predominant acceptance of AC over DC. In this section, we will explore the technical principles, applications and computations associated with single phase transformers. We will touch on some fundamental characteristics of three phase transformers, as well.

Let's begin by exploring the construction of a simple, single phase, AC transformer based on our first introduction of magnetic flux in Chapter 1. We will build upon the toroidal core example we introduced in Chapter 1 by adding an AC voltage source to the conductor coiled onto the left side of the core, as shown in Figure 3.7, and by winding N_s number of turns of a conductor on the right side of the same toroidal core. This converts the toroidal core into a basic transformer with N_p number of turns coiled on the AC source side and N_s turns on the side that is connected to the load.

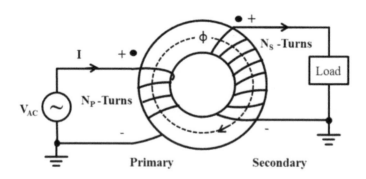

Figure 3.7

Primary and Secondary Side of a Transformer: The side or segment of the transformer that is connected to the power source is referred to as the *primary* side. The segment of the transformer that is connected to the load is called the *secondary* side. Note that the toroidal transformer depiction in Figure 3.7 above is simplified for voltage transformation principle illustration purpose. In actual, practical and commercial applications, the primary and secondary windings of a *toroidal core transformer* are concentric, with the core in the center as shown in the Figure 3.8 below.

Figure 3.8: Toroidal core transformer by Omegatron®

Even though we used a toroidal core in this basic transformer example, we could have used a more common rectangular or square core which would make this transformer resemble an ordinary transformer as shown in Figure 3.9 below.

Figure 3.9: Ideal transformer with complete primary and secondary circuits.

As we examine Figures 3.7 and 3.9, we notice that one of the voltage source (V_{AC}) terminals is grounded. This permits the other terminal to serve as the higher potential, or "hot," terminal. Hence, the current flows from left to right, or in the clockwise direction through the primary circuit of the transformer. Applying the right hand rule, as explained in Chapter 1, would result in a clockwise magnetic flux ϕ through the core. On the secondary side, through converse application of the right hand rule, the clockwise core flux ϕ results in the induction of current emerging from the top left side of the secondary circuit. This induced current flows in a clockwise fashion through the load as shown in Figure 3.9. Note that the direction of flow of AC current, on the primary or secondary sides of the transformer, signifies the ***direction of flow of AC power or AC energy***, from the source to the load.

The Dot Convention for Transformers: The dots shown, in Figures 3.7 and 3.9, on the primary and secondary sides of the transformer, indicate the ***direction of each winding relative to the others***. Voltages measured at the dot end of each winding are in phase. In other words, ***they have the same angle***. By convention, the current flowing into the ***dot end of a primary coil*** will result in current flowing out of the ***dot end of a secondary coil***.

Ideal vs. Real Transformers: A thorough and comprehensive study of "*real*" transformers is complex. Core and winding losses in most ***real*** transformers constitute a small percentage—less than 5%—of the total power transformed. Therefore, for simplicity, transformer circuit analyses are often conducted on the premise that they are "***ideal***" and "***lossless***." This means that the power fed into an ideal transformer, on the primary side, is equal to the power put out on the secondary side. In this text, we will limit the scope of our discussion to ideal transformers, with just a brief introduction to the concept of core and winding losses associated with real transformers.

An ideal transformer can be modeled as shown in Figure 3.10.

The two parallel vertical lines, in the middle of the ideal transformer diagram shown in Figure 3.10, represent the assumption that the core of an ideal transformer consists of a ferromagnetic or magnetic material.

Some of the formulas that govern relationships between various basic operational parameters of an ideal transformer are as follows:

$$\text{Turns Ratio} = a = \frac{N_P}{N_S} \qquad\qquad \textbf{Eq. 3.12}$$

$$V_s = \frac{V_p}{a} = \left(\frac{N_s}{N_p}\right) \cdot V_p \qquad\qquad \textbf{Eq. 3.13}$$

$$I_s = a.I_p = \left(\frac{N_p}{N_s}\right) \cdot I_p \qquad\qquad \textbf{Eq. 3.14}$$

Figure 3.10: Ideal transformer model

$$\left|S_{1-\Phi}\right| = V_s.I_s = V_p.I_p \qquad\qquad \textbf{Eq. 3.15}$$

Where,

N_p =Number of winding turns on the primary side of the transformer
N_s =Number of winding turns on the secondary side
V_p =Voltage applied on the primary side of the transformer
V_s =Voltage on the secondary side of the transformer
I_p =Current flowing through the primary side of the transformer
I_s =Current flowing through the secondary side of the transformer
$S_{1-\phi}$= Single phase AC power being transmitted by the transformer

Since turns ratio **a** represents a *ratio* of the primary turns to secondary turns, it is unit-less. Voltages and currents referred to, typically, in transformer circuit analysis and specifications, are in RMS form. Also, when transformer voltages and currents are measured using voltmeters and clamp-on ammeters, respectively, the resulting measurements are in RMS form. This is the reasons why many brands of voltmeters and ammeters are labeled as *True RMS Voltmeters* and *True RMS Ammeters*, respectively. In Eq. 3.15, $S_{1-\phi}$ represents the single phase AC power; or, to be more accurate, AC apparent power, *S*. Apparent power is discussed, in greater detail, in Chapter 4. Practical significance of the ideal transformer equations is illustrated through Example 3.4

Example 3.4

The primary of the transformer shown in Figure 3.9 is fed from a 120 VAC source. As apparent from the circuit diagram, the transformer has four (4) turns on the primary and eight (8) turns on the secondary. The current measured on the primary, using a clamp-on ammeter, is 2A. Determine the following un-

known parameters:
  (a) Turns ratio
  (b) Secondary current, I_s
  (c) Secondary voltage, V_s

Solution:

The scenario described in this example can be illustrated pictorially by annotating the circuit diagram Figure 3.9 as follows:

(a) According to ideal transformer equation Eq. 3.12,

$$\text{Turns Ratio} = a = \frac{N_P}{N_S}$$

As given, $N_p = 4$ and $N_s = 8$.

$$\therefore \text{Turns Ratio} = \left(\frac{4}{8}\right) = \frac{1}{2} \text{ or } 1:2$$

(b) According to ideal transformer equation Eq. 3.14,

$$I_s = a.I_p = \left(\frac{N_p}{N_s}\right).I_p$$

Since the turns ratio was computed as ½ in part (a), and primary current I_p is given as 2A in the problem statement,

$$I_s = \left(\frac{1}{2}\right).I_p = \left(\frac{1}{2}\right).(2A) = 1A$$

(c) According to ideal transformer equation Eq. 3.13,

$$V_s = \left(\frac{V_p}{a}\right)$$

Since the turns ratio was computed as 1:2 or ½ in part (a), and primary voltage V_p is given as 120V in the problem statement,

$$V_s = \left(\frac{V_p}{a}\right) = \left(\frac{120V}{1/2}\right) = 240V$$

In Example 3.4, the number of secondary turns, N_S, is twice the number of primary turns, N_p; resulting in the secondary voltage that is twice the magnitude of the primary voltage. This configuration is called a "***step up***" transformer configuration. In other words, the transformer is being used to boost or step up the primary voltage. Conversely, as shown in Figure 3.11, if the primary windings are greater than the secondary windings, the transformer would constitute a "step down" transformer. Such configuration is used to reduce incoming utility or source voltage to a desired lower voltage that is compatible with the driven load or equipment. See self-assessment problem 4, at the end of this chapter, for a practical illustration of a step down transformer scenario and associated analysis.

Figure 3.11: A Step down transformer configuration

The ideal transformer model shown in Figure 3.11 can be expanded to include impedance on the primary side of the transformer and the impedance of the load. If those two impedances are introduced into the discussion, as is the case in practical AC electrical systems, the AC transformer circuit would be represented by the circuits shown in Figures 3.12 (a) and (b).

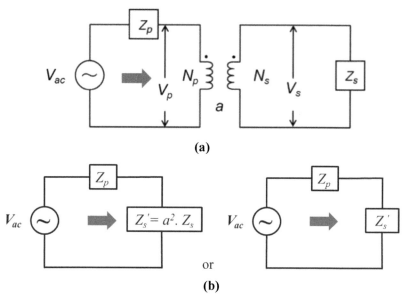

Figure 3.12: (a) Ideal transformer model with impedance consideration **(b)** Alternate representations of an ideal transformer equivalent circuit.

Note that the left to right arrows denote the flow of AC power, from the source to the load. The explanation of the nomenclature used in Figure 3.12 (a) and (b) is as follows:

V_{ac} =AC voltage source
V_p = Voltage on the primary side of the transformer
V_s = Voltage on the secondary side of the transformer
a = Turns ratio = N_p/N_s
Z_p = Impedance between the AC voltage source and the primary coil of the transformer.
Z_s = Impedance of the load connected to the secondary side of the transformer.
Z'_s = Impedance of the load as seen from the primary side of the transformer = $a^2.Z_s$

Impedances $\mathbf{Z_p}$ and $\mathbf{Z_s}$ do not represent the impedances of the transformer itself. In other words, $\mathbf{Z_p}$ represents the combined impedance of all circuit elements connected to the primary side of the transformer and $\mathbf{Z_s}$ represents the load impedance or the combined impedance of all circuit elements connected to the secondary side of the transformer. Simplification of the transformer circuit involved in the transition from Figure 3.12 (a) to Figure 3.12 (b) can be followed through the basic steps listed below:

1) Combine the impedances of all circuit elements connected between the primary side of the transformer and the AC voltage source. This combined impedance is represented by Zp.

2) Combine the impedances of all circuit elements connected between the secondary side of the transformer and the load. This combined impedance is represented by Zs.

3) The combined secondary impedance Zs is reflected to the primary side— or appears from the primary side of transformer—as Z's. The symbol Z's is read as "Z" sub "s" prime. The relationship between Zs, Z's and the turns ratio a is represented by the Eq. 3.16 below:

$$Z'_s = a^2 . Z_s \qquad\qquad \text{Eq. 3.16}$$

4) The original transformer circuit with numerous elements can then be shown in a simplified or condensed form—with only three elements: Vac, Zp and Z's—as shown in Figure 3.12 (b).

Example 3.5
Calculate the equivalent impedance as seen from the vantage point of the AC source V_{ac} in the circuit shown below. The transformer in the circuit is assumed to be ideal. The values of the primary and secondary circuit elements are: $X_{Lp} = 2\Omega$, $R_p = 4\Omega$, $R_s = 10\Omega$, $X_{Ls} = 5\Omega$, $N_p = 100$, $N_s = 200$.

Solution:
Note that, for the sake of simplicity, the primary and secondary inductors are given in form of their reactances "X_L" instead of actual inductance values,

in form of Henries. As explained earlier in this chapter, in the process of reducing the AC circuits to an equivalent impedance form, the individual inductive and capacitive reactances are combined in form of their respective impedance, or $\mathbf{Z}$, contributions. In other words, $\mathbf{X_L}$s are represented by $\mathbf{Z_L}$s and $\mathbf{X_C}$s are represented by $\mathbf{Z_C}$s.

Employing steps (1) through (4) from the section above results in the following computation and the equivalent circuit:

On the primary side: $\mathbf{Z_{Lp}} = j\mathbf{X_{Lp}} = j2\Omega$
    Therefore, $\mathbf{Z_P} = \mathbf{R_p} + \mathbf{Z_{Lp}} = 4 + j2\Omega$

On the secondary side: $\mathbf{Z_{Ls}} = j\mathbf{X_{Ls}} = j5\Omega$
    Therefore, $\mathbf{Z_s} = \mathbf{R_s} + \mathbf{Z_{Ls}} = 10 + j5\Omega$

a = Turns ratio = $\mathbf{N_p}/\mathbf{N_s}$ = 100/200 = 1/2
$\mathbf{Z'_s} = a^2.\mathbf{Z_s}$
$\mathbf{Z'_s} = (1/2)^2.\,(10 + j5\Omega) = 2.5 + j1.25\Omega$
$\mathbf{Z_{eq}} = \mathbf{Z_p} + \mathbf{Z'_s} = (4 + j2\Omega) + (2.5 + j1.25\Omega) = \mathbf{6.5 + j3.25\Omega}$

Or,

And, combination of the two impedances, as supported by the math above in:

The final equivalent impedance Z_{eq} can be represented in rectangular complex form as:

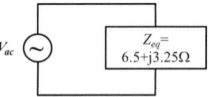

The Zeq derived in the rectangular for above can be stated in polar or phasor form as:

$$Z_{eq} = 6.5 + j3.25\Omega = 7.27\angle 26.6°$$

This conversion from rectangular to phasor form can be accomplished through a scientific calculator, with complex math feature or, as illustrated earlier in this chapter, through application of Pythagorean Theorem and trigonometry.

COMMON SINGLE PHASE AC TRANSFORMER APPLICATIONS

Three common single phase transformer configurations are described in this section, accompanied by their respective schematics.

Generic step-up or step-down single phase transformer application: Even though the single phase transformer schematic shown in Figure 3.13 represents greater number of turns on the primary side versus the secondary side—implying that it is a step down transformer—this transformer could be wired or connected in a reverse fashion, resulting in a step up voltage transformation function.

Figures 3.13: Generic step-down single phase transformer configuration

Step down voltage transformation—residential power distribution: Single phase transformer schematic shown in Figure 3.14 represents step down voltage transformation typically utilized in residential power distribution applications. As shown in the schematic, typically, 7200 volts—in phase (φ) to neutral three phase Y power supply configuration—are fed to the primary of a single phase power transformer. The secondary of the residential power distribution transformer is split or "tapped" into two segments, with the center tap connected to a neutral or ground. Such secondary configuration allows electricians to utilize both 120 volt, φ—neutral, segments for multiple 120 volt loads at a particular residence. On the other hand, for the 240 volt loads, such as clothes dryers, water heaters, HVAC systems and electric ranges, the electricians connect the line side of the respective load breakers across terminals X_1 and X_2, phase to phase.

As shown in Figure 3.14, the 7200 volts presented to the primary of this residential power distribution transformer are stepped down to the secondary voltage of 240 volts. Both terminals X1 and X2 are energized or "hot." That is why, if you examine a typical residential breaker panel, you will note that the breakers for loads such as HVAC systems, dryers, water heaters and ranges are installed in duplex or dual formation. In other words, two breakers are connected per circuit; one on each energized phase. See the breaker interlocking bar—mechanically interconnecting the two breakers together for simultaneous operation—in Figure 3.15. This interlocking or duplexing feature permits simultaneous breaking/operation of both energized phases when the breaker trips or is turned off. The schematic segment of Figure 3.15 shows the electrical interface and location of the breakers in two phase systems.

Figures 3.14: Step down, residential power distribution transformer

Figures 3.15: Step down, residential power distribution transformer with load

Load to source, power isolation transformer: In most industrial and many commercial facilities, equipment such as arc welders, intermittent motor loads and other types of switching loads tend to feedback electric "spikes" or "noise" to the supply line or supply bus. Such undesirable noise and spikes on the supply line or feeder are subsequently seen by all loads sharing the same supply line. Such switching of electromagnetic circuits also results in generation of electromagnetic radiation. This electromagnetic radiation, sometimes referred to as EMI, or electromagnetic interference, can have an adverse effect on sensitive electronic equipment such as computers, TVs, audio/video stereo systems, VFDs or variable frequency drives, PLCs (Programmable Logic Controllers), PCs (Personal Computers) and other types of electronic equipment. The electrical noise on the utility or power supply lines and EMI often tend to cause trips, faults and computational errors in sensitive electronic equipment.

There are various methods for attenuating or eliminating the electrical noise on the power supply lines and power distribution systems. Some such approaches include the application of capacitors and electrical filters that serve as "sponges" or noise absorbers. Another common method for protecting against electrical noise consists of an isolation transformer. See Figure 3.16 for a schematic of a typical, single phase isolation transformer. A distinguishing characteristic of an isolation transformer is that it's primary to secondary winding turns ration is 1:1. In other words, in isolation transformers, the number of turns on the primary side equals the number of turns on the secondary side. So, the primary and secondary voltages are the same. The net gain is the "cleansing" of voltage or current achieved through conversion electrical current to magnetic flux ϕ on the primary side and subsequent conversion of the flux to electrical AC current in the load or secondary side of the transformer. This flux to current and current to flux conversion of depicted in Figure 3.11.

Isolation
Transformer

Figures 3.16: Load to source, 1:1, power isolation transformer.

AUTOTRANSFORMERS

An autotransformer is a type of transformer that requires fewer windings and a smaller core. Therefore, it offers an economical and lighter means for converting voltage from one level to another as compared with its, regular, two winding counterparts. Practical, cost effective, application of an autotransformer is, typically, limited to voltage ratio of about 3:1.

Figure 3.17: Autotransformer

A typical autotransformer has one winding and four electrical connections, referred to as taps. In autotransformers, a portion of the winding serves as the primary as well as the secondary. See Figure 3.17 where the winding spanning from terminals **X1** to **X2** serves as the primary as well as the secondary.

VOLTAGE REGULATION, VOLTAGE REGULATORS AND BUCK BOOST TRANSFORMERS

Envision a scenario where you own and manage a small factory that operates on an 8 to 5 day shift. Assume that all power is turned off at the main breaker at the end of the day. As the first person to arrive at the facility in the morning, in an effort to assess the power quality of your facility, you measure the voltage at the main breaker before any electrical equipment is turned on. Let's assume that the voltage you measure is 480 V_{RMS}. If you measure the voltage at that main breaker, once again, after turning on all of the routine load, as well as, your production machinery, you will most likely read some lower voltage, say 460 V_{RMS}. This reduction in voltage, as load is added to the power distribution system, when quantified as a percentage, is referred to as voltage regulation. As the load on a power distribution system is increased, the current being pulled or demanded, from the source, by the load, increases. Since the ultimate mechanical source of electrical energy has definite or limited capacity, or limited and relatively constant power, P, as the demand for current **I** rises, the voltage sags, proportionally. This can be explained on the basis of Eq. 3.17 and 3.18.

For DC systems, P = V.I, or,

$$I = \frac{P}{V} \quad \text{or,} \quad V = \frac{P}{I} \qquad\qquad \textbf{Eq. 3.17}$$

In AC systems, **S,** AC Apparent Power = **V.I,** or,

$$|I| = \frac{|S|}{|V|} = |V| = \frac{|S|}{|I|} \qquad\qquad \textbf{Eq. 3.18}$$

As more current is demanded by loads, more current flows through DC supplies and AC power transformers. This results in a voltage drop on the load side of the DC power supply or AC power transformer, as the load approaches the maximum power capacity of the system. For some voltage regulation can be best understood and applied in its mathematical form as stated in Eq. 3.19.

$$\text{Voltage Regulation in \%} = \left[\frac{V_{NL} - V_{FL}}{V_{FL}} \right] \times 100 \qquad \textbf{Eq. 3.19}$$

Where,
V_{NL} = voltage measured at the source under no load condition
V_{FL} = voltage measured at the source under full load condition

Note that Eq. 3.19 may appear in a slightly different form in some texts, as stated in Eq. 3.20.

$$\text{Voltage Regulation in \%} = \left[\frac{V_{NL} - V_{Rated}}{V_{Rated}} \right] \times 100 \qquad \textbf{Eq. 3.20}$$

In Eq. 3.20, V_{Rated} is considered synonymous with the full load voltage, V_{FL}. In other words, the rated voltage is assumed to be the voltage when the system is operating at full load.

As apparent from Eq. 3.19, in an ideal "utopian" scenario, voltage regulation would approach "0." In other words, in an ideal situation, the voltage magnitude at the source would stay constant as the load escalates from no load to rated or full load level. A source that offers a low or negligible regulation is sometimes referred to, by electricians or electrical engineers, as a "*stiff*" source. In a typical power distribution system, a voltage regulation of 1% would be considered as acceptable over input voltage variations of +10% to –20%. Voltage regulation tends to be larger (or worse) in inductive load systems as compared with resistive load systems. Power distribution systems with larger voltage regulation are sometimes referred to as exhibiting '*loose*' voltage regulation.

Example 3.6

Assume that the second (rated or full load) voltage measured in the voltage regulation scenario captured in the first paragraph of this section is 460 V_{RMS}. Determine the voltage regulation.

Solution:

V_{NL} = 480 V, given
V_{FL} = 460 V, given

Apply Eq. 3.9:

$$\text{Voltage Regulation in \%} = \left[\frac{V_{NL} - V_{FL}}{V_{FL}}\right] \times 100$$

$$\text{Voltage Regulation in \%} = \left[\frac{480 - 460}{460}\right] \times 100 = 4.35\%$$

Loose or unacceptable voltage regulation can be remedied through various means, three of such methods are listed below:

1) Application of buck-boost voltage regulating transformers
2) Application of voltage regulating transformers
3) Application of ferroresonant, magnetic saturation based, transformers

We will elaborate on the more common buck-boost voltage regulating transformer in this text. A schematic diagram of a buck-boost voltage regulating transformer is shown in Figure 3.18. This type of voltage regulator is built around an auto-transformer. It has a movable contact that makes electrical connection with available coil taps.

A buck-boost autotransformer type voltage regulator can raise voltage (boost) or lower voltage (buck) as Necessary. These regulators are, typically, rated +/- 10%; meaning that they are capable of maintaining the output (or secondary) voltage within +/- 1% in response to +/- 10% variation of the input (or primary) voltage.

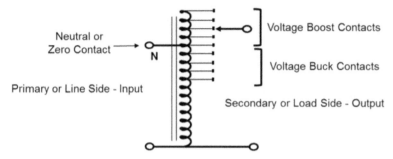

Figures 3.18: Buck-boost voltage regulating autotransformers

THREE PHASE AC

While single phase AC is based on one energized phase (or conductor) and a ground or neutral, three phase AC systems consist of three energized phases (or conductors). Some three phase AC systems consist of three energized conductors and a grounded neutral. Note that electrical engineers and electricians often refer to the energized conductors as "*hot*" conductors. When measuring voltages in single phase or three phase systems, the energized or "hot" terminal or conductor is touched with the red (anode) probe of the voltmeter and the black (cathode) probe is connected to the neutral or ground terminal. See Figure 3.19 below for a contrast between a single phase system and a three phase system. The single phase circuit, shown in Figure 3.19 (a), is powered by a 120 volt AC source. This single phase AC source is assumed to have an angle of 0°. Therefore, complete vector or phasor representation of the source voltage would be 120∠0° VAC. The load in this single phase AC circuit is represented by impedance "**Z**."

On the other hand, Figure 3.19 (b) represents a three phase AC circuit, with a three phase AC source supplying power to three phase AC load. Note that the three phases in the three phase AC system are identified as phases A, B and C. The three phase load consists of: (a) impedance Z_A, connected between phase A and the neutral, (b) impedance Z_B, connected between phase B and the neutral, and (c) impedance Z_C, connected between phase C and the neutral. In many common three phase AC load scenarios—such as three phase motors— the impedance values for each of the three phases are equal. When the three phases present equal impedance, the three phase load is said to be balanced. In other words:

$$Z_A = Z_B = Z_C = Z$$

Unequal phase impedances result in an imbalanced load condition, unequal line currents, and in some cases, unequal line voltages. Under most normal circumstances, the neutral current I_N in Y-loads is negligible. However, under unbalanced load conditions, I_N is non-zero. Therefore, neutral line current I_N is often monitored by installing a current transformer (CT) on the neutral leg of the three phase system. See additional discussion on current transformers and their application in the next section. Due to the fact that the line fuses or breakers protecting the three phase systems are sized typically with the assumption of balanced phase loading, when loads phases become unbalanced, currents in one or more of the three phases can exceed the fuse or breaker trip threshold; thus, resulting in the clearing or "blowing" of the fuse or tripping of the breaker.

Figure 3.19: (a) Single phase AC system. **(b)** Three phase AC system, Y-configuration.

THREE PHASE AC SYSTEMS VS. SINGLE PHASE AC SYSTEMS

The advantages of three phase AC systems over their single phase AC counterparts are as follows:

a) Three phase circuits and power distribution systems are more efficient than single phase systems. In other words, three phase power distribution systems can deliver the same magnitude of power with fewer and smaller conductors.

b) Unlike single phase induction motors, three phase AC induction motors do not require additional starting windings.

c) Unlike single phase induction motors, three phase AC induction motors provide uniform torque. Single phase motors often tend to deliver pulsating torque.

d) When rectifying AC current to DC current, three phase AC rectification yields smoother, relatively ripple free, DC.

CURRENT TRANSFORMERS

Current transformers are transformers that are used to measure or sample currents in AC electrical circuits. In measurement applications, a current transformer, abbreviated as "CT," is connected to metering instrumentation to display the magnitude of AC current through an AC ammeter or some other form of annunciation, such as a computer based HMI, Human Machine Interface, system. As shown in Figure 3.20 below, the conductor whose current is to be measured, is routed through the CT. When current, **I**, flows through the conductor, it produces magnetic flux—with magnetic flux density **B**—in the cylindrical core of the CT. This magnetic flux, as explained in Chapter 2, initiates the flow of current through the windings of the CT. The current I_{CT} induced in the windings of the CT is a proportional representation of the current **I** flowing through the conductor being monitored. The proportional relationship between **I** and I_{CT} can be represented, mathematically, by Eq. 3.21.

$$I_{CT} = \frac{I}{N} \qquad\qquad \textbf{Eq. 3.21}$$

Current transformer

Magnetic flux density in Wb/m²

Ammeter

Coil with N-turns

Current carrying conductor

Figure 3.20: Current transformer operating principle. By: Beizl, April 15, 2009.

When CTs are used to measure phase or line currents in three phase AC systems, they are installed in a fashion depicted in Figure 3.21. The CT installation photographed and shown in Figure 3.21 pertains to a situation where only the three energized phases need to be metered; a CT on the neutral is, therefore, not needed.

Current carrying three phases

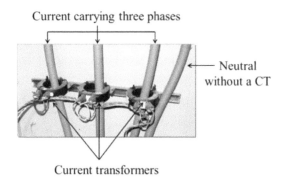

Neutral
without a CT

Current transformers

Figure 3.21: Current transformers used as part of metering equipment for three phase 400Amp electricity. By: Ali, December 13, 2004.

While majority of CT applications pertain to metering or display of AC current, in form or another, CTs are often applied for control purposes. When CTs are included in electrical power distribution systems for control purposes, they are terminated at—or connected to—the input/output (I/O) blocks of a control system, such as, a Programmable Logic Controller (PLC) or a Direct Digital Control (DDC) System. In such control applications, the current induced in the CT windings is often scaled or transduced to a low voltage signal at the I/O blocks. In many control systems, the PLC or DDCs CPU (Central Processing Unit) make control decisions on the basis of the scaled signal, available at the CT I/O block, by continuous monitoring and comparison of the real time current signal against "**High**" or "**Low**" set points. These "**High**" or "**Low**" set points are embedded in the program or algorithm of the PLC or DDC; where, control programs, typically, cycle at the rate of approximately once per 20 milliseconds. The scan rates are a function of the complexity of the program and the number of inputs and outputs included in the architecture of the overall control system. Tripping of breakers, shutting down of equipment, turning on of alarms and horns are a few examples of controls functions executed on the basis of current sensed by CTs. Schematics in Figures 3.22 (a) and (b) show the distinction between CTs being applied in single phase AC scenarios versus three phase AC scenarios.

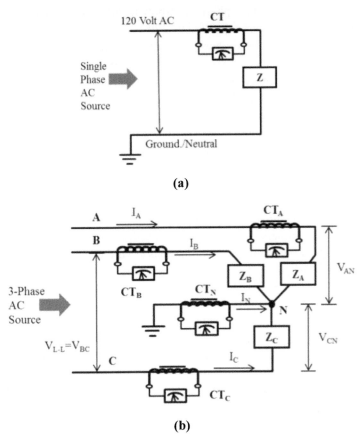

(a)

(b)

Figure 3.22: (a) Current transformer applied to a single phase AC circuit, for metering purposes. (b) Current transformers applied to a three phase AC circuit, for metering purposes.

THREE PHASE AC TRANSFORMERS

Three phase transformers, like the basic single phase transformers, serve the purpose of transforming voltage from one magnitude to another. Like single phase transformers, three phase transformers can be used as isolation transformers, where the primary and secondary voltages are the same and the main purpose of the transformer is to insulate and isolate a sensitive load from electrical noise and voltage fluctuations prevalent on the AC supply line. Three phase transformers' intrinsic and operational parameters, such as impedance, voltage and current—similar to their single phase counterparts—are represented by complex numbers and vectors. Upon close examination of Figure 3.19

(b) it is apparent that in three phase transformers, the voltages and currents for each of the three phases are separated by 120°.

Three identical single-phase transformers can be connected to form a 3-phase bank. Primary and secondary sides of such a bank of three phase transformers can be connected in Star (Y)-Delta (Δ), Star (Y)-Star (Y) or Delta (Δ)-Delta (Δ) combinations. Various possible three phase transformer configurations are shown in Figures 3.23 through 3.30.

When determining the voltages and currents in a 3-phase transformer, one must employ the line and phase relationship of star or delta connections along with the ratio of transformation between the coupled windings as explained in the various transformer configurations below.

Three Phase Δ-Δ Transformer Configuration

Most transformers, unless otherwise specified, are delivered to customers or installation sites without final terminations or "pre-configuration;" meaning, unless otherwise specified, they are not prewired for Δ-Δ, Δ-Y, Y-Δ, or Y-Y application. One could view an "un-configured" three phase transformer as a combination of three single phase transformers. As such, a set of three individual single phase transformers could be wired or configured as Δ-Δ, Δ-Y, Y-Δ, or Y-Y. If three individual transformers are connected together as depicted in Figure 3.23 (a), the resulting configuration would be a Δ-Δ three phase transformer. Where the points labeled X1 through X6 and H1 through H3 are terminals, wire termination points, or physical connection points. The schematic version of this Δ-Δ transformer would be represented by Figure 3.23 (b). The schematic, as usual, conveys the functional information about the transformer. In other words, the schematic illustrates how the voltage and current get transformed between the primary and secondary of the transformer.

(a)

(b)

Figure 3.23: Δ-Δ, or Delta-Delta, 3-phase transformer configuration. (a) wiring diagram, (b) schematic

As shown in Figure 3.23 (b), key mathematical relationships between the phase voltages and line voltages, and phase currents and line currents are as follows:

$$V_{P\text{-}Pri} = V_{L\text{-}Pri} = V \qquad\qquad \text{Eq. 3.22}$$

$$V_{P\text{-}Sec} = V_{L\text{-}Sec} \qquad\qquad \text{Eq. 3.23}$$

$$I_{P\text{-}Pri} = \frac{I_{L\text{-}Pri}}{\sqrt{3}} = \frac{I}{\sqrt{3}} \qquad\qquad \text{Eq. 3.24}$$

$$I_{L\text{-}Sec} = \sqrt{3}.I_{P\text{-}Sec} = a.I \qquad\qquad \text{Eq. 3.25}$$

Where,

$V_{P\text{-}Pri}$ = Primary phase voltage = **V**
$V_{L\text{-}Pri}$ = Primary line voltage = **V**
$V_{P\text{-}Sec}$ = Secondary phase voltage
$V_{L\text{-}Sec}$ = Secondary line voltage

I = Primary line current
$I_{P\text{-}Pri}$ = Primary phase current
$I_{P\text{-}Sec}$ = Secondary phase current
$I_{L\text{-}Pri}$ = Primary line current = **I**
$I_{L\text{-}Sec}$ = Secondary line current

For the sake of simplicity, and in congruence with the context and scope of this text, we will limit the discussion in this three phase transformer section to the magnitudes of currents and voltages. Of course, when full vector or phasor analyses of voltages and currents are required, the angles of voltages and currents must be considered in conjunction with the respective magnitudes.

As illustrated in Figure 3.24, the line voltage and line current transformations for Δ-Δ three phase transformers are governed by the following equations:

$$\frac{V_{L\text{-Sec}}}{V_{L\text{-Pri}}} = \frac{1}{a}$$

or, $V_{L\text{-Sec}} = \frac{V_{L\text{-Pri}}}{a} = \frac{V}{a}$

Eq. 3.26

$$\frac{I_{L\text{-Sec}}}{I_{L\text{-Pri}}} = a$$

or, $I_{L\text{-Sec}} = a.I_{L\text{-Pri}} = a.I$

Eq. 3.27

Figure 3.24: Δ-Δ, or Delta-Delta, voltage and current transformation

Three Phase Δ-Y Transformer Configuration

The wiring diagram for a three phase Δ-Y transformer is depicted in Figure 3.25(a). In this diagram, terminals H1 through H3 represent the primary "Δ" connection points and terminals X0 through X3 represent the secondary "Y" connection points. In the wiring diagram, in accordance with electrical convention, the solid, circular, black dot represents an internal connection that determines the Δ or Y configuration of the transformer primary or secondary. The wires connecting the solid black dot terminals are, sometimes, referred to as "jumpers." The hollow, circular, rings represent the terminals where external field connections are terminated. The schematic version of this Δ-Y transformer is shown in Figure 3.25 (b).

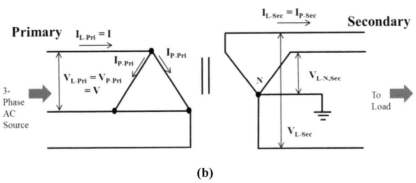

(b)

Figure 3.25: (a) Δ-Y, or Delta-Wye, transformer wiring diagram, **(b)** Δ-Y, transformer schematic

As illustrated in Figure 3.25 (b), key mathematical relationships between the phase voltages and line voltages, and phase currents and line currents are as follows:

$$V_{P\text{-}Pri} = V_{L\text{-}Pri} \qquad\qquad \text{Eq. 3.28}$$

$$V_{L\text{-}Sec} = V_{P\text{-}Sec} \cdot \sqrt{3} \qquad\qquad \text{Eq. 3.29}$$

$$I_{P\text{-}Pri} = \frac{I_{L\text{-}Pri}}{\sqrt{3}} = \frac{I}{\sqrt{3}} \qquad\qquad \text{Eq. 3.30}$$

$$I_{P\text{-}Sec} = I_{L\text{-}Sec} \qquad\qquad \text{Eq. 3.31}$$

Where,

V_{P-Pri} = Primary phase voltage = **V**
V_{L-Pri} = Primary line voltage = $V_{Line-Delta}$ = **V**
V_{P-Sec} = Secondary phase voltage
V_{L-Sec} = Secondary line voltage = V_{Line-Y}

I_{L-Pri} = Primary line current = **I**
I_{P-Pri} = Primary phase current
I_{P-Sec} = Secondary phase current = I_{Line-Y}
I_{L-Sec} = Secondary line current = I_{Line-Y}

For a Δ-Y three phase transformers, as illustrated in Figure 3.26, the voltage and current transformations can be assessed using the following equations:

$$\frac{V_{Line-Y}}{V_{Line-Delta}} = \frac{\sqrt{3}}{a}$$

$$\text{or, } V_{Line-Y} = \frac{\sqrt{3}}{a} \cdot V_{Line-Delta} = \frac{\sqrt{3}}{a} \cdot V$$

Eq. 3.32

$$\frac{I_{Line-Y}}{I_{Line-Delta}} = \frac{a}{\sqrt{3}}$$

$$\text{or, } I_{Line-Y} = \frac{a}{\sqrt{3}} \cdot I_{Line-Delta} = \frac{a}{\sqrt{3}} \cdot I$$

Eq. 3.33

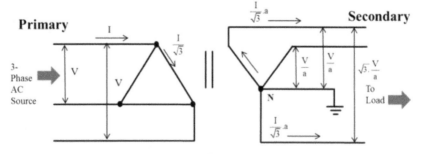

Figure 3.26: Δ-Y, or Delta-star, 3-phase voltage and current transformation

Three Phase Y-Δ Transformer Configuration

The wiring diagram of a three phase Y- Δ transformer is shown in Figure 3.27 (a); where the points labeled H0 through H3 represent the "Y" primary terminals and points X1 through X3 represent the Δ secondary terminals. The schematic version of this Y-Δ transformer is represented by Figure 3.27 (b).

(a)

(b)

Figure 3.27: (a) Y-Δ, or Wye-Delta, transformer wiring diagram, **(b) Y-Δ**, transformer schematic.

As illustrated in Figure 3.27 (b), in a Y-Δ three phase transformer, the phase and line voltages; and phase and line currents can be calculated using the following equations:

$$V_{P\text{-Pri}} = V_{L\text{-N, Pri}} = \frac{V_{L\text{-Pri}}}{\sqrt{3}} = \frac{V}{\sqrt{3}}$$

Eq. 3.34

$$V_{L\text{-Sec}} = V_{P\text{-Sec}}$$

Eq. 3.35

$$I_{P\text{-Pri}} = I_{L\text{-Pri}}$$

Eq. 3.36

$$I_{L\text{-Sec}} = \sqrt{3}.I_{P\text{- Sec}}$$

Eq. 3.37

As illustrated in Figure 3.28, key mathematical formulae for transformations between the primary and secondary voltages and primary and secondary currents are follows:

$$V_{Line\text{ - Delta}} = \frac{V_{Line\text{ - Y}}}{a\sqrt{3}}$$

or, $V_{Line\text{ - Delta}} = \dfrac{V}{a\sqrt{3}}$

Eq. 3.38

$$I_{Line\text{ - Delta}} = a\sqrt{3}.I_{Line\text{ - Y}}$$

or, $I_{Line\text{ - Delta}} = a\sqrt{3}.I$

Eq. 3.39

Where,
$V_{P\text{-Pri}}$ = Primary phase voltage
$V_{L\text{-Pri}}$ = Primary line voltage = $V = V_{Line\text{-Y}}$
$V_{P\text{-Sec}}$ = Secondary, Δ, phase voltage
$V_{L\text{-Sec}}$ = Secondary, Δ, line voltage

$I_{L\text{-Pri}}$ = Primary line current = $I = I_{Line\text{-Y}}$
$I_{P\text{-Pri}}$ = Primary phase current
$I_{P\text{-Sec}}$ = Secondary, Δ, phase current
$I_{L\text{-Sec}}$ = Secondary, Δ, line current

Three Phase Y-Y Transformer Configuration

The wiring diagram of three phase Y- Y transformer is shown in Figure 3.29 (a); where the points labeled H0 through H3 represent the "Y" primary terminals and points X0 through X3 represent the Y secondary terminals. Note

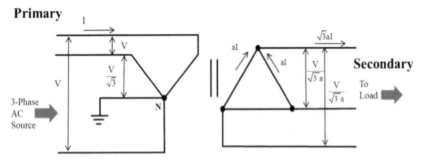

Figure 3.28: Y-Δ, or Star-Delta, 3-phase voltage and current transformation schematic

that the neutrals on primary and secondary windings of Y-Y transformers may or may not be grounded. As shown in Figures 3.29 (a) and (b), the Y-Y transformer, in this specific example, has grounded neutrals on the primary **and** secondary sides. The decision to ground the neutrals, or to leave them *"floating,"* is premised on specific concerns associated with noise, harmonics and load balancing. Schematic version of this Y-Y transformer is depicted in Figure 3.29 (b). The term *"floating,"* in the electrical domain implies isolation of a circuit or point from a verified low resistance ground. Floating electrical circuits could belong to an AC or DC system; regardless, such points would exhibit voltages above ground, i.e. +/- 5 VDC, +/- 10 VDC, +/- 24 VDC, 14 VAC, 24 VAC, 110 VAC, etc.

(a)

(b)

Figure 3.29: (a) Y-Y, or Star-star, 3-phase transformer wiring diagram **(b)** Y-Y, or Star-Star, 3-Phase transformer schematic

For a typical Y-Y three phase transformer, as depicted in Figure 3.29 (b), mathematical relationships between the phase and line voltages and phase and line currents are as follows:

$$V_{P\text{-}Pri} = V_{L\text{-}N,\, Pri} = \frac{V_{L\text{-}Pri}}{\sqrt{3}} = \frac{V}{\sqrt{3}}$$ Eq. 3.40

$$V_{L\text{-}Sec} = \sqrt{3}V_{P\text{-}Sec} = \sqrt{3}V_{L\text{-}N,\, Sec}$$ Eq. 3.41

$$I_{P\text{-}Pri} = I_{L\text{-}Pri} = I$$ Eq. 3.42

$$I_{L\text{-}Sec} = I_{P\text{-}Sec}$$ Eq. 3.43

As illustrated in Figure 3.30, key mathematical formulas for transformations between the primary and secondary voltages and primary and secondary currents are follows:

$$V_{L\text{-}sec} = \frac{V_{L\text{-}Pri}}{a} = \frac{V}{a}$$ Eq. 3.44

$$I_{L\text{-}Sec} = a.I_{L\text{-}Pri} = a.I$$ Eq. 3.45

Where,

$V = V_{\text{L-Pri}}$ = Primary line voltage

$V_{\text{L-N, Pri}}$ = Primary line to neutral voltage = Primary phase voltage

$V_{\text{L-N, Sec}}$ = $V_{\text{P-Sec}}$ = Secondary phase voltage, or, line to neutral secondary phase voltage

$V_{\text{L-Sec}}$ = Secondary line voltage

I = Primary line current = $I_{\text{L-Pri}}$

$I_{\text{P-Pri}}$ = Primary phase current

$I_{\text{P-Sec}}$ = Secondary phase current

$I_{\text{L-Sec}}$ = Secondary line current

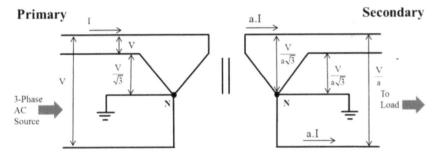

Figure 3.30: Y-Y, or Star-star, voltage and current transformation

Example 3.7

Consider the power distribution system shown in the schematic below. Determine the following unknown parameters on the, **Y**, load side of the transformer given that the turns ratio is 2:1:

a) $|I_{\text{L-Sec}}|$ = Magnitude of load or secondary line current

b) $|I_{\text{P-Sec}}|$ = Magnitude of secondary phase current or load phase current

c) $|V_{\text{P-Pri}}|$ = Magnitude of phase voltage on the source or primary side of the transformer

d) $|V_{\text{L-Sec}}|$ = Magnitude of line voltage on the load or secondary side of the transformer

e) $|V_{\text{P-Sec}}|$ = Magnitude of phase voltage on the load or secondary side of the transformer

f) $|V_{\text{L-N, Sec}}|$ = Magnitude of line to neutral voltage on the load or secondary side of the transformer

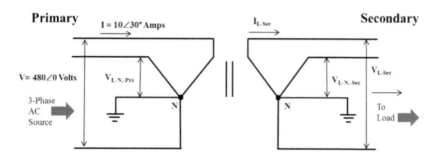

Solution:

a) Using Eq. 3.42

$$I_{P-Pri} = I_{L-Pri} = 10\angle 30° \text{ Amps}$$

$$\left|I_{L-sec}\right| = a.\left|I_{L-Pri}\right| = a.\left|I\right| = (2).\left|10\angle 30°\right| = (2).10 \text{ Amps} = 20 \text{ Amps}$$

b) According to Eq. 3.43 and as computed in part (a):

$$I_{P-Sec} = I_{L-Sec} = 20\angle 30° \text{ Amp}$$

$$\therefore \left|I_{P-Sec}\right| = 20 \text{ Amps}$$

c) According to Eq. 3.40:

$$\left|V_{P-Pri}\right| = \left|V_{L-N,\,Pri}\right| = \frac{\left|V_{L-Pri}\right|}{\sqrt{3}} = \frac{\left|V\right|}{\sqrt{3}} = \frac{\left|480\angle 0°\right|}{\sqrt{3}} = 277V$$

d) According to Eq. 3.41:

$$\left|V_{L-sec}\right| = \frac{\left|V_{L-Pri}\right|}{a} = \frac{\left|V\right|}{a} = \frac{\left|480\angle 0°\right|}{2} = \frac{480}{2} = 240 \text{ Volts}$$

e) As computed in part (d), |VL-Sec| = 240V. Then, using Eq. 3.41:

$$V_{L-Sec} = \sqrt{3}V_{P-Sec} = \sqrt{3}V_{L-N,\,Sec}$$

$$\therefore \left|V_{P-Sec}\right| = \left|V_{L-N,\,Sec}\right| = \frac{\left|V_{L-Sec}\right|}{\sqrt{3}}$$

$$\text{and, } \left|V_{P-Sec}\right| = \frac{240}{\sqrt{3}} = 139 \text{ V}$$

$$V_{\text{P-Sec}} = V_{\text{L-N, Sec}} = \frac{V_{\text{L-Sec}}}{\sqrt{3}}$$

And, by rearranging Eq. 3.41

f)   According to the rearranged version of Eq. 3.27, in part (e):

$$V_{\text{p-Sec}} = V_{\text{L-N, Sec}} \text{ and,}$$

and, $\left|V_{\text{P-Sec}}\right| = \left|V_{\text{L-N, Sec}}\right| = \dfrac{240}{\sqrt{3}} = 139 \text{ V}$

Self-Assessment Problems & Questions—Chapter 3

1.   A plating tank with an effective resistance of 100 ohm is connected to the output of a full-wave rectifier. The AC supply voltage is $340V_{\text{peak}}$. Determine the amount of time, in hours, it would take to perform 0.075 faradays worth of electroplating?

2.   Determine the source current I_{rms} in the AC circuit below.

3.   Calculate the impedance Z_{EQ} as seen by the AC voltage source in the circuit below:

4.   A single phase 1 kVA resistive load, designed to operate at 240 VAC, has to be powered by a 480 VAC source. A transformer is applied as shown in the diagram below. Answer the following questions associated with this scenario:

(a) Would the transformer be connected in a "**step up**" configuration or a "**step down**" configuration?

(b) When installing the transformer, what turns ratio, **a**, should it be connected for?

(c) What would be the secondary current, I_s, when the load is operating at full capacity?

(d) What would be the primary current at full load?

5.   Calculate the equivalent impedance as seen from the vantage point of the AC source V_{ac} in the circuit shown below. The transformer in the circuit is assumed to be ideal. The values of the primary and secondary circuit elements are: $X_{lp} = 1\Omega$, $R_p = 4\Omega$, $R_s = 10\Omega$, $X_{Ls} = 5\Omega$, $X_{Cs} = 10\Omega$, $N_p = 100$, $N_s = 200$.

6.   The no load voltage at the main switch yard of a manufacturing facility is 13,400 V_{AC}. The voltage regulation of the main switch yard is 4%. What is the rated full load voltage that is most likely to be measured on the load side of the main switch yard?

7.   Consider the power distribution system shown in the schematic below. Determine the following unknown parameters on the, **Y**, load side of the transformer given that the turns ratio is 2:1:

a) $|I_{L\text{-Sec}}|$ = Magnitude of load or secondary line current
b) $|I_{P\text{-Sec}}|$ = Magnitude of secondary phase current or load phase current
c) $|V_{P\text{-Pri}}|$ = Magnitude of Phase voltage on the source or primary side of the transformer

d) $|V_{L\text{-Sec}}|$ = Magnitude of Line voltage on the load or secondary side of the transformer

e) $|V_{P\text{-Sec}}|$ = Magnitude of Phase voltage on the load or secondary side of the transformer

f) $|V_{L\text{-N, Sec}}|$ = Magnitude of Line to neutral voltage on the load or secondary side of the transformer

Chapter 4

DC and AC Power

Introduction

This chapter begins with an introduction to concepts of energy, DC power, and DC system efficiency. DC power and efficiency discussion sets the stage for the more complex concepts of single phase AC power and AC energy. And, discussion of single phase AC serves as a suitable segue for consideration of the more complex subject of three phase AC power. Significance of efficiency and energy will be explained within the context DC power and AC power. Once the basic understanding of AC and DC power is clear, we will be better poised to examine the flow of power and energy from the point of generation to the point of consumption. The flow of power and energy from some common sources such as steam and water to electricity are described, with illustrations in form of example problems and self-assessment problems. The flow of power from utility to pumps is discussed in the "Wire to Water" section.

POWER

Power can be defined in multiple ways. Power can be defined as the rate of performance of work, rate of generation of energy, and even as rate of application of torque. This is congruent with the fact that work, energy and torque are equivalent. All three can be quantified or measured—in the Metric or SI system—as N-m (Newton-meters) or J (Joules). In the US unit realm, work, energy and torque can be measured in ft-lbf, in-lbf, Btu (British thermal unit), Hp-hr (horsepower-hour), etc. Since power is a derivative of energy with respect to time, its units in the SI or metric realm are N-m/s (Newton- meters/second) or J/s (Joules/second). And, the units for power in the US domain are ft-lbf/s (foot-pound force/second), in-lbf/s (in-pound force/second), Btu/s (British thermal units/second) and hp (horsepower).

The power and energy relationships stated above gain practical significance when stated mathematically as follows:

$$E = W = \tau \qquad\qquad\qquad \textbf{Eq. 4.1}$$

Where, **E** represents energy, **W** represents work and τ denotes torque.
Two common forms of energy in the mechanical and civil realm are as follows:

$$\text{Potential Energy} = mgh \qquad\qquad \textbf{Eq. 4.2}$$

$$\text{Kinetic Energy} = \frac{1}{2}mv^2 \qquad\qquad \textbf{Eq. 4.3}$$

$$\text{Work} = F \times d \qquad\qquad\qquad \textbf{Eq. 4.4}$$

$$\tau = F \times l \qquad\qquad\qquad \textbf{Eq. 4.5}$$

Where,

m = mass of an object, in kg (SI units system)
g = Acceleration due to gravity = 9.8m/s² (SI units system)
h = elevation of the object, in meters
v = Velocity in m/s
F = Force in N (Newtons)
d = Distance, in meters, over which the object is pushed or pulled by force, F
l = Moment arm, or perpendicular distance, in meters, from the fulcrum, at which the force is applied

In order to illustrate that potential energy, kinetic energy, work and torque can ultimately be measured in joules and kWh (unit for electrical energy), let's expand Equations 4.2 through 4.5 into N-m and Joules:

$$\text{Potential Energy} = mgh = (kg)\left(\frac{m}{s^2}\right)(m) = \text{N-m} = J$$

$$\text{Kinetic Energy} = \frac{1}{2}mv^2 = \frac{1}{2}(kg)\left(\frac{m}{s}\right)^2 = \frac{1}{2}(kg)\left(\frac{m}{s^2}\right)(m)$$

$$= \frac{1}{2}\text{N-m} = \frac{1}{2}J$$

Work $= F \times d = N\text{-}m = J$

$\tau = F \times l = N\text{-}m = J$

Since the four forms of energy correspond and correlate to the basic unit of energy, Joule, the basic unit of electrical power can be derived as:

$$\text{Power} = \frac{\text{Energy}}{\text{Time}} = \left(\frac{J}{s}\right) = \text{Watt};$$

and $1{,}000\,\text{Watts} = 1\text{kW}$, or 1 kilo-Watt

Conversely,

$$\text{Energy} = (\text{Power}).(\text{Time}) = (W).(s) = \left(\frac{J}{s}\right)(s) = J$$

Or, 1kWh of Energy $= (1000W).(1h)$

$$= (1000)\left(\frac{J}{s}\right)(3600s) = 3600\,\text{kJ}$$

Now, in order to demonstrate the interchangeability between electrical energy and hydrocarbon fuel (i.e. fuel oil, natural gas, propane, gasoline, etc.), let's extrapolate the conversion of 1 kWh into the Btus and dekatherms:

Since $1055\,J = 1$ Btu,

$$1\text{kWh of Energy} = 3600\,\text{kJ} = 3{,}600{,}000\,J = \left(\frac{3{,}600{,}000\,J}{1055\,J/\text{Btu}}\right)$$

$$= 3{,}412\,\text{Btu}$$

And since $1{,}000{,}000\,\text{Btu} = 1\text{MMBtu} = 1$ DT,

$1\text{kWh of Energy} = 3600\,\text{kJ} = 3{,}412\,\text{Btu} = 0.003412\,\text{MMBtu}$

$$= 0.003412\,\text{DT (Dekatherm)}$$

Following the energy and power unit conversion methods illustrated above, one can perform operating energy cost comparison between alternative pieces of equipment.

Example 4.1

Which of the following two water heaters would cost the least to operate, on annual cost basis, under the given assumptions?

A. **Electric Water Heater:**
    Estimated annual energy consumption: **5000 kWh**
    Efficiency: **100%**
    Cost Rate: **$0.10/kWh**

B. **Natural Gas Water Heater:**
    Estimated annual energy consumption: **Same as the electric water heater**
    Efficiency: **100%**
    Cost Rate: **$10.87/DT**

Solution:

Since the cost rate for electrical energy and the annual electrical energy consumption for the electric water heater are given:

Total annual cost for operating the *electric water heater* **(5000 kWh). ($0.10/kWh) = $500**

The annual energy consumption by the *gas water heater*, assumed to be the same as the electric water heater = **5000 kWh**

Then, the annual energy consumption by the gas water heater, in DT or MMBtu would be = **(5000 kWh).(3412Btu/kWh)(1DT/1,000,000Btu) = 17.06 DT.**

Since the natural gas cost rate is given as **$10.87/DT**, the annual operating cost for the gas water heater would be = **(17.06 DT).($10.87/DT) = $185.44.**

Answer: The gas water heater would cost substantially less to operate than the electric water heater.

Electric Motor Horsepower Required to Move/Convey Mass[*]

Conveyance of material through powered conveyors, aside from the selection and design of other pieces of equipment, requires sizing and specification of conveyor motors. Two simple formulas that can be used to assess the brake

---

[*]Thermodynamics Made Simple for Energy Engineers, By: S. Bobby Rauf)

horsepower required to move *"loose"* (i.e. aggregate bulk materials like sand, flour, grain, sugar, etc.) or discrete *"unit"* mass (i.e. solid objects like steel parts, fabricated mechanical components, wooden objects, etc., are listed below:

$$P = F.\Delta v \qquad\qquad\qquad \textbf{Eq. 4.6}$$

$$F = \dot{m}\, \Delta v \qquad\qquad\qquad \textbf{Eq. 4.7}$$

Where,
- **P** = Power in J/s or Watts = brake horsepower required at the motor shaft.
- **F** = Linear force required to move the conveyor (or conveyor belt), loaded with the aggregate or unit mass to be moved
- **Δv** = Change in the velocity of aggregate or unit mass being conveyed, in the direction of the applied force and in the direction the material is being moved.
- **ṁ** = Mass flow rate.

Application of this method for assessing the size of a conveyor motor is illustrated through Example 4.2

Example 4.2

Sand drops at the rate of 20,000 kg/min onto a conveyor belt moving with a velocity of 2.0 m/s.
(a) What force is required to keep the belt moving?
(b) What is the minimum motor size that should be specified for this application?

Solution:
(a) $F = \dot{m}.\Delta v = (20,000 \text{ kg/min}).(2.0 \text{ m/s} - 0 \text{ m/s}).(1/60 \text{ min/s})$
$$= 666.67 \text{ kg.m/s2 or, } F = 666.67 \text{ N}$$

(b) Power $= P = F. v = (666.67 \text{ N}) . (2 \text{ m/s})$
$$= 1333.3 \text{ N-m/s} = 1333.3 \text{ J/s} = 1333.3 \text{ Watts}$$
Since 746 W = 1 hp, minimum motor horsepower required:
$$= 1333.3/746 = 1.79 \text{ hp}$$
∴ Specify a 2 hp (standard size) motor

DC Power

Power in direct current realm is equivalent to power in the mechanical realm, for most practical purposes. DC electrical power is also referred to as the "real" power. DC power is called real because, unlike AC power, it can be transformed entirely into work or other forms of energy, i.e. heat energy, potential energy, kinetic energy, etc. Such transformation can be assessed through power and energy relationships discussed earlier in this chapter. Eq. 4.8 below represents one way to correlate DC, real, power to mechanical work performed.

$$W = Energy = P.t \qquad \textbf{Eq. 4.8}$$

Where,

    W = Mechanical work performed under DC electrical power, measured in Joules (or N-m).

    P = DC electrical power, in Watts (or J/s).

    t = DC power application duration, in seconds (s)

DC power can be defined mathematically as stipulated in Eq. 4.9.

$$P = V.I \qquad \textbf{Eq. 4.9}$$

Where,

    P = DC electrical power, in Watts (or J/s)

    V = DC voltage measured in volts (V)

    I = DC current measured in amps (A)

We can quantify mechanical work in terms of voltage, current and time by substituting Eq. 4.9 into Eq. 4.8:

$$W = Energy = V.I.t \qquad \textbf{Eq. 4.10}$$

In the SI or Metric unit system, DC power or "real" power is traditionally measured in Watts, kW, MW, GW, TW (10^{12} W). Where k = 1000, M = 1,000,000, G = 1 billion and T = 1 trillion.

Some of the more common power conversion factors that are used to convert between SI System and US system of units are listed below:

1.055 kJ/s = 1.055 kW = 1 Btu/s

One hp = 746 Watts = 746 J/s = 746 N-m/s = 0.746 kW = 550 ft-lbf/sec

In the SI or Metric unit system, DC energy or "real" energy is traditionally measured in Wh, kWh, MWh, GWh, TWh (10^{12} Wh).

Some mainstream conversion factors that can be used to convert electrical energy units within the SI realm or between the SI and US realms are referenced below:

1000 kW x 1h = 1 MWh
1 Btu = 1055 J = 1.055 kJ
1 Btu = 778 ft-lbf
1 hp x 1 hour = 1 hp-hour

Example 4.3

An automobile is parked with parking lights left on for one hour. The automobile battery is rated 12 V_{DC} and the lamps are incandescent. If all of the parking lamps are drawing a combined current of 4A, what is the total energy consumed by the parking lights, in form of heat and light?

Solution:

Apply Eq. 4.10:

Energy = V.I.t = (12V).(4A).(1hr) = (48W).(3600 s)

$$= \left(48\frac{J}{s}\right)(3600s) = 172,800 \text{ J}$$

SINGLE PHASE AC POWER

AC, single and three phase, was introduced in Chapter 3. In Chapter 3, we also discussed the differences between DC, single phase AC and three phase AC. In this chapter, in addition to the DC power discussion in the foregoing section, we will focus mostly on single phase and three phase power. While single phase AC power is, primarily, power or rate of performance of work, it differs from DC power, real power or mechanical power in the fact that it is not devoted *entirely* to performance of work or conversion into other forms of energy. In most cases, a percentage of AC power is "sequestered" in form of electromagnetic energy and power. This percentage of AC power is not available for performance of mechanical work or conversion to other forms of energy; instead, it is dedicated toward charging and discharging inductances and capacitances in the AC electrical system. Yet, the utility or the power company must generate and supply the *entire* AC power to sustain *all* customers regardless of the size of inductances and capacitances in their systems.

The term "single phase AC" implies that the AC power source consists of one energized or higher electrical potential conductor or terminal. The other conductor, wire or terminal serves as a neutral and is typically connected to building or power distribution system ground. Mechanical engineers, with thermodynamic background, could view the energized line or terminal as the "heat source" and could apply the analogy of "heat sink" to the neutral or ground. This analogy is premised on the fact that, for heat engines to perform work through steam turbines, the superheated steam must have a heat source and a heat sink to traverse between.

As introduced briefly in Chapter 3, total AC power is called Apparent Power and is denoted by **S**. Note that the apparent power symbol **S** is bold faced. The bold font signifies that fact that apparent power is a *vector* and can be "completely" represented by a *complex number* that consists of a *magnitude* and an *angle*.

Fundamental mathematical definition of single phase AC power is represented in terms of Eq. 4.11 and Eq. 4.12. The magnitude and angle of single phase AC apparent power can be can be determined through Equations 4.13 and 4.14, respectively. The vector values of single phase AC voltage and current can be calculated using Equations 4.15 and 4.16, respectively. The magnitudes of single phase AC voltage and current can be calculated using Equations 4.17 and 4.18, respectively. Values of P, Q and |S| can also be calculated in terms of $\mathbf{V_{RMS}}$, **R**, **X** and **Z** as shown in equations 4.19, 4.20 and 4.21.

$$\bar{S} = \bar{V}.\bar{I}^{*} \qquad\qquad\qquad \text{Eq. 4.11}$$

$$\bar{S} = P + jQ \qquad\qquad\qquad \text{Eq. 4.12}$$

$$|S| = |V|.|I| \qquad\qquad\qquad \text{Eq. 4.13}$$

$$\text{Angle of } S = \theta_{s} = \text{Tan}^{-1}\left(\frac{Q}{P}\right) \qquad\qquad \text{Eq. 4.14}$$

$$\bar{V} = \frac{\bar{S}}{\bar{I}^{*}} \qquad\qquad\qquad \text{Eq. 4.15}$$

$$\bar{I}^{*} = \frac{\bar{S}}{\bar{V}} \qquad\qquad\qquad \text{Eq. 4.16}$$

$$|V| = \frac{|S|}{|I|} \qquad\qquad\qquad \text{Eq. 4.17}$$

$$|I| = \frac{|S|}{|V|} \qquad\qquad\qquad \text{Eq. 4.18}$$

$$P = \frac{V_{RMS}{}^2}{R}$$

Eq. 4.19

$$Q = \frac{V_{RMS}{}^2}{X}$$

Eq. 4.20

$$|S| = \frac{|V_{RMS}|^2}{|Z|}$$

Eq. 4.21

Where,

$\bar{S}$ = Complex, vector, representation of single phase AC power, measured in VA, or Volt-Amperes, kVA or MVA

$\bar{V}$ = Complex, vector, representation of single phase AC voltage. measured in volts

$\bar{I}$ = Complex, vector, representation of single phase AC current. measured in amperes

$|V|$ = Magnitude or absolute value of single phase AC voltage, in volts

$|V_{RMS}|$ = Magnitude of the RMS value of single phase AC voltage

$|I|$ = Magnitude or absolute value of single phase AC current, in amps

$|S|$ = Magnitude or absolute value of single phase AC apparent power, measured in VA, or Volt-Amperes, kVA or MVA

θ = Angle of apparent power **S**

P = Real power, or real component of apparent power **S**, measured in Watts, kW or MW.

Q = Reactive power, or imaginary component of apparent power **S**, measured in VARs, or, Volt-Amperes Reactive, kVAR or MVAR.

j = Unit vector, with a magnitude of "1" or unity, portending an angle of 90° with respect to the x-axis. As explained in Chapter 3, unit vector **j** can be written in polar or phasor form as $1\angle 90°$.

X = Reactance, in Ω

R = Resistance in Ω

$|Z|$ = Magnitude or absolute value of AC impedance in Ω

Since apparent power **S** is represented as a vector with horizontal component "P" and vertical component "Q," in Eq. 4.12, an alternate method for calculating the magnitude of apparent power S (vector) can derived on the Pythagorean Theorem as stipulated by Eq. 4.22. The angle of apparent power vector **S** could still be computed using Eq. 4.14.

$$|S| = \text{Magnitude of AC Apparent Power} = \sqrt{P^2 + Q^2}$$ **Eq. 4.22**

Power Factor: A concept that is inherently important in most analysis and consideration of AC power is "Power Factor." While power factor is discussed, in detail, in Chapter 5, as a prelude to that detailed discussion and to complete the discussion of single phase AC power, mathematical definition of power factor is stated below in form of equations 4.23 and 4.24.

$$PF = \text{Power Factor} = \frac{|P|}{|S|}$$ **Eq. 4.23**

$$PF = \text{Power Factor} = Cos(\theta_v - \theta_I)$$ **Eq. 4.24**

Where,

$|P|$ = Magnitude of the real component of AC apparent power

$|S|$ = Magnitude of the oveall AC apparent power

θ_v = Angle of AC voltage
θ_I = Angle of AC current

Example 4.4
   A 156Sin377t sinusoidal voltage is connected across a load consisting of a parallel combination of a 10 Ω resistor and a 5Ω inductive reactance.
(a)   Determine the real power dissipated by the resistor.
(b)   Determine the reactive power sequestered in a **5Ω** parallel reactance.
(c)   Calculate the total apparent power delivered to this parallel R and X circuit by the AC voltage source.

Solution:

The circuit diagram for this scenario would be as depicted below:

$V_{ac} = 156 \, Sin(377t)$ $R = 10\Omega$ $X = 5\Omega$

a) We can apply Eq. 4.19 to determine the power dissipated or consumed in the 10Ω resistor. However, we must first derive the VRMS from the given AC voltage of 156Sin377t. This is due to the fact that, by convention, coefficient 156 stated in the give AC voltage function of 156Sin400t is the peak or maximum voltage, Vm.\As discussed in Chapter 3 and stipulated by Eq. 3.3:

$$V_m = \sqrt{2} V_{RMS} \quad \text{or} \quad V_{RMS} = \frac{V_m}{\sqrt{2}}$$

$$\therefore V_{RMS} = \frac{156}{\sqrt{2}} = 110.3V$$

Then, according to Eq. 4.19:

$$P = \frac{V_{RMS}^2}{R} = \frac{(110.3)^2}{10} = 1217 \text{ Watts}$$

b) Apply Eq. 4.20 to determine the reactive power sequestered in the 5Ω parallel reactance.

$$Q = \frac{V_{RMS}^2}{X} = \frac{(110.3)^2}{5} = \frac{12,166}{5} = 2433 \text{ VAR}$$

c) Apply Eq. 4.12 to calculate the total apparent power S delivered to this parallel R and X circuit by the AC voltage source.

$$\bar{S} = P + jQ = 1217 + j2433 = 2720\angle 63.4°\Omega$$

Note: We can also apply Eq. 4.22 to verify the magnitude of the total apparent power S delivered to this parallel R and X circuit:

$$|S| = \text{Magnitude of AC Apparent Power} = \sqrt{P^2 + Q^2}$$
$$= \sqrt{1217^2 + 2433^2} = 2720 \text{ VA}$$

Note that jQ reactive power entity is entered into the apparent power calculation as +jQ because of the fact that inductance in the given AC circuit results in positive impedance contribution or "+jX."

Ancillary: Reader is encouraged to verify the apparent power of 2720 VA by applying equation 4.21. **Hint:** The Z, in this case must be computed through parallel combination of R and Z_L as shown below:

$$|Z| = \left| \frac{R.Z_L}{R + Z_L} \right| \qquad\qquad \textbf{Eq. 4.25}$$

Example 4.5
    The AC circuit shown below depicts a simplified, single phase, one-line diagram of a "special purpose" power generating station. Assume that there is no voltage drop between the generator and the power distribution system. The line current is measured to be 1,000 amps, rms. Calculate the following if the power factor is known to be 0.9:

a)   Magnitude of the apparent power presented to the power distribution system.
b)   Magnitude of the real power presented to the power distribution system.

Solution:

a) Magnitude of the apparent power presented to the power distribution system:

$$|S|_{1-\Phi} = \sqrt{(P)^2 + (Q)^2} = |V.I*|$$

Since we are interested in the magnitude of the single phase AC power:

$$|S|_{1-\Phi} = |V.I*| = |V|.|I|$$

Where, V and I are RMS values. Note that, by convention, the 156V in the given AC voltage function is V_P, or the peak value.

$$V_{rms} = \frac{V_p}{\sqrt{2}} = \frac{156}{\sqrt{2}} = 110 \text{ V}$$

$$
\begin{aligned}
|S|_{1-\Phi} &= |V.I*| = |V|.|I| \\
&= (110 \text{ V}).(1,000 \text{ amps}) = 110,000 \text{VA} \\
&= 110 \text{ kVA}
\end{aligned}
$$

b) Magnitude of the real power presented to the power distribution system can be determined by rearranging and applying 4.23 as follows:

$$PF = \text{Power Factor} = \frac{|P|}{|S|} \qquad \qquad \textbf{Eq. 4.23}$$

$$|P| = |S|.PF = (110 \text{ kVA}).(0.9) = 99.34 \text{ kW}$$

THREE PHASE AC POWER

As explained in Chapter 3, while single phase AC offered substantial advantages over DC in terms of transmission of power from power generating stations to consumers, three phase AC took electrical power transmission to an even higher level of efficiency and effectiveness. Having explored the concepts, principles analytical techniques, equations and conventions associated with single phase AC, in the last section, we are now adequately prepared to explore basic yet practical concepts, principles analytical techniques, equations and conventions associated with three phase AC. Because conveyance and transformation of electrical energy into mechanical and other forms of energy is more effective and efficient in three phase configuration than single phase configuration, larger loads and large transfers of electrical power and energy are mostly conducted in three phase formation. Therefore, we find most large motors and other types of loads, with a few exceptions, in industrial and commercial facilities, to be three phase.

Analytical techniques, mathematical formulas, schematics and other types of three phase drawings tend to be more complex than their single phase counterparts. Due to the profundity and complexity of three phase AC, we will limit our discussion of three phase AC power, in this section, to the basic level. Efficient and effective application of three phase AC requires that the load, on each of the three phases be as equal as possible. When the loads on each of the three phases are equal, three phase AC system is said to be balanced. Basic three phase power formulas are listed below:

$$\bar{S}_{3-\phi} = \text{Three phase apparent power} = \sqrt{3}\bar{V}_{L\text{-}L}\bar{I}_L^{\,*} \qquad \textbf{Eq. 4.27}$$

$$\left|S_{3-\phi}\right| = \sqrt{3}\left|V_{L-L}\right|.\left|I_L^{\,*}\right| \qquad \textbf{Eq. 4.28}$$

$$\bar{S}_{3-\phi} = 3.\,\bar{S}_{1-\phi} \qquad \textbf{Eq. 4.29}$$

$$\left|\bar{S}_{3-\phi}\right| = 3\left|\bar{S}_{1-\phi}\right| = 3\left(\frac{V_{L\text{-}L}.I_L}{\sqrt{3}}\right) = \sqrt{3}V_{L\text{-}L}.I_L, \text{ for three} \qquad \textbf{Eq. 4.30}$$

phase Y and Δ circuits.

Where,

$\bar{S}_{3-\phi}$ = Complex, vector, representation of three phase AC power, measured in VA, or Volt-Amperes, kVA or MVA.

$\left|\bar{S}_{3-\phi}\right|$ = Magnitude of three phase AC power, measured in VA, or Volt-Amperes, kVA or MVA.

$\bar{S}_{1-\phi}$ = Complex, vector, representation of one of the three phases of AC power, measured in VA, or Volt-Amperes, kVA or MVA.

$\left| \vec{S}_{1-\phi} \right|$ = Magnitude of one of the three phases of AC power, measured in VA, or Volt-Amperes, kVA or MVA.

$\vec{V}_{L-L}$ = Complex, vector, representation of line to line, or phase to phase, RMS voltage, measured in volts.

$\left| \vec{V}_{L-L} \right|$ = Magnitude of the Line to Line, or Phase to Phase, RMS voltage, measured in volts.

$\vec{I}_{L}$ = Complex, vector, representation of line RMS current, with magnitude measured in amperes, typically, using a clamp-on ammeter or a CT, Current Transformer.

$\vec{I}_{L}^{*}$ = Conjugate of vector, representation of line RMS current. Sign of the angle of the conjugate current is opposite of the sign of the original current.

$\left| \vec{I}_{L}^{*} \right| = \left| \vec{I}_{L} \right|$ = Magnitude of the RMS value of the line current.

Example 4.6

The AC circuit shown below depicts a three phase, one-line schematic of a hydroelectric power generating station, modeled after the Three Gorges Dam, China. Assume that there is no voltage drop between the generator and the primary side of the transmission system transformer. The line current is measured to be **22,453 amps**, RMS. Calculate the following if the power factor is known to be **0.9**:

a) Magnitude of the apparent power presented to the transmission lines.
b) Magnitude of the real power presented to the transmission lines.

Solution:

(a) Magnitude of the apparent power presented to the transmission lines:
   Note that the AC voltage function is specified in RMS form and not in

"peak" or "maximum" form. Therefore, we don't need to derive the RMS voltage. The line current is given in RMS form, as well.

According to Eq. 4.30:

$$\left|\bar{S}_{3-\phi}\right| = 3\left|\bar{S}_{1-\phi}\right| = 3\left(\frac{V_{L-L}.I_L}{\sqrt{3}}\right) = \sqrt{3}V_{L-L}.I_L, \text{ for three}$$

phase Y and Δ circuits.

Therefore,

$$\left|\bar{S}_{3-\phi}\right| = \sqrt{3}V_{L-L}.I_L = \sqrt{3}(20,000V).(22,450A) = 777,690,813VA$$
$$= 777,690 \text{ kVA} = 777.69 \text{ MVA}$$

(b) Magnitude of the real power presented to the transmission lines can be determined by rearranging and using Eq. 4.23:

$$PF = \text{Power Factor} = \frac{|P|}{|S|}$$

or, $|P| = PF.|S| = (0.9).(777.69MVA) = 700MW$

EFFICIENCY

Efficiency is defined, generally, as the ratio of output to input. The output and input, in general, could be in form of power, energy, torque or work. Efficiency assumes a more specific definition when considered in the context of a specific form of energy, work or power. The concept of efficiency, when applied in the electrical engineering domain, typically, involves power or energy.

In electrical engineering, when *power* is the subject of analysis, efficiency is defined as follows:

$$\text{Efficiency in percent} = \eta = \frac{\text{Output Power}}{\text{Input Power}} \times 100 \qquad \textbf{Eq. 4.31}$$

Where,

η (Eta) is a universal symbol for efficiency

In electrical engineering, when energy is the subject of analysis, efficiency is defined as follows:

$$\text{Efficiency in percent} = \eta = \frac{\text{Output Energy}}{\text{Input Energy}} \times 100 \qquad \textbf{Eq. 4.32}$$

Although work is not used as commonly in the computation of efficiency, in the electromechanical realm, where applicable, the overall system efficiency calculation, based on work, could be stated as follows:

$$\text{Efficiency (\%)} = \eta = \frac{\text{Work Performed by the Electromechanical System}}{\text{Input Energy}} \times 100 \qquad \textbf{Eq. 4.33}$$

As obvious from the definitions of efficiency above, since energy cannot be created, efficiency cannot exceed "1" or 100%. The decimal result for efficiency is often converted to, and stated as, a percentage value.

In the following section, we will explore the relationship between power and efficiency in steam, mechanical and electrical systems, and develop better understanding of the flow of power in steam type electrical power generating systems.

POWER CONVERSION FROM STEAM TO ELECTRICAL FORM

The power delivered by steam to the turbine blades, call it P_{Steam}, in a simplified—no heat loss, no kinetic head loss, no potential head loss and zero frictional head loss—scenario can be represented by the mathematical relationship stated in form of **Eq. 4.34**. In the context of flow of energy from steam to electricity, functional relationship between electrical power, $P_{Electrical}$, generator efficiency $\eta_{Generator}$, steam turbine efficiency $\eta_{Turbine}$, and P_{Steam} can be expressed in form of **Eq. 4.35**.

$$^1 P_{Steam} = (h_i - h_f) \cdot \dot{m} \qquad\qquad\qquad \textbf{Eq. 4.34}$$

$$^2 P_{Electrical} = (P_{Steam}) \cdot (\eta_{Turbine}) \cdot (\eta_{Generator}) \qquad\qquad \textbf{Eq. 4.35}$$

See the power flow diagram depicted in Figure 4.1. Even though this diagram refers to the flow of power, in conformance with Eq. 4.32, it applies just as well to flow of energy. This diagram shows the harnessing of power and energy contained in superheated steam—with high enthalpy or heat content—and their conversion to mechanical brake horsepower in the steam turbine. The brake horsepower thus imparted onto the turbine shaft is conveyed to the electrical power generator, typically, through direct coupling of the turbine and generator shafts. The generator, subsequently, converts the brake horsepower and mechanical energy into electrical power (Watts, kW, MW, etc.) and energy (Wh, KWh, MWh, etc.).

However, as annotated in Figure 4.1, the transformation of power and energy from steam to electricity is de-rated first in the turbine and later in the generator by respective efficiencies.

Figure 4.1: Steam to wire power and energy flow—steam power generation system.

[1,2]*Thermodynamics Made Simple for Energy Engineers,* By: S. Bobby Rauf

The flow of power and energy from—steam to electricity—as depicted in Figure 4.1, is also referred to as "steam to wire" flow of power and energy. Of course, the electrical power galvanized by the generator is presented to the power distribution grid via necessary switchgear and transformers.

Figure 4.2: Water to wire power and energy flow.

POWER CONVERSION FROM WATER TO ELECTRICAL FORM

In hydroelectric power plants, power and energy is transferred from water to electrical form, via hydraulic turbines. This flow of power and energy—referred to as "**Water to Wire**" flow of power—is illustrated in Figure 4.2. The power carried by the water to the turbine blades is called "**Water Horse Power**," or **WHP**; sometimes, also referred to as *hydraulic horsepower* or *fluid horse power*. As explained in the next section, when energy or power flows from electrical power source to the fluid, through a hydraulic pump, the **WHP** is considered equivalent to—or referred to as—P_P, the *pump horse power*.

The transformation of power from hydraulic form to electrical can be computed through equations similar to the ones listed below:

$$WHP = \frac{(h_A)(\gamma)\left(\dot{v}\right)}{550} \qquad \text{Eq. 4.36}$$

$$\text{WHP} = \frac{(\Delta P)\left(\overset{\bullet}{V}\right)}{550}$$ Eq. 4.37

Where,

WHP = Water horse power, imparted the water onto the turbine, in hp.

h_A = Net head added to the water, in ft, by virtue of the height of water in the reservoir.

γ = Specific weight or weight density = 62.4 lbf/ft^3

ΔP = Differential pressure, across the turbine, in pounds per ft^2

$\overset{\bullet}{V}$ = Volumetric flow rate of water flowing through the turbine, measured in ft^3/sec

In the context of flow of energy from water to electricity, functional relationship between electrical power, $P_{Electrical}$, generator efficiency $\eta_{Generator}$, turbine efficiency $\eta_{Turbine}$, and **WHP** can be expressed in form of **Eq. 4.38.**

$$P_{Electrical} = (WHP) \cdot (\eta_{Turbine}) \cdot (\eta_{Generator})$$ Eq. 4.38

See the power flow diagram depicted in Figure 4.2. Even though this diagram refers to the flow of power, in conformance with Eq. 4.38, it applies just as well to flow of energy. This diagram shows the harnessing of power and energy contained in water—including potential head, kinetic head pressure heads—and their conversion to mechanical brake horsepower in the turbine. The brake horsepower thus imparted onto the turbine shaft is conveyed to the electrical power generator, typically, through direct coupling of the turbine and generator shafts. The generator, subsequently, converts the brake horsepower and mechanical energy into electrical power (Watts, kW, MW, etc.) and energy (Wh, KWh, MWh, etc.).

However, as annotated in Figure 4.2 and supported by Eq. 4.38, the transformation of power and energy from water to electricity is first depreciated in the turbine and later in the generator by respective efficiency values of the turbine and generator.

The flow of power and energy from—water to electricity—as depicted in Figure 4.2, is also referred to as "***water to wire***" flow of power and energy. Electrical power generated by the generator is routed to the power grid through the necessary switchgear and transformers.

POWER CONVERSION IN HYDRAULIC PUMP SYSTEMS—
WIRE TO WATER TRANSFORMATION OF POWER AND ENERGY

Hydraulic pumps, driven by electric motors, represent the exact converse of the hydroelectric power generating systems. The flow of power in electric pump applications follows the path depicted in Figure 4.3. Equations 4.36 and 4.37 are still, mathematically, pertinent. However, as annotated in form of equations 4.39 and 4.40, the WHP equations are interpreted in a context that is opposite of equations 4.36 and 4.37.

$$\text{WHP} = \frac{(\Delta P) \left(\dot{V} \right)}{550} \qquad \qquad \text{Eq. 4.39}$$

$$\text{WHP} = \frac{(\Delta P) \left(\dot{V} \right)}{550} \qquad \qquad \text{Eq. 4.40}$$

$$\text{WHP} = (P_{\text{Electrical}}) \cdot (\eta_{\text{Motor}}) \cdot (\eta_{\text{Pump}}) \qquad \qquad \text{Eq. 4.41}$$

Where,

WHP = P_P = Water horse power imparted by the pump and motor system onto the water, measured in hp. Also referred to as "pump horse power" or "fluid horse power."

EHP $\rightarrow$ **BHP** $\rightarrow$ **WHP**

Motor Efficiency, η_{motor}

Pump Efficiency, η_{pump}

Electrical Power from Utility

HP, Delivered by the Utility to the Motor

HP Del. By Utility = BHP / η_{motor}

BHP, Delivered by the Motor to the Pump

BHP = WHP / η_{pump}

WHP, or P_P Delivered by the Pump to the Fluid

Figure 4.3: Wire to water power flow—hydraulic pump system discussion on equations 4.39, 4.40 and 4.41, cont'd:

h_A = Net head added by the pump and motor system onto the water, measured in ft.

γ = Specific weight or weight density = 62.4 lbf/ft^3.

ΔP = Differential pressure, across the pump, measured in pounds per ft^2.

$\dot{V}$ = Volumetric flow rate of water through the motor and pump system, measured in ft^3/sec.

$P_{Electrical}$ = Electrical power drawn by the motor, from the power source or utility.

η_{Motor} = Motor efficiency.

η_{Pump} = Pump efficiency.

As evident from Figure 4.3, Eq. 4.39, Eq. 4.40 and Eq. 4.41, in electric motor and hydraulic pump systems, power and energy is transferred from an electrical power source to a hydraulic or fluid system. Because in this scenario the energy stems from the power grid—transported by power transmission wires—and terminates into the fluid, it is referred to as **wire to water** flow of energy and power.

In electric motor and pump systems—as annotated in Figure 4.3 and supported by Eq. 4.41—during the transformation of power and energy from electricity to water, energy is first depreciated in the motor and later in the pump by respective efficiency values of the motor and the pump.

Some of the efficiency, energy flow and power flow principles and methods described above are illustrated through Examples 4.7 and 4.8 and self-assessment problems at the end of the chapter.

Example 4.7

As an energy engineer, you are charged with the task to estimate the heat content or enthalpy, h_i, of the superheated steam that must be fed to a steam turbine in order to supply **10 MW** (Megawatt) of electrical power to the electrical grid. Assume that there is no heat loss in the turbine system and that difference between the enthalpies on the entrance and exit ends of the turbine is converted completely into work, minus the inefficiency of the turbine. All of the data available and pertinent to this project is listed below:

— Electrical Power Generator Efficiency: 90%
— Steam Turbine Efficiency: 70%
— Mass flow rate for steam, ṁ: 25 kg/s (55 lbm/s)
— Estimated exit enthalpy, hf, of the steam: 2875 kJ/kg (1239 Btu/lbm)

Solution

Strategy: In order to determine the estimated enthalpy, h_i, of the incoming steam, we need to start with the stated output **(10 MW)** of the generator and work our way upstream to derive the energy delivered to the vanes of the turbine. The assumption that there is no heat loss in the turbine system and that the difference between the enthalpies on the entrance and exit ends of the turbine is converted completely into work, minus the inefficiency of the turbine, implies that the energy delivered by the steam is equal to the net energy delivered to the turbine vanes. Also, note that net energy delivered to the turbine vanes is reduced or de-rated according to the given efficiency of the turbine.

Solution in SI/Metric Units

Since, $1 J/s = 1 W$ and $1 kJ/s = 1 kW$,

$$\textbf{Power output of the generator} = \textbf{10 MW} = \textbf{10,000 kW}$$
$$= \textbf{10,000 kJ/s}$$

Brake horsepower delivered by the turbine to the generator, through the turbine shaft is determined as follows:

$$\textbf{BHP} = \frac{\textbf{Generator Output}}{\textbf{Generator Efficiency}}$$

$$= \frac{10,000 kJ/s}{0.9} = \textbf{1.11 x } 10^4 \textbf{ kJ/s or 11,111 kJ/s}$$

Power delivered by the steam to the turbine vanes is determined as follows:

$$\textbf{P}_{\textbf{steam}} = \frac{\textbf{BHP}}{\textbf{Turbine Efficiency}}$$

$$= \frac{1.11 \text{ x } 10^4 kJ/s}{0.7} = \textbf{1.5873 x } 10^4 \textbf{ kJ/s or 15,873 kJ/s}$$

Of course, we could obtain the same result, in one step, by rearranging and applying Eq. 4.35 as follows:

$$\textbf{P}_{\textbf{steam}} = \frac{\textbf{P}_{\textbf{Electrical}}}{\left(\eta_{\textbf{Turbine}}\right).\left(\eta_{\textbf{Generator}}\right)} = \frac{\textbf{10,000 kJ/s}}{\left(\textbf{0.7}\right).\left(\textbf{0.9}\right)} = \textbf{15,873 kJ/s}$$

Since the difference in the turbine entrance and exit enthalpies, in this scenario, is equal to the energy delivered to the turbine vanes:

$$P_{steam} = (h_i - h_f) \cdot \dot{m}$$

Eq. 4.34

$$P_{steam} = 15{,}873 \text{ kJ/s} = \left(h_i - 2875 \text{ kJ/kg}\right) \cdot 25 \text{ kg/s}$$

$$h_i = \frac{15{,}873 \text{ kJ/s}}{25 \text{ kg/s}} + 2875 \text{ kJ/kg}$$

$$h_i = 3{,}509 \text{ kJ/kg}$$

Solution in US/Imperial Units

Power output of the generator = 10 MW = 10,000 kW = 10,000 kJ/s
Since 1.055 kJ = 1.0 Btu,

Power output of the generator = (10,000 kJ/s) . (1/1.055 kJ/Btu)

= **9,479 Btu/s**

Brake horsepower delivered by the turbine to the generator:

$$BHP = \frac{\textbf{Generator Output}}{\textbf{Generator Efficiency}} = \frac{9{,}479 \text{ BTU/s}}{0.9} = \mathbf{10{,}532 \text{ BTU/s}}$$

Power delivered by the steam to the turbine vanes is determined as follows:

$$P_{steam} = \frac{BHP}{\text{Turbine Efficiency}} = \frac{10{,}532 \text{ BTU/s}}{0.7} = \mathbf{15{,}046 \text{ BTU/s}}$$

Since the difference in the turbine entrance and exit enthalpies, once again, is equal to the energy delivered to the turbine vanes:

$$P_{steam} = (h_i - h_f) \cdot \dot{m}$$

Eq. 4.34

$$15{,}046 \text{ BTU/s} = \left(h_i - 1239 \text{ BTU/lbm}\right) \cdot \left(55 \text{ lbm/s}\right)$$

$$h_i = \frac{15{,}046 \text{ BTU/s}}{55 \text{ lbm/s}} + 1239 \text{ BTU/lbm}$$

$$= \mathbf{1512 \text{ BTU/lbm}}$$

Verification: The accuracy of the initial enthalpy, h_i, value of **1512 Btu/lbm** can be verified by simple unit conversion to the metric form as follows:

Since 1.055 kJ = 1.0 Btu, or, there are 0.9479 Btu per kJ

$$\frac{1512 \text{ Btu/lbm}}{0.9479 \text{ Btu / kJ}} \times 2.2046 \text{ lb/kg} \Rightarrow 3,517 \text{ kJ/kg}$$

$$\approx 3,509 \text{ kJ/kg (as computed in SI units at the outset)}$$

Example 4.8

Pressures on the intake and exit ends of a turbine are measured to be 50 psia and 10 psia, respectively. The volumetric flow rate for the fluid (water) is 120 ft³/sec. See the diagram below. The turbine is driving an electric generator with nameplate efficiency of 90%. Calculate the following:

a)   Water horsepower delivered by the water to the turbine.
b)   Brake horsepower delivered by the turbine to the shaft driving the generator. The efficiency of the turbine is 65%.
c)   Maximum power generated, in kWs, by this hydroelectric power generating system.

Solution
Given:

P_{in} = 50 psia
P_{out} = 10 psia
ΔP = 40 psia

ΔP Conv. to psf $= 40$ lb/in^2 x 144 in^2/ft$^2 = 5{,}760$ psf

$\dot{V}$ = Volumetric Flow Rate $= 120$ cu $-$ ft / sec

Turbine Efficicncy $= 65\%$
Generator Efficiency $= 90\%$

a)   According to Eq. 4.32, water horsepower delivered by the water to the turbine can be stated as:

$$\text{WHP} = P_p = \frac{(\Delta P).(\dot{V})}{550},$$ provided the differential pressure, ΔP, is expressed
in psf, or pounds per square foot

$$\therefore \; P_p = \frac{(5760 \text{ lbf / ft}^2).(120 \text{ ft}^3 \text{ / sec })}{550} = 1257 \text{ hp}$$

b)   Brake horsepower delivered by the turbine to the generator—via the shaft—at a turbine efficiency, η_t, of 65%:
   BHP $=$ WHP x η_t
         $= 1257$ hp x 0.65
         $=$ **817 hp**

c)   Maximum power generated, in kWs, by this hydroelectric power generating system, with the generator efficiency, η_g, of 90%:
   Electrical Power Generated (**EHP***), in kW $=$ BHP x η_g x 0.746 kW/hp
         $= 817$ hp x 0.9 x 0.746 kW/hp
         $= 549$ kW

Example 4.9
   A pump is to be installed to supply a maximum of 200 ft^3/sec at a gage pressure (ΔP) of 8,000 psf. Determine the size of the motor required and the size (kVA) of the transformer for this system. Assume that the efficiency of the pump is 85% and that the nameplate efficiency of the "custom" designed special motor is 92%. Power factor of the transformer is 0.8. The frictional and other hydraulic losses are negligible.

---

*Electric Horse Power, abbreviated as EHP, is loosely used to refer to generated electrical power measured in Horse Power, hp or kW.

Solution

Solution strategy in this case would be to use Eq. 4.32 to compute the WHP. Then, the amounts of real power "**P**" drawn by the motor and delivered by the motor would be computed based on the given efficiencies of the pump and the motor. Once the real power drawn by the motor is computed, we can use the power factor of the transformer to compute the kVA, or apparent power rating "**S**," of the transformer.

Given:

$\Delta P = 8000$ psf
$V =$ Volumetric Flow Rate $= 200$ cu-ft/sec
Pump Efficiency $= 85 \%$
Motor Efficiency $= 92 \%$
Power factor of the transformer $= 0.8$

According to Eq. 4.32, water horsepower delivered by the pump to the water can be stated as:

$$\text{WHP} = P_p = \frac{(\Delta P).(\dot{V})}{550}, \text{ provided the differential pressure, } \Delta P, \text{ is expressed}$$

in psf, or pounds per square foot

$$\therefore P_p = \frac{(8000 \text{ lbf}/\text{ft}^2).(200 \text{ ft}^3/\text{sec})}{550} = 2{,}909 \text{ hp}$$

According to the wire to water (hydraulic pump) power flow diagram in Figure 4.3:

$$\text{BHP delivered by the motor to the pump must be } = \frac{\text{WHP}}{\eta_p}$$

$$= \frac{2909}{0.85} \text{ hp} = 3{,}422 \text{ hp } \{\text{Custom, non-standard, size}\}$$

$\therefore$ the size of this custom motor would be $3,422$ **hp**.

Once again, according to the wire to water power flow diagram in Figure 4.3:

$$\text{EHP drawn by the motor, from the transformer} = \frac{\text{BHP}}{\eta_{\text{motor}}}$$

$$= \frac{3{,}422}{0.92} \text{ hp} = 3{,}720 \text{ hp}$$

Or, EHP, in kW $= (3,720$ **hp**$).(0.746\text{kW/hp}) = 2{,}775$ kW

Then, according to Eq. 4.23, the apparent power, **S** (kVA) rating of the feeder transformer would be:

$$PF = \text{Power Factor} = \frac{|P|}{|S|}$$

$$\text{Or, } |S| = \frac{|P|}{PF} = \frac{2,775\text{kW}}{0.8} = 3,469 \text{ kVA}$$

Self-assessment Problems & Questions—Chapter 4

1. Consider a hydroelectric reservoir where water is flowing through the turbine at the rate of 1100 ft³/sec. See the diagram below. The turbine exit point is 700 ft lower than the elevation of the reservoir water surface. The turbine efficiency is 90% and the total frictional head loss through the penstock shaft and turbine system is 52 ft.
   a) Calculate the power output of the turbine in MWs.
   b) If the efficiency of the Electric Power Generator is 92%, what would the electric power output be for this hydroelectric power generating system?

2. Which of the following two water heaters would cost the least to operate, on annual cost basis, under the given assumptions?
    A. Electric Water Heater:
        Estimated annual energy required to heat the water: **9000 kWh**
        Efficiency: **95%**
        Cost Rate: **$0.10/kWh**
    B. Natural Gas Water Heater:
        Estimated annual energy required to heat the water: Same as the Electric water heater
        Efficiency: **98%**
        Cost Rate: **$10.87/DT**

3. A computer manufacturing company is testing a prototype for the amount of heat it dissipates as wasted energy over a 10 hour period of operation. The computer is powered by a 24V DC power supply and is designed to draw 3A of current. Determine the total energy dissipated in Btu.

4. In response to a significant near miss incident and midair fire on a new commercial jet aircraft, a governmental agency is performing forensic analysis on the type of Lithium Ion aircraft battery suspected to be the root cause. Estimate the amount of *current* involved in the suspected fault on the basis of the following forensic data:
    — Total energy released in the catastrophic failure of the battery: 866 kJ
    — Estimated duration of fault: 2 seconds
    — Rated voltage of the battery: 3.7 VDC

5. A **156Sin377t** sinusoidal voltage is connected across a load consisting of a parallel combination of a **20 Ω** resistor and a 10Ω capacitive reactance.
    (a) Determine the real power dissipated by the resistor.
    (b) Determine the reactive power stored in a **10Ω** parallel capacitive reactance. (c) Calculate the total apparent power delivered to this parallel R and X circuit by the AC voltage source.

6. A **156Sin400t** sinusoidal voltage is connected across an unknown resistive load. If the power dissipated in the resistor is 1000 Watts, what is the resistance of the resistive load?

7. The AC circuit shown below depicts a three phase, one-line, schematic of a hydroelectric power generating station. Assume that there is no voltage

drop between the generator and the primary side of the transmission system transformer. The line current is indicated by an EMS system to be **10 kA, RMS**. Calculate the following if the power factor is known to be **0.95**:

a) Magnitude of the apparent power presented to the transmission lines.
b) Magnitude of the real power presented to the transmission lines.
c) The RMS line to neutral voltage at the source.

8.   A pump is to be installed on the ground floor of a commercial building to supply 200 ft³/sec of water up to an elevation of 100 ft. Determine the minimum size of the motor for this application. Assume that the efficiency of the pump is 80%. The weight density of water $\gamma = 62.4$ lbf/ft³.

Chapter 5

Power Factor

Introduction

Power factor was briefly introduced in Chapter 4 in order to gain complete and comprehensive understanding of the concepts, principles and application of AC power. Power factor stems from the fact that AC current lags behind or leads the AC voltage in AC circuits or systems that are predominantly inductive. Conversely, AC current leads the AC voltage in AC circuits or systems that are predominantly capacitive. If an AC circuit is purely resistive, or if the inductive reactance is completely offset by the capacitive reactance in the circuit, the voltage and current would be completely coincident, resulting in the perfect "utopian" power factor of, as shown in Figure 5.1.

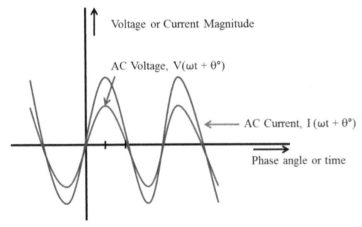

Figure 5.1: Coincident AC voltage and current and perfect power factor.

In AC electrical systems where inductive reactance, X_L, exceeds X_C, AC current and voltage will not be coincident. The current will lag behind the voltage with respect to time and phase angle; resulting in a power factor that is less than "1" or 100%. See Figures 5.2 and 5.3. The graph shown in Figure 5.2 shows AC voltage and current plotted as a function of time, while Figure 5.3 depicts AC voltage and current plotted as a function of phase angle. Upon closer examination one can discern the fact that, in both of these graphs, the current function (or graph) *follows* the voltage function in the rise and fall below the horizontal (abscissa) axis. This would constitute a lagging power factor scenario, explained in greater detail in later in this chapter. At this point, note that the "*leading*" and "*lagging*" qualifier associated with the term power factor is always assessed from current's vantage point. In other words, it is always the AC current that is lagging behind, leading or in synch with the AC voltage.

One way to remember that the current lags behind voltage in predominantly inductive circuits and that the current leads the voltage in predominantly capacitive circuits is through a well-known mnemonic:

ELI the ICE man
–   The **L** in **ELI** means *inductance*. The **E** (voltage) comes first, then the **I** (current) lags behind. Inductive reactance produces a lagging power factor

–   The **C** in **ICE** means *capacitance*. The **I** (current) comes first (leads) then the **E** (voltage) comes later. Capacitive reactance produces a leading power factor.

In this chapter, we will explore the physical aspects of power factor in greater detail, and we will illustrate the difference between lagging power factor and leading power factor. We will explain and demonstrate potential economic benefit that can be derived from improvement of power factor in AC systems. We will explain the consequence of low power factor on the energy productivity*, energy cost and the life of electrical equipment. We will conclude the power factor discussion with the introduction of conventional alternatives for power factor improvement.

POWER FACTOR
Power factor of an AC system may be defined "qualitatively" as the capacity of an AC system to covert transmitted or delivered apparent power to actual work or other forms of energy. The mathematical or "quantitative" definition of power factor—as introduced in Chapter 4—is represented by equations 5.1 and 5.2.

$$PF = \text{Power Factor} = \frac{|P|}{|S|} \qquad\qquad \textbf{Eq. 5.1}$$

Finance and Accounting for Energy Engineers, By S. Bobby Rauf.

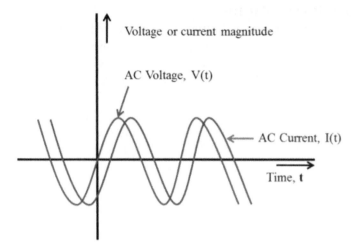

Figure 5.2: Lagging power factor graph as a function of time.

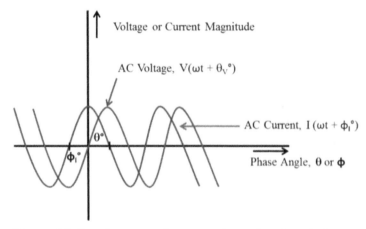

Figure 5.3: Lagging power factor graph as a function of phase angle.

$$PF = \text{Power Factor} = Cos(\theta_V - \theta_I) \qquad \textbf{Eq. 5.2}$$

Where,

$|P|$ = Magnitude of the real component of AC apparent power

$|S|$ = Magnitude of the overall AC apparent power

θ_V = Angle of AC voltage

θ_I = Angle of AC current, also denoted as ϕ_I

Lagging Power Factor

Since loads in most industrial, commercial and institutional facilities consist of inductive equipment such as motors, transformers, solenoids, etc., the inductive reactance, X_L, in such facilities, exceeds any capacitive reactance, X_C, that might be present. When AC voltage is applied to such predominantly inductive systems, the resulting AC current lags behind the voltage. If this lag were measured in terms of phase angle, the voltage angle θ_V would be greater than the current angle θ_I, the angular difference would be positive and would be quantified as ($\theta_V - \theta_I$). The cosine of this angular difference would represent the lagging power factor of the AC system and would be a stated in form of Eq. 5.2. Alternatively, power factor or the "efficiency" of conversion of AC apparent power, **S**, to real power **P** (rate of performance of work, or delivery of torque and energy) could be represented by the ratio of the magnitudes of the real power **P** and apparent power **S**, as stipulated by Eq. 5.1.

Note: Since cosine of any angle that lies in the 1st or 4th quadrants of the Cartesian coordinates is positive, power factor would be a positive number—declining with increasing angular difference—whether or not the current lags behind the voltage.

Example 5.1:
In an AC system, the voltage angle $\theta_V = 65°$ and the current angle $\phi_I = 20°$.
   a)   Calculate and define the power factor
   b)   Calculate and define the power factor if the voltage and current angles are reversed.
Solution:
   a) $Cos(65° - 20°) = Cos(45°) = 0.707$; or 70.7% lagging
   b) $Cos(20° - 65°) = Cos(-45°) = Cos(+45°) = 0.707$; or 70.7% leading

Lagging Power Factor from Impedance Perspective

A basic AC circuit with a lagging power factor, as depicted in Figure 5.4, would consist of resistance and inductance, represented by **R** and X_L, respectively. As shown in Figure 5.4, the AC voltage source, V_{ac}, drives AC current **I** through the total resistance **R** and the total inductance **L** in the AC system. Customarily, **R** would represent not only the resistive loads in the overall system but also the inherent resistance of the power distribution system, i.e., resistance of the conductors and other current carrying devices. The total inductance in the system, **L**, could be an aggregate of the inductances of all inductors in the system, i.e., motors, solenoids, transformers, etc. The total inductance, **L**, could be represented by its inductive reactance X_L, as shown in Figure 5.4, or by its impedance contribution, Z_L as annotated in Figure 5.5. While, in practice, there may be some capacitance and capacitive reactance in the system, for the sake of simplicity, it is assumed that this AC system is predominantly inductive, X_L >> X_C, or X_C is negligible.

Figure 5.4: AC circuit with a lagging power factor

One salient effect of the dominant inductive reactance X_L in this circuit is that it causes the current to lag behind the voltage. The current lags behind the voltage in terms of ***time*** as well as the ***angle*** as depicted graphically in the Figures 5.5 (a) and (b).

While Figure 5.4 illustrates the physical—circuit element—aspect of a predominantly inductive AC electrical system and Figure 5.5 highlights the resulting discordance between the voltage and current waveforms, Figure 5.6 helps us visualize the formation of a, less than perfect, non-unity power factor in terms of vectors.

The combination of the resistance and the inductance in the circuit to form the total impedance of the circuit is depicted in form of vectors **R** and Z_L and **Z**, respectively, in Figure 5.6 (a). Note that the impedance contribution of inductance in the circuit is represented by the vector Z_L pointing straight up at an angle of 90°. As discussed earlier in this text, impedance contribution Z_L by the inductance **L**, or inductive reactance X_L in an AC system, is denoted in complex (rectangular) number form as jX_L. Where, unit vector "**j**" is represented in polar/phasor form as $1∠90°$, and is responsible for assigning the 90° angle to Z_L.

As always, resistance **R** possesses an angle of zero; therefore, it is represented by a horizontal vector along the x-axis as shown in Figure 5.6 (a).

Vector addition of vectors Z_L and **R** results in the resultant vector **Z**. Where, **Z** represents the total or equivalent impedance of the predominantly inductive AC circuit. As ostensible from the Figure 5.5 (a), magnitude of **Z**, denoted as |**Z** |, is a geometric hypotenuse of the triangle formed by component

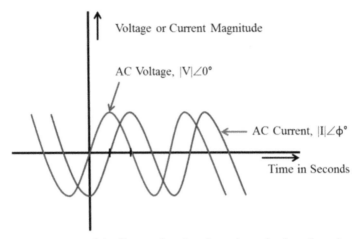

(a) Current lagging the voltage in time domain.

(b) Current lagging the voltage in phase angle domain.

Figure 5.5: AC circuit with a lagging power factor; current lagging the voltage in phase angle and time domains.

vectors **R** and $\mathbf{Z_L}$. Therefore, as annotated in Figure 5.5 (a), **Z** can be represented in polar form as $|\mathbf{Z}| \angle \theta°$, or in rectangular form as $\mathbf{R + jX_L}$.

One of the methods for assessment of power factor requires the knowledge of AC voltage and AC current in polar or phasor form. The voltage is assumed to be at an angle of **0°** and as shown in Figure 5.4, represented in vector/

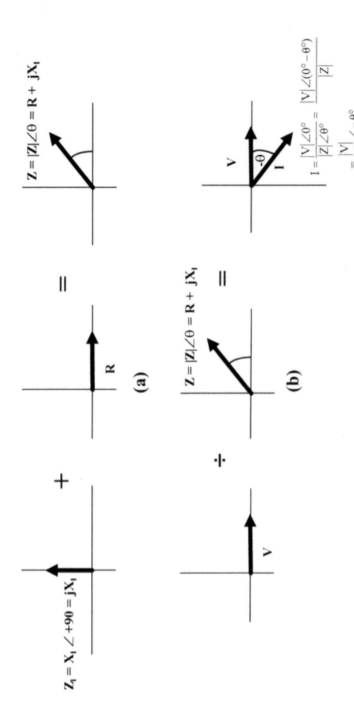

Figure 5.6: Graphical analysis of lagging power factor

polar form as $|V| \angle 0°$. The current is not known. The current can be determined by applying ohms law as follows:

$$I = \frac{|V| \angle 0°}{|Z| \angle \theta°} = \frac{|V| \angle (0° - \theta°)}{|Z|}$$

$$= \frac{|V|}{|Z|} \angle - \theta°$$

Then, by definition and in accordance with Eq. 5.2,

$$PF = \text{Power Factor} = \text{Cos}(\theta_V - \theta_I)$$

Since, in this lagging power factor case,

$$\theta_V = 0°$$
And, $\phi_I = \theta_I = -\theta°$
$$PF = \text{Power Factor} = \text{Cos}(\theta_V - \theta_I) = \text{Cos}\{0 - (\theta°)\}$$
Or,   Power Factor = Cosine of the angle of the impedance, Z.
$$PF = \text{Cos}(\theta_Z) \hspace{4cm} \textbf{Eq. 5.3}$$

Example 5.2
    A $110 \angle 0°$ voltage is applied to load consisting of a 10Ω resistance and 10Ω inductive reactance, in series. Calculate the power factor as seen by the source.

Ancillary: Verify the result using Eq. 5.1.

Solution:
    The circuit diagram for this system would appear as follows:

Eq. 5.3, derived above, can be applied to determine the power factor if the angle of the total impedance were known. The total impedance can be assessed, in polar/phasor form, using a scientific calculator or through the angle computation formula covered in the preceding chapters.

$$\left|\theta_Z\right| = \text{Tan}^{-1}\left(\frac{R}{\left|X_L\right|}\right) = \text{Tan}^{-1}\left(\frac{10\Omega}{10\Omega}\right) = \text{Tan}^{-1}(1) = 45°$$

$\therefore$ Power Factor $= PF = \text{Cos}(\theta_Z) = \text{Cos}(45°) = 0.707$,

or 70.7% lagging.

Ancillary: In order to verify the power factor computed above, as obvious from Eq. 5.1, we need to find the real power component **P** and the total apparent power **S** dissipated in the given circuit.

Apparent power S can be calculated using Eq. 4.11 introduced in Chapter 4.

$$\vec{S} = \vec{V}.\vec{I}^{\,*} \qquad\qquad\qquad \textbf{Eq. 4.11}$$

However, we must calculate the phasor value of the AC current first.

$$I = \frac{110\angle 0°}{\left|Z\right|\angle\theta_Z{}°}$$

$$\left|Z\right| = \sqrt{R^2 + X^2_L} = \sqrt{10^2 + 10^2} = \sqrt{200} = 14.14$$

$$\therefore I = \frac{110\angle 0°}{14.14\angle 45°} = 7.78\angle -45°$$

Note: The negative angle of the current, with the voltage angle being 0°, signifies the fact that the current is *lagging* behind the voltage, and therefore the power factor will be a *lagging* power factor.

S = V.I* = (110∠0°).(7.78∠45°) = 856∠45°
Note: "*" stands for conjugation of the original current angle
Then, P = |S|.Cos(45°) = |856|.Cos(45°) = 605W
And, Q = |S|.Sin(45°) = |856|.Sin(45°) = 605W {Redundant}

$$\therefore PF = \frac{P}{\left|S\right|} = \frac{605}{856} \quad 0.707 \text{ or } 70.7\% \text{ lagging}$$

LEADING POWER FACTOR FROM IMPEDANCE PERSPECTIVE

A basic AC circuit with a leading power factor, as depicted in Figure 5.7, would consist of resistance and capacitance, represented by **R** and X_C, respectively. As shown in Figure 5.7, the AC voltage source, V_{ac}, drives AC current **I** through the total resistance **R** and the total capacitance **C** in the AC system. The resistance, **R,** would represent not only the resistive loads in the overall system but also the inherent resistance of the power distribution system, i.e., resistance of the conductors and other current carrying devices. The total capacitance in the system, **C,** could be an aggregate of the capacitances in the system, transmission line capacitance, etc. Leading power factor can also be caused by machines (i.e. certain power generators) that might be operating at a leading power factor, computer switched mode power supplies, etc. The total capacitance, **C,** is represented by its capacitive reactance X_C, as shown in Figure 5.7, or by its impedance contribution, Z_C as annotated in Figure 5.9. While, in practice, there may be some inductance and inductive reactance in the system, for the sake of simplicity, it is assumed that this AC system is predominantly capacitive, $X_C \gg X_L$, or X_L is negligible.

Figure 5.7: AC circuit with a leading power factor

One distinct effect of the dominant capacitive reactance X_C in this circuit is that it causes the current to lead the voltage. The current leads the voltage in terms of *time* as well as the *angle*, as depicted graphically in the Figure 5.8 (a) and (b).

While Figure 5.7 illustrates the physical—circuit element—aspect of a predominantly capacitive AC electrical system and Figure 5.8 highlights the resulting discordance between the voltage and current waveforms, Figure 5.9 depicts the formation of a, less than perfect, non-unity power factor in terms of vectors.

The combination of the resistance and the capacitance in the circuit to form the total impedance of the circuit is depicted in form of vectors **R** and Z_C **and Z**, respectively, in Figure 5.9 (a). Note that the impedance contribution of capacitance in the circuit is represented by the vector Z_C pointing directly downward at an angle of **-90°**. As discussed earlier in this text, impedance contribution Z_C by the capacitance **C** or capacitive reactance X_C in an AC system

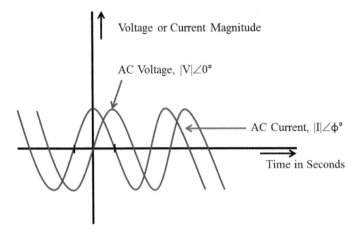

(a) Current leading the voltage in time domain.

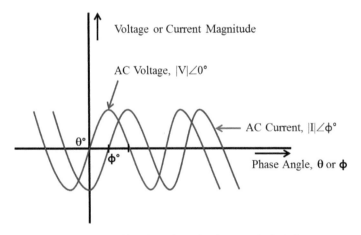

(b) Current leading the voltage in phase angle domain.

Figure 5.8: AC circuit with a leading power factor; current leading the voltage in phase angle and time domains

is denoted in complex (rectangular) number form as $-j\mathbf{X_C}$. Where, unit vector "$\mathbf{-j}$" is represented in polar/phasor form as $\mathbf{1\angle{-90°}}$, and is responsible for assigning the $\mathbf{-90°}$ angle to $\mathbf{Z_C}$.

Resistance $\mathbf{R}$ portends an angle of zero; therefore, it is represented by a horizontal vector along the x-axis as shown in Figure 5.9 (a).

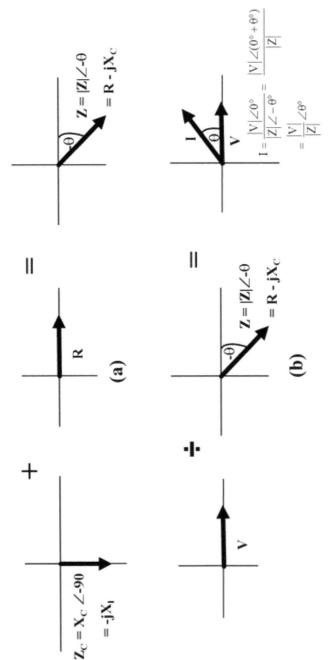

Figure 5.9: Graphical analysis of leading power factor

Vector addition of vectors $\mathbf{Z_C}$ and $\mathbf{R}$ results in the resultant vector $\mathbf{Z}$. Where, $\mathbf{Z}$ represents the total or equivalent impedance of the predominantly capacitive AC circuit. As obvious from the Figure 5.9 (a), magnitude of $\mathbf{Z}$, denoted as $|\mathbf{Z}|$, is a geometric hypotenuse of the triangle formed by component vectors $\mathbf{R}$ and $\mathbf{Z_C}$. Therefore, as annotated in Figure 5.9 (a), $\mathbf{Z}$ can be represented in polar form as $|\mathbf{Z}|\angle\text{-}\theta°$, or in rectangular form as $\mathbf{R} - \mathbf{jX_C}$.

One approach for assessment of power factor requires the knowledge of AC voltage and AC current in polar or phasor form. The voltage is assumed to be at an angle of $\mathbf{0°}$ and as shown in Figure 5.7, represented in vector/polar form as $|\mathbf{V}|\angle\mathbf{0°}$. The current is not known. The current—as with the lagging power factor scenario—can be determined by applying ohms law as follows:

$$I = \frac{|V|\angle 0°}{|Z|\angle -\theta°} = \frac{|V|\angle(0° + \theta°)}{|Z|}$$

$$= \frac{|V|}{|Z|}\angle\theta°$$

Then, by definition and in accordance with Eq. 5.2,

$$PF = \text{Power Factor} = \text{Cos}(\theta_V - \theta_I)$$

In this leading power factor case,

$\theta_V = \mathbf{0°}$
And, $\mathbf{\phi_I} = + \theta°$, or just $\theta°$

$$PF = \text{Power Factor} = \text{Cos}(\theta_V - \theta_I) = \text{Cos}(0 - (\theta°))$$
$$= \text{Cos}(-\theta_I) = \text{Cos}(\theta°)$$

Or, Power Factor = Cosine of the angle of the impedance, Z.
$PF = \text{Cos}(\theta_Z)$, which is the same as in the lagging
power factor scenario **Eq. 5.3**

Example 5.3

A $110\angle 0°$ voltage is applied to load consisting of a 10Ω resistance and 10Ω capacitive reactance, in series. Calculate the power factor as seen by the source.

Ancillary: Verify the result using Eq. 5.1.

Solution:

The circuit diagram for this system would appear as follows:

Eq. 5.3, derived above, can be applied to determine the power factor if the angle of the total impedance were known. The total impedance can be assessed, in polar/phasor form, using a scientific calculator or through the angle computation formula covered in the preceding chapters.

$$\left|\theta_z\right| = \text{Tan}^{-1}\left(\frac{R}{\left|X_L\right|}\right) = \text{Tan}^{-1}\left(\frac{10\Omega}{10\Omega}\right) = \text{Tan}^{-1}(1) = 45°$$

As shown in Figure 5.9 (b), **Z**, in the leading power factor scenario lies in the fourth quadrant, therefore, θ_Z = -45°.

∴ Power Factor = PF − Cos (θ_Z) − Cos (-45°) = 0.707, or 70.7% leading.

Note: Because power factor computation is based on "Cosine" of impedance angle, power factor remained the same as we transitioned from 10Ω inductive reactance system in Example 5.2 to a 10Ω capacitive reactance system in Example 5.3; it simply transformed from being lagging to leading.

Ancillary: In order to verify the power factor computed above, as obvious from Eq. 5.1, we need to find the real power component **P** and the total apparent power **S** dissipated in the given circuit.

Apparent power **S** can be calculated using Eq. 4.1 introduced in Chapter 4.

$$\vec{S} = \vec{V}.\vec{I}^{\;*} \hspace{4cm} \textbf{Eq. 4.11}$$

However, as in Example 5.2, we must calculate the phasor value of the AC current first.

$$I = \frac{110\angle 0°}{|Z|\angle -\theta_z°}$$

$$|Z| = \sqrt{R^2+X^2_L} = \sqrt{10^2+10^2} = \sqrt{200} = 14.14$$

$$\therefore \; I = \frac{110\angle 0°}{14.14\angle -45°} = 7.78\angle +45° \; \text{or} \; 7.78\angle 45°$$

Note: The positive angle of the current, with the voltage angle being 0°, signifies the fact that the current is *leading* the voltage, and therefore the power factor will be a *leading* power factor.

S = V.I* = (110∠0°).(7.78 ∠ – 45°) = 856∠ – 45°

Note: "*" stands for conjugation of the original current angle

Then, P = |S|.Cos(-45°) = |856|.Cos(-45°) = -605W

And, Q = |S|.Sin(-45°) = |856|.Sin(-45°) = -605W {Redundant}

Or, S = P + jQ = 605 + j(-605) = 605 - j605 VA

$$\therefore \; PF = \frac{P}{|S|} = \frac{605}{856} \quad 0.707 \; \text{or} \; 70.7\% \; \text{lagging}$$

Example 5.4

An $X_L = 10\Omega$ capacitive reactance worth of capacitance is added in series with the series resistor and inductive reactance AC circuit from Example 5.2.

    (a) Draw the circuit diagram for the new circuit.

    (b) Determine the net impedance of the new circuit.

    (c) Calculate the new power factor.

Solution:

(a)

(b)   Since all elements in the modified circuit are connected in series, the total or net impedance in this AC circuit would be as follows:

$$Z = R + jX_L - jX_C = 10\Omega + j10\ \Omega - j10\ \Omega = 10\ \Omega,$$
$$\text{or, } Z = 10\Omega\ \angle 0°\Omega$$

(c)   Since, in this case, the inductive reactance is completely offset by the capacitive reactance, the net or total impedance is composed solely of pure resistance of 10Ω. As discussed earlier in this chapter and Chapter 4, power factor in circuits that are purely resistive or have "0" net reactance, is "1" or 100%.

      This can be proven by applying Eq. 5.3 derived earlier, and recognizing the fact that θ_Z, the angle of total impedance Z, as derived in part (b) is $0°$:

Power Factor = Cosine of the angle of the impedance, Z.
    Or, PF = $Cos(\theta_Z) = Cos(0) = 1$

As obvious by now, in essence, Example 5.4 shows us how capacitive reactance can be added to improve or correct power factor in a predominantly inductive AC circuit. This serves as an appropriate segue into the next section that pertains to the Power Factor Correction.

POWER FACTOR CORRECTION

There are multiple methods for correcting power factor. Some of the more common approaches for correcting power factor are as follows:

    a)   Addition of power factor correcting capacitors
    b)   Addition of power factor correcting inductors
    c)   Frequency manipulation

The power factor correction method chosen depends on the composition or design of the AC system. In general, if the power factor is less than "**1**," or 100%, due to the predominance of inductive reactance, $\mathbf{X_L}$, capacitance or capacitive reactance, $\mathbf{X_C}$, must be added in series or shunt to raise the power factor closer to 100%. In fact, Example 5.4—where 10Ω of $\mathbf{X_L}$ is offset by the addition of 10Ω of $\mathbf{X_C}$—is a classic illustration of this approach. Frequency manipulation method is applicable when, in addition to resistance, the both inductive and capacitive reactances are present. As described later in this chapter, when the power factor is less than 100% due to the dominance of capacitance in an AC electrical system, inductive reactance may be introduced into the circuit to offset the capacitive reactance and improve the system's power factor.

Addition of capacitance or capacitive reactance

Since power factor correcting capacitors, essentially, impact the AC circuits or systems through addition of capacitive reactance X_C, or $Z_C = -jX_C$, their addition to a predominantly inductive circuit yields different results depending upon whether they are connected in series with the load or in parallel (or shunt). Example 5.4 represents a scenario where power factor correcting capacitance is added in series. Comprehensive discussion on the distinction between addition of power factor correcting capacitors in parallel versus series is beyond the scope of this text. Recognizing the fact that the difference in parallel versus series addition of capacitors is mainly due to the difference in parallel versus series addition of impedances, in general, should suffice.

Power factor correcting capacitors can also be viewed as devices used to reduce current in power systems. Capacitors do so by providing a leading current to loads that require lagging current. In other words, capacitors provide "off- setting" current for inductive loads.

Another side benefit in the use of a series capacitor is in the fact that they can reduce voltage drop, especially in motors starting situations and other high impulse loads. This can be attributed to the fact that capacitors, as discussed earlier in this text, don't permit instantaneous drop or rise in voltage.

There are multiple approaches for assessment of capacitance required for specific improvement of power factor. One method relies on the knowledge of existing and desired reactive powers, Q_1 and Q_2, respectively; the operating voltage and the frequency. The formula for determination of the required per phase capacitance as follows:

$$C = \frac{(Q_1 - Q_2)}{2\pi f V^2} \qquad\qquad \textbf{Eq. 5.4}$$

Where,

    C = Capacitance, in Farads, required to reduce the reactive power from Q_1 to Q_2 per phase

    Q_1 = Initial, higher, reactive power, in VARs per phase

    Q_2 = Improved, lower, reactive power, in VARs, per phase

    V = Voltage, in Volts

    f = Frequency, in Hz

This formula is premised on the following mathematical definitions of reactive power Q:

$$|Q| = \frac{V^2}{X_C} = V^2 2\pi fC \qquad\qquad \textbf{Eq. 5.5}$$

Or, for inductive reactance:

$$|Q| = \frac{V^2}{X_C} = \frac{V^2}{2\pi fC} \qquad\qquad \textbf{Eq. 5.6}$$

An electrical engineer may use Eq. 5.4 to determine the capacitance required to improve the power factor, by a specific amount, in lieu of the apparent power (S) reduction approach. The apparent power based power factor calculation method is illustrated, later, through Example 5.7.

Example 5.5

The HMI (Human Machine Interface) monitor of an EMS system, monitoring a chilled water pump motor, is showing that the motor is drawing a reactive power, Q_1, of 50 kVARs. Funding has been allocated as a part of a DSM (Demand Side Management) energy conservation project to improve the power factor of the motor branch circuit such that the reactive power is reduced to Q_2 of 10 kVARs. The branch circuit is operating at 240 V_{RMS}. The system frequency is 60 Hz. Calculate the amount of capacitance that must be added, per phase, to achieve the power factor improvement project objective.

Solution:
Apply Eq. 5.4:

$$C = \frac{(Q_1 - Q_2)}{2\pi fV_2} = \frac{(50\ kVAR - 10\ kVAR}{2\pi(60Hz)(240V)^2}$$

$$= \frac{(50,000\ VAR - 10,000\ VAR)}{2\pi(60Hz)(240V)^2} = 0.001843F = 1.84\ mF$$

Example 5.6

Power Factor Improvement and Cost Savings: An air compressor station is consuming 2,000 kW at a power factor of 0.7. The utility company charges a $5.00/kVA per month as penalty for poor power factor. What would be the annual, pre-tax, savings if capacitors could be installed and power factor improved to 0.9?

Solution:

Since the objective is to assess the cost savings on the basis of apparent power S (kVA) reduction, we must determine the apparent power S_1 being drawn by the air compressor motor at the existing power factor of 0.7 (70%), and the apparent power S_2 at the desired power factor of 0.9 (90%). We can compute S_1 and S_2 by rearranging and applying Eq. 5.1.

$$PF = \text{Power Factor} = \frac{|P|}{|S|} \qquad\qquad \textbf{Eq. 5.1}$$

Rearrange Eq. 5.1:

$$\text{Magnitude of Apparent Power } |S| = \frac{|P|}{PF}$$

Therefore,

$$\text{Magnitude of existing apparent power } |S_1| = \frac{|P|}{PF_1} = \frac{2{,}000 \text{ kW}}{0.7}$$

$$= 2857 \text{ kVA}$$

$$\text{Magnitude of desired apparent power } |S_2| = \frac{|P|}{PF_2} = \frac{2{,}000 \text{ kW}}{0.9}$$

$$= 2222 \text{ kVA}$$

Note that real power drawn, which is a function of the actual work performed or torque generated, remains unchanged at 2,000 kW.

$$= (2857\text{kVA} - 2222\text{kVA}) \cdot \left(\frac{\$5}{\text{kVA - Month}}\right) \cdot (12 \text{ Months/Year})$$

$$= \$38{,}095$$

Automatic Switching PF Correcting Capacitor System

Imagine a plant engineering project that entails installation of a large, 500 hp, air compressor. The 500 hp rating of the air compressor is premised on the size of the motor. In other words, the 500 hp rating, in most cases would be real power (**P**) delivered by the motor to the compressor. However, as we discussed earlier, motors add ***inductive reactance*** to electrical system. So, the branch circuit the motor pertains to, in this facility, would have a non-unity, or less than 100% power factor, say 80%. Sizeable loads, such as large air compressors, therefore, degrade the aggregate power factor of the entire electrical system. On new installations, such as the one being considered here, degradation of power factor can be prevented by including automatic switching power factor correcting capacitor banks such as the one shown in Figure 5.10. Automatic switching power factor correcting capacitor banks constitute a more optimal and effective approach for automatic introduction and removal of capacitance to maintain the system power factor at an optimal level.

The controls in an automated power factor correcting system connect or disconnect capacitors as the inductive load appears and disappears from the overall electrical circuit. This ensures that the power factor correcting capacitors—and the associated capacitive reactance—drop out of the circuit when the compressor motor load (and associated inductive reactance) switches off. ***Important point***: If the power factor capacitors were to stay connected to the compressor motor branch circuit when the motor is off, the circuit would become predominantly capacitive, as X_C would exceed X_L, resulting in a low leading power factor. In other words, if the capacitor bank is not disconnected when the compressor based inductive reactance, X_L, drops out, it would result in a ***capacitive reactance "overdose."***

As annotated in Figure 5.10, there are six key functional components in the pictured automatic power factor correcting system:

Figure 5.10: Automated power factor correcting capacitor bank.

1.   Control contactor or relay: for connecting and disconnecting of the capacitors.

2.   Terminal Strip or Connection Points: for interface to the motor circuit, sensors, and other field connections.

3.   Slow blow fuses: for overload protection.

4.   Inrush limiting contactors: for limiting the motor inrush current.

5.   Capacitors (single-phase or three-phase units, delta-connection: power factor correcting capacitors (X_C).

6.   Control transformer: for obtaining control voltage, i.e. 110 VAC, from the existing power source, operating at power circuit voltage, i.e. 480 VAC.

**General information on power factor and
the application of power factor correcting capacitors:**

What is a suitable power factor goal for industrial, commercial and institutional facilities? It would be difficult to present a specific and perfect answer to this question. While one could find enough data to support the claim that the overall average power factor in the United States is approximately 82%, the best answer varies from situation to situation, and industry to industry. For example, on one end of the spectrum, one might observe power factors approaching 90% to 97% in the institutional and commercial structures, while on the low end of the power factor spectrum it is not uncommon to come across power factors to the tune of 50% or lower in certain foundry, plastics and chemical industries, where the melting processes employ electrical inductive heating.

What is the financial return and payback associated with power factor correction projects? Some utilities base demand charges on kVA demanded by the customer facility. In such cases, an approximate 10% demand (kVA) reduction can be realized if a power-factor-correction capacitor bank is installed to improve the power factor from 82% to 90%. Such and investment has a typical payback of, approximately, two (2) years. Note, however, that the payback periods, on power factor correction based energy projects—or for that matter, on any energy project—will be *shorter* for consumers located in higher energy cost regions, i.e. the North Eastern United States, the Western states and Hawaii. Conversely, the payback periods will be longer in lower energy cost regions, i.e. the South East, the Dakota's, etc. See Finance and Accounting for Energy Engineers, by S. Bobby Rauf.

Application, installation, operation and maintenance of power factor correcting capacitors: Application of power factor correcting capacitors should be preceded by proper analysis and design. Improper application of power factor correcting capacitors can damage the electrical system. Improper application of power factor correcting capacitors can cause fuses to blow and can amplify system harmonics. After installation, capacitor banks must be included in the facility's Preventive Maintenance Program. Loose connections and inherent defects, when ignored, can lead to unanticipated failures, faults, downtimes and arc flash hazards*.

Higher power factor means lower apparent power, S (in kVA), demanded and lower "unproductive" current in the electrical system. This "unproductive" current is wasted, as heat, as it flows through conductors, protective devices, power monitoring devices, transformers and loads. As we know, in general, operation of equipment above normal temperatures results in shortening of the life of equipment. Case in point and analogy: overheated machines, such as automobile engines, tend to cease and fail when overloaded or operated at abnormally high temperatures. Therefore, improving the power factor not only results in improved energy efficiency but it also extends service life, and reduces maintenance cost of equipment.

If inductive reactance (X_L, in) and reactive power (Q, in VARs) are encountered through a distribution transformer, application of PF capacitor is recommended at the distribution bus feeding the transformer. When the inductive reactance (X_L, in Ω) and reactive power (Q, in VARs) are contributed by large motors, the greatest PF corrective effect and energy loss reduction is realized when the capacitors are installed at each motor load (on line side), or between the motor starters and the motors. However, the initial cost for a number of small banks of capacitors would tend to be higher than the initial cost of a single, large, bank of capacitors, catering to several motor loads. In such cases, the higher initial cost must be weighed against gains in energy cost and PF correction benefits. Systems that experience large load swings, throughout the day or month, are good candidates for automatic switching correction systems. As explained earlier, these systems, automatically, correct the PF as load variations occur.

Addition of inductance or inductive reactance to correct for leading power factor

If the power factor is less than "**1**" due to the predominance of capacitance or capacitive reactance, X_C, inductance or inductive reactance, X_L, must

---

*NFPA® 70E, 2015

be added in series or shunt to raise the power factor closer to 100%.

In some AC electrical systems, such as the power transmission lines, the total impedance as seen from the utility side is indeed predominantly capacitive. In such cases, **inductance** must be added to improve the power factor.

Addition of power factor correcting inductive reactance, and analytical methods to correct power factor in predominantly capacitive electrical systems, is illustrated in self- assessment problems 2 and 3.

Frequency manipulation approach for correction of power factor

While pursuit of a perfect **unity** power factor in low power factor situations is not always feasible or practical, for simplicity, we will limit our discussion associated with frequency based power factor correction to scenarios where power factor is, indeed, raised to 1 or 100%. As stated earlier, frequency manipulation method pertains to AC electrical systems where the both inductive and capacitive reactances are present, in addition to resistance.

Consider the AC circuit shown in Figure 5.11 with an existing power factor that is less than 1.

Figure 5.11: An RLC (Resister, Inductor and Capacitor) circuit.

In order for this circuit's power factor to be unity or 100%, and the circuit must be purely resistive. In other words, $\mathbf{Z_{Eq}} = \mathbf{R}$ and $\mathbf{Z_C}$ must cancel or offset $\mathbf{Z_L}$. So, when inductive reactance $\mathbf{X_L}$ and capacitive reactance $\mathbf{X_C}$ are fixed, the electrical frequencies "f" and "ω" may be changed such that the magnitudes of $\mathbf{Z_C}$ and $\mathbf{Z_L}$ become equal and cancel each other. The frequency at which $|\mathbf{Z_C}|$ and $|\mathbf{Z_L}|$ equate is called **resonant frequency, f_0 or ω_0**. An electrical circuit operating at resonance frequency is called a **resonant circuit**. The derivation and mathematical proof for resonance frequency is as follows:

In order for Z_L and Z_C to cancel each other when added together in a series combination: $Z_L = -Z_C$.

Since $Z_L = j2\pi fL$ and $Z_C = -jX_C = -j\dfrac{1}{2\pi fC}$

$$j2\pi fL = -\left(-j\dfrac{1}{2\pi fC}\right)$$

Or, $X_L = X_C$

$$2\pi fL = \dfrac{1}{2\pi fC} \text{ and, } f^2 = \dfrac{1}{(2\pi)^2 LC}$$

And,

$$f = f_o = \dfrac{1}{2\pi\sqrt{LC}}, \text{ or, since } 2\pi f = \omega$$

$$\omega = \omega_o = \dfrac{1}{\sqrt{LC}}$$

Where,

L = Inductance in Henry, or H
C = Capacitance in Farads, or F
ω = Angular frequency, in general, in radians/sec.
ω_0 = Resonance angular frequency, in radians/sec.
f = Electrical frequency, in general, in cycles/sec, hertz, or Hz; 60Hz in the US.
f_0 = Resonance electrical frequency, in hertz.
Z_C = Impedance contribution by capacitance in an AC circuit, in Ω.
Z_L = Impedance contribution by inductance in an AC circuit, in Ω.
X_C = Reactance due to capacitance in an AC circuit, in Ω.
X_L = Reactance due to inductance in an AC circuit, in Ω.

As obvious, the frequency method for correcting power factor is feasible only in circumstances where the electrical frequency can be varied through the use of electromechanical systems such as electrical power generators, or via electronic variable frequency drives.

Example 5.7

The output of a variable frequency drive, as shown in the circuit below, is $157\mathrm{Sin}\omega t$. The VFD output is currently set at 60Hz. This drive is connected to a resistive load, capacitive reactance, and an inductive reactance.

(a)   What should be the new frequency setting to attain a power factor of 1, or 100%?

(b)   What is the resonance frequency for this AC system?

(c)   Could a power factor other than unity be attained through variation of frequency?

Solution

(a)   As explained earlier in this section, the power factor of an AC circuit, consisting of inductive and capacitive reactance, peaks to the maximum value of unity, or 100%, at resonance frequency, f_0. However, in this case, because the inductance and capacitance are represented as inductive and capacitive reactances, respectively, we cannot apply Eq. 5.8, directly, to compute f_0. We must convert reactances X_C and X_B to corresponding capacitance, C, and inductance, L, values.

$$\text{Since } X_L = 2\pi f L, \; L = \frac{X_L}{2\pi f}, \text{ or, } L = \frac{10\Omega}{2\pi(60hz)} = 0.02654 \text{ H}$$

And,

$$X_C = \frac{1}{2\pi f C}, \; C = \frac{1}{2\pi f X_C}, \text{ or, } C = \frac{1}{2\pi(60hz)(30\Omega)} = 88.46 \times 10^{-6} \text{F}$$

$$f = f_0 = \frac{1}{2\pi\sqrt{LC}} = \frac{1}{2(3.14)\sqrt{(0.02654).(88.46 \times 10^{-6})}} \qquad \text{Eq. 5.8}$$

$$\therefore \quad f = f_0 = 104hz$$

(b)   As explained earlier, resonance frequency *is* the AC frequency at which the total capacitive reactance in an AC circuit cancels or offsets the total inductive reactance. Therefore, resonance frequency, f_0, for the given AC circuit is the frequency calculated in part (a). In other words, $f_0 = 104Hz$.

(c)   The power factor does, indeed vary with frequency, with all other entities held constant. For instance, with all parameter held constant, if the frequency is lowered to 30Hz, the power factor drops down to, approximately, 96%.

Ancillary: Reader is encouraged to prove this. Also see self-assessment problem 4.

Chapter 5
Self-assessment Problems & Questions

1.   Determine the power factor of the circuit shown below, as seen by the AC source.

2.   Assume that the circuit depicted below represents one phase of a special power transmission line. Determine the power factor of the circuit shown below, as seen by the AC source.

3.   If the power factor in problem 2 is less than 1.0, how much capacitance or inductance must be added in series to raise the power factor to unity?

4.   The output of a variable frequency drive, as shown in the circuit below, is 157Sinωt. The VFD output is currently set at 50Hz. This drive is connected to a resistive load, capacitive reactance, and an inductive reactance.
      (a) What should be the new frequency setting to attain a power factor of 1, or 100%.

(b) What is the existing power factor, at 50 Hz?

(c) What would be the power factor if all circuit elements remain unchanged and the VFD frequency is lowered to 30Hz.

$R = 30\,\Omega$ $X_C = 30\Omega$

$157Sin\omega t$

$X_L = 10\Omega$

I

5. The HMI (Human Machine Interface) monitor of an Automated HVAC system, monitoring an air washer supply fan motor is indicating a reactive power, Q_1, of 60 kVARs. This system is located in the United Kingdom, where the AC frequency is 50 Hz. Determine the amount of capacitance that must be added to improve the power factor of the motor branch circuit such that the reactive power is reduced to Q_2 of 20 kVARs. The branch circuit is operating at 240 V_{RMS}.

6. Power Factor Improvement and Cost Savings: In conjunction with the local utility company DSM program, a manufacturing plant is being offered $2 per kVA for improvement in power factor from 0.75 to 0.85. The plant is operating at its contract level of 30 MW. Determine estimated annual pre-tax revenue if the plant accepts the offer.

7. The output of a variable frequency drive, as shown in the circuit below, is 157Sinωt. The VFD output is currently set at 60Hz. This drive is connected to a resistive load, capacitive reactance, inductive reactance and a "black box" load, Z_B, of 10Ω∠45°. What should be the new frequency setting to attain a power factor of 1, or 100%?

$R = 40\,\Omega$ $X_C = 30\Omega$

$157Sin\omega t$

$X_L = 10\Omega$

$10\Omega\angle45°$

I

Chapter 6

Demand, Load Factor, Service Factor and Electrical Power Bill Computation

Introduction

In this short chapter we will begin with the discussion on fundamental concepts of electrical demand and peak demand. Introduction to the concepts of demand and peak demand should poise us well for investigation of the concept of load factor, its role in power quality considerations and electrical energy cost reduction through peak shaving. Later in the chapter, we will explain the concept of service factor as it applies to electrical equipment. This chapter concludes with discussion and analysis of large industrial power bill computation. The bill calculation examples are designed to help the reader develop skills and acumen to analyze and understand important components of large power bills, in their efforts to identify cost reduction opportunities. As with other electrical concepts presented in this text, we will substantiate our discussion with analogies, mathematical and analytical models, as applicable.

DEMAND

In the electrical power distribution and energy realms, generally, the term "demand" implies electrical power demanded by electrical loads. The term "power" could be construed as *apparent power*, **S**, measured in VA, kVA, MVA, etc., or it could be interpreted as real power, **P**, measured in W, kW, MW, etc. Many electrical power utilities tend to apply the term *demand* in the context of *real power*, **P**, demanded from the grid. However, some electrical power utilities use the term demand to signify *apparent power* demanded from the grid, measured in kVA. When demand is known to represent real power in kWs or MWs, average demand is computed in accordance with Eq. 6.1.

Average Demand, in kW

$$= \frac{\text{Energy (kWh or MWh) consumed during the billing Month}}{\text{Total number of hours in the month}} \qquad \textbf{Eq. 6.1}$$

PEAK DEMAND

The term peak demand has two common or mainstream interpretations. The first interpretation is associated, mostly, with electrical power load profiling, such as the real power, P (kW), load profile or graph included among the various power monitoring screens available through EMS, Energy Monitoring Systems*, or BMS, Building Management Systems. The graph in Figure 6.1 is a screen capture of load profile screen in an EMS system. The graph depicts a plot of real power load (in kWs) plotted as a function of time. The darker line spanning from the left to right in the graph represents the *actual power demand (in kW),* recorded, by the hour, over a 24-hour period. The lighter line in the graph represents *demand forecasted* on the basis of the actual load profile measured and recorded by the EMS system, over the long term. The forecasted demand plays an essential role in energy conservation related peak shaving programs†. As obvious, demand forecast premised on actual, measured, load profile—for a specific facility—would tend to be more reliable than general load profiles based on theoretical models. This constitutes an important justification for EMS systems.

The second interpretation of the term peak demand is associated more directly with the computation of electrical power bills for large industrial and commercial consumers of electricity. The peak demand, in the billing context, is computed by most power companies on the basis of demand intervals.

Some utilities (power companies) base their peak demand computation on the basis of 30 minute intervals and others base the demand computation on the basis of 15 minute intervals. We will limit our consideration of the peak demand interval to 30 minute intervals.

Under one of the more common large power consumer contract rate schedules, referred to as the OPT, Time of the Day, or Time of Use Rate Schedule, the power company captures and records the energy consumed by the consumer, in kWh, over half hour intervals. Each energy data point thus captured is converted to the corresponding demand value, in kW, as stated under Eq. 6.2.

$$\text{Demand} = \text{Power in kW} = \frac{\text{Energy in kWh}}{\text{Time in Hours}} = \frac{\text{Energy in kWh}}{1/2 \text{ Hour}} \qquad \textbf{Eq. 6.2}$$

---

*†*Finance and Accounting for Energy Engineers* by S. Bobby Rauf

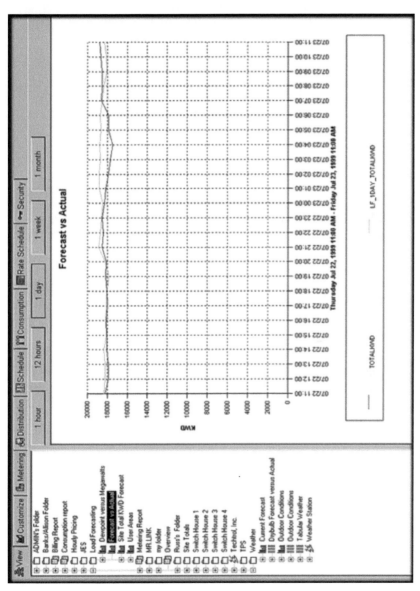

Figure 6.1: Actual vs. forecasted power demand (kW)

This could amount to, approximately, 1440 data points for a 30 day bill-ing month, with each data point representing the demand (kW) for a specific 30 minute interval during the billing month. The utility provider (power company) selects the highest demand data point, during the declared "on-peak" periods and labels it **Peak Demand** for the month. See additional information on peak demand under the bill calculation section.

LOAD FACTOR

The most common and general definition of load factor is that it is the ratio of *average power* to *peak demand*. This definition of load factor can be translated into the Eq. 6.3.

$$\text{Load Factor} = \frac{\text{Average Demand for the Month, in kW or MW}}{\text{Peak Demand for the Billing Month, in kW or MW}} \qquad \textbf{Eq. 6.3}$$

Equation 6.3 can be modified to define load factor (LF) in terms of **Energy Usage (in kWh)** and **Demand (in kW).**

$$\text{Load Factor} = \frac{\text{kWh used in the Billing Period)/}{(\text{Hrs in Billing Period})}}{\text{Peak Demand in kWh}} \qquad \textbf{Eq. 6.4}$$

A closer look at Eq. 6.3 and Eq. 6.4 highlights the following facts:

1) High peak demands (large denominator) result in low load factors. Low load factors result in higher energy cost due to high peak demand and, often, due to low load factor triggered penalties.

2) Higher load factors signify better power management, better power quality, higher energy productivity and lower energy cost.

3) High load factor implies relatively leveled and controlled demand (kWs), short peaks with low amplitude, accomplished in some cases through "peak shaving" and targeted peak avoidance.

A load factor of unity or "1" indicates that demand peaks are negligible resulting in steady and relatively leveled power (kW) demand. Since most fa-cilities don't operate 24 hours a day, and since the demand (kW) varies during a 24 hour period and the billing month, load factor—similar to power factor—in most practical applications is usually less than one "1."

Low load factor is a good indicator of the cost savings potential in shifting some electric loads to off-peak hours, thus reducing on-peak demand.

In a utopian situation, theoretically, if the load factor of a facility is already almost unity, further demand reduction can be accomplished through replacement of existing *low efficiency* equipment with *higher efficiency* equipment

Example 6.1

The EMS System at an automotive plastic component manufacturing plant that operates around the clock, 365 days a year, is displaying following electrical power data:

Billing Days in the Current Month: 30

On Peak Energy Consumption: 4,320,000 kWh

Off Peak Energy Consumption: 17,280,000 kWh

Highest 30 Minute Energy Recorded: 17,500 kWh

Calculate the following assuming that this facility is on OPT, Time of Use, contract with 30 minutes demand interval:

(a) Average demand.

(b) Peak demand.

(c) The load factor for the current month.

Solution:

(a) Average demand can be calculated by applying Eq. 6.1 as follows:

Average Demand, in kW

$$= \frac{\text{Energy (kWh or MWh) consumed during the billing month}}{\text{Total number of hours in the billing month}}$$

$$= \frac{\text{On-peak Energy} + \text{Off-peak Energy Consumption}}{\text{Total number of hours in the billing month}}$$

$$= \frac{4,320,000 \text{ kWh} + 17,280,000 \text{ kWh}}{(24 \text{ hrs/day}) \cdot (30 \text{ days/month})}$$

$$= 30,000 \text{ kW or } 30 \text{ MW}$$

(b) Peak demand can be calculated by applying Eq. 6.2 to the 30 minute interval during which the highest energy consumption is recorded:

$$\text{Demand} = \text{Power in kW} = \frac{\text{Energy in kWh}}{\text{Time in hours}} = \frac{\text{Energy in kWh}}{1/2 \text{ hour}}$$

$$= \frac{17{,}500 \text{ kWh}}{0.5 \text{ h}} = 35{,}000 \text{ kW or 35 MW}$$

(c)   Load factor can be calculated by applying Eq. 6.3 as follows:

$$\text{Load Factor} = \frac{\text{Average Demand for the Month, in kW or MW}}{\text{Peak Demand for the Billing Month, in kW or MW}} \qquad \textbf{Eq. 6.4}$$

Using the Average Demand, calculated in part (a) as 30 MW and the Peak Demand calculated in part (b) as 35 MW:

$$\text{Load Factor} = \frac{30 \text{ MW}}{35 \text{ MW}} = 0.8571 \text{ or } 85.71\%$$

SERVICE FACTOR

Service factor of electrical equipment such as motors, transformers, switchgear, etc. can be defined as the ratio of load the equipment can sustain continuously and the load rating of that equipment. Another way to view service factor is that it is a ratio of "safe" operating load to standard (nameplate) load. Service factor is typically expressed in decimal. Occasionally, the decimal value of load factor is presented in percentage form. The formula for service fact is stated as Eq. 6.5 below.

$$\text{Service Factor} = \frac{\text{Safe or Continuous Load, in kW, kVA or hp}}{\text{Nameplate rating of equipment, in kW, kVA or hp}} \qquad \textbf{Eq. 6.5}$$

Example 6.2

A 20-hp motor has been tested by a motor manufacturer to safely and continuously sustain a load of 25 hp. What service factor should the manufacturer include on the nameplate of the motor?

Solution:

$$\text{Service Factor} = \frac{\text{Safe or Continuous Load, in kW, kVA or hp}}{\text{Nameplate rating of equipment, in kW, kVA or hp}}$$

$$= \frac{\text{Safe Operating Load}}{\text{Full Load Rating of the Motor}} = \frac{25 \text{ HP}}{20 \text{ hp}} = 1.25$$

COMPUTATION OF LARGE INDUSTRIAL OR COMMERCIAL ELECTRICAL POWER BILLS

Electrical bill calculation for large electrical power consumers is different and more complex than the assessment of monthly residential bills. As supported by the electric bill shown in Figure 6.2, residential electrical bill of an average size American home in the Southeast can be calculated using Eq. 6.6.

Baseline Charge = (Present Reading - Prev. Reading) . (Rate in #kWh)**Eq. 6.6**

Total Bill = Baseline Charge + Special Riders + Taxes   **Eq. 6.7**

Example 6.3

If the "present reading" in the residential bill depicted in Figure 6.2 were 85552, what would be the baseline bill for the month? The energy cost rate for this property remains unchanged, at ¢11.9/kWh.. **(Hint: The baseline charge does not include riders or taxes.)**

Solution:

According to Eq. 6.6,
Baseline Charge = (85552 - 84552) . ($0.119/kWh)
      = (1000) . ($0.119/kW) = $119

Case Study 6.1: EMS Based Electrical Power Bill— Large Industrial Power Consumer*

This case study is modeled on a scenario with large industrial customer in the southeastern region of the United States. The rates in $/kW or $/kWh and the fixed charges are assumed to represent that time frame, during non-summer months. This consumer is assumed to have a 45 MW (megawatt) contract with the power company. This facility is a 24/7, 365-day operation and a significant portion of its load can be shifted from on-peak periods to off-peak periods, during its daily operations. This overall service contract of 45 MW is split into the following two components:

A 30 MW, uninterruptible, power service, on OPT, Time of Use Schedule.

Finance and Accounting for Energy Engineers, by S. Bobby Rauf

Service From: FEB 26 to MAR 26 (28 Days)   Your next scheduled meter reading will occur between   APR 26 and MAY 01

PREVIOUS BILL AMOUNT	PAYMENTS (-)	NEW CHARGES (+)	ADJUSTMENTS (+ OR -)	AMOUNT DUE (=)
$51.94	$51.94	$47.35	$0.00	$47.35

METER NUMBER	METER READINGS: PREVIOUS	PRESENT	MULTI-PLIER	TOTAL USAGE	RATE SCHEDULE DESCRIPTION	AMOUNT
431400	84552	84938	1	386 KWH	RS - Residential Service	45.75
					Renewable Energy Rider	.22
					Sales Tax	1.38

Amount Due   47.35
To be drafted on or after   Apr 06 , 2013

Electricity Usage	This Month	Last Year
Total KWH	386	N/A
Days	28	N/A
AVG KWH per Day	14	N/A
AVG Cost per Day	$1.63	N/A

Our records indicate your telephone number is ???-???-???? . If this is incorrect, please follow the instructions on the back of the bill

A late payment charge of 1.0 % will be added to any past due utility balance not paid within 25 days of the bill date.

Figure 6.2: Residential electrical power bill

A.   A 15 MW, HP, Hourly Pricing Schedule. This schedule applies to incremental, interruptible, load and was offered to this customer as a part of DSM (demand-side management) program.

The monthly electrical bill computation for this facility is performed automatically by an existing EMS, Energy Monitoring System. A screen print of one such bill is shown in Figure 6.3.

The calculation methods associated with the various line items in this bill are explained through a specific bill calculation scenario described and analyzed in Case Study 6.2.

Case Study 6.2—Electrical Power Baseline Parameters and Bill Calculation—Large Industrial Power Consumer*

In an effort to illustrate the interpretation and arithmetic behind the various components of an OPT based electrical power bill, let us consider data from a specific billing period, recorded and processed in Spreadsheet 6.1. This spreadsheet shows all of the pertinent measured data, derived billing parameters, standard charges, tiered demand charge rates, energy charge rates, and computed line item charges for the billing month. In the following sections we will highlight the arithmetic behind the derived or extended line item charges.

Baseline Billing Determinants: The on peak billing demand and the billing demand are typically the same. This segment of the bill represents derived and measured data. The measured portion is in form of the energy consumption (kWh) measured in 15- or 30-minute intervals; which are then divided by 0.25 hour or 0.5 hour, respectively. This yields demand (kW) for each of the intervals for the entire billing month. The utility company then selects the highest of these demands—registered during the On-Peak periods—as the peak demand or billing demand for the month. The other billing determinants consist of the total energy usage during on-peak and off periods. The consumed energy data is in measured and recorded for the billing month, in kWh.

Basic Facilities Charge: This charge could be considered to represent administrative cost associated with the generation and processing of the bill. This charge stays, relatively, constant over time.

Extra Facilities Charge: This charge is a means for the utility to recoup its cost embedded in providing "extra facilities" to the customer. In this case

---

**Finance and Accounting for Energy Engineers*, by S. Bobby Rauf

HMLS - playground
HOURLY PRICING
SITE TOTALS
UTILITIES
CHILLERS
COMPRESSORS
GRU
PNG
POWER BILL
　Power Bill HP
　Power Bill OPT
WTP
WEATHER

BASELINE BILLING DETERMINANTS

BILLING DEMAND			25,594 KW
ON-PEAK BILLING DEMAND			25,594 KW
ON-PEAK ENERGY			3,696,021 KWH
OFF-PEAK ENERGY			15,138,552 KWH

BASELINE CHARGES

BASIC FACILITIES CHARGE			$36.07
ON-PEAK BILLING DEMAND CHARGE			
FOR THE FIRST 2,000 KW	2,000 X(	$7.64 per KW) =	$15,280.00
FOR THE NEXT 3,000 KW	3,000 X(	$6.54 per KW) =	$19,620.00
FOR ALL OVER 5,000 KW	20,594 X(	$5.43 per KW) =	$113,997.42
ECONOMY DEMAND CHARGE	0 X(	$1.030000 per KW) =	$0.00
ON-PEAK ENERGY CHARGE	3,696,021 X(	$0.042393 per KWH) =	$156,685.42
OFF-PEAK ENERGY CHARGE	15,138,552 X(	$0.021039 per KWH) =	$318,500.00
TOTAL BASELINE CHARGE		($0.033135 per KWH)	$624,082.83

HP BILLING DETERMINANTS

ACTUAL DEMAND			39,132 KW

HP CHARGES

INCREMENTAL DEMAND CHARGE	13,138 X(	$0.260000 per KW) =	$3,415.88
NEW LOAD X AVG HOURLY PRICE	7,845,899 X(	$0.021672 per KWH) =	$170,032.72
TOTAL HP CHARGE		($0.022107 per KWH)	$173,448.60
TOTAL CHARGE		($0.029892 per KWH)	$797,531.44

Figure 6.3: Industrial electrical power bill—EMS system based.

study, the extra facilities consisted of a set of redundant transmission lines installed by the power company to enhance reliability of the power service to this customer. In some cases, this type of charge is associated with the upgrade of main switch yard step down transformers, regulators, separate metering, etc.

On-peak Billing Demand Charge: The overall billing demand of 26,000 kW is tiered into three segments: first 2,000 kW, next 3,000 kW and the remaining 21,000 kW. Each of these demand tiers are applied respective charges, in $/kW, as shown in Spreadsheet 6.1, to yield tiered line item demand charges. These tiered line item demand charges are later added to other billing line item charges.

Economy Demand Charge: This charge is triggered when the highest integrated demand during off-peak periods exceeds the highest integrated demand recorded during the on-peak periods. The latter also serving as the peak demand for the billing month. When economy demand charge is triggered on the premise described above, the difference between the two demands is labeled as the economy demand. The economy demand figure thus derived is multiplied with the stated $1.03 rate multiplier to compute the economy demand charge for the month. Since, in this case study, the peak demand is assumed to be greater than the demands set during the off-peak periods, economy demand is ***not*** triggered and, therefore, the economy demand charge is zero.

Energy Charge: This charge accounts for the actual (measured) energy consumed during the billing month. This charge comprises of two components as shown in Spreadsheet 9.3; the on-peak energy charge and the off-peak energy charge. As apparent from examination of the spreadsheet, the on-peak rate ($0.042/kWh) is almost twice the magnitude of the off-peak rate ($0.021/kWh).

Total Baseline Charge: The total baseline charge is simply the sum of all line items calculated or identified to that point; $650,787.12, in this case.

HP, Hourly Pricing, Billing Determinants: This portion of the bill represents the demand and energy charges associated with the load that fall under the Hourly Pricing contract. The hourly pricing demand was measured and billed at 14,000 kW, in this case. The energy consumed under the hourly pricing contract was 8,805,800 kWh.

HP Charges: Special HP rates are applied to the recorded demand and energy under HP contract, yielding respective line item charges as shown on Spreadsheet 6.1.

Spreadsheet 6.1—Large industrial electric bill calculation

Description	Demand and Energy Parameters	Rates	Line Item Charge
Baseline Billing Determinants			
Billing Demand	26,000 kW		
On-Peak Billing Demand	26,000 kW		
On-Peak Energy Usage	4,334,573 kWh		
Off-Peak Energy Usage	14,500,000 kWh		
Baseline Charges			
Basic Facilities Charge		$36.07	$36.07
Extra Facilities Charge		$13,000	$13,000
On-Peak Billing Demand Charge			
For the First 2,000 kW	2,000 x	$7.64 per kW =	$15,280.00
For the Next 3,000 kW	3,000 x	$6.54 per kW =	$19,620.00
For Dem. Over 5,000 kW	21,000 x	$5.43 per kW =	$114,030

Total Bill for the Month: The last line of the bill represents the sum of Total Baseline Charge and the HP Charge, amounting to the total bill of **$845,266.42** for the month.

Spreadsheet 6.1—Large industrial electric bill calculation

Economy Demand Charge	0 x	$1.03 per kW =	$0
On-Peak Energy Usage	4,334,573 x	$0.042393 Per kWh =	$183,755.55
Off-Peak Energy Usage	14,500,000 x	$0.021039 Per kWh =	$305,065.50
Total Baseline Charge		$0.033135 Per kWh	**$650,787.12**
HP Billing Determinants			
Actual Demand	40,000 kW		
HP Charges			
Incremental Demand Charge	14,000 kW x	$0.26 / kW =	$3,640.00
New Load x Avg. Hourly Price	8,805,800 x	$0.021672 Per kWh =	$190,839.30
HP Subtotal			**$194,479.3**
Total Bill for the Month			**$845,266.42**

Self-assessment Problems & Questions—Chapter 6

1. The BMS System at a truck assembly plant, that operates 365 days a year, is displaying following electrical power and energy consumption data:
   — Billing Days in the Current Month: 30
   — On Peak Energy Consumption: 2,880,000 kWh
   — Off Peak Energy Consumption: 11,520,000 kWh
   — Three highest 30 minute energy usages for the billing month are (i) 12,500 kWh, (ii) 12,300 kWh and (iii) 12,290 kWh.

— Assuming this facility is on OPT, Time of Use, contract with 30 minutes demand interval, determine the following:
   (a) Average demand.
   (b) Peak demand.
   (c) The load factor for the current month.
   (d) Average annual demand

2. A 200 kVA transformer has been tested by the manufacturer to safely and continuously sustain a load of 230 kVA. What service factor should the manufacturer include on the nameplate of this transformer?

3. A 5 hp single phase AC motor, rated at a service factor of 1.10, is being tested at maximum safe load, powered by 230 V_{AC} source. Determine the amount of current drawn by this motor, from the power source if the motor efficiency is 90% and the power factor is 0.85.

4. Consecutive electrical power meter readings at a home in Hawaii are listed below. Determine the total electrical power bill for the month of this residence if the flat $/kWh cost rate is ¢21/kWh. The renewable energy rider is $15 and the energy sales tax rate is 4%.
   Previous reading: 45000
   Current or present reading: 46000

5. If the peak demand in Case Study 6.2 is reduced by 10% through implementation of peak shaving measures, what would be the baseline cost for the demand portion of the bill?

Chapter 7

Electric Motors and Generators

Introduction

Generators and motors are, primarily, rotating machines. The rotating machines are called *motors* when they *consume* electrical energy, and are referred to as *generators* when they *produce* electrical energy. In practical applications, while DC machines are almost always single phase, AC machines can be single phase or three phase. In this chapter we will explore fundamental operating principles and concepts associated with DC and AC motors and generators. The electromagnetics principles behind the operation of generators and motors will be illustrated through simplified electrical diagrams. Basic principles and equations, governing important and practical functions and operational parameters of motors and generators, will be introduced. Common calculations involving electric motors will be illustrated. Concept of induction motor slip is explained and associated calculations are covered. Roles that slip and frequency play in determination of the motor shaft speed are illustrated through example problems. Significance of certain common classifications of motors is explained. Interpretation of a common motor nameplate is discussed.

DC GENERATOR

A direct current generator, also referred to as a *dynamo*, is an electromagnetic device designed to convert mechanical energy, or mechanical power—namely, brake horsepower—to electrical energy or electrical power. The electrical energy and power developed in DC dynamos consists of DC current and DC voltage. A DC generator is, fundamentally, an AC generator. The feature that differentiates a DC generator's function and output from an AC generator is a "commutator." Common commutator consists of two rings as shown in Figure 7.1. As depicted in Figure 7.1, the current, **I**, and associated voltage, are generated in the dynamo by virtue of the motion, movement or rotation of the coil windings within a magnetic field set up by permanent magnets or, in some cases, by "field windings." Of course, the rotation of the winding coils is caused by force, torque, work producing system or energy source such as steam turbines, hydroelectric turbines, hydraulic pumps, etc. The end result is conversion of mechanical energy into electrical energy.

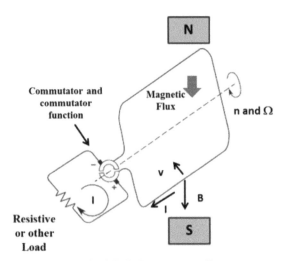

Figure 7.1: A DC Generator or Dynamo.

As shown in Figure 7.1, and as stipulated by Eq. 7.1, Eq. 7.2, and Eq. 7.3, key variables, the interaction of which, results in the production of voltage across the windings are:

— Magnetic flux density **B**,
— Number of turns, **N** the coil consists of,
— The area of cross-section **A**,
— The electrical frequency, **ω**, at which the armature is rotated by external work producing force,
— Rotational speed of the armature, **Ω**,
— The number of poles, **p**.

$$V(t) = V_m \text{Sin}\omega t = \omega NAB \sin \omega t \qquad \textbf{Eq. 7.1}$$

Voltage generated by the dynamo, or DC generator, can be expressed in terms of the rotational speed of the armature, Ω (in rads/sec), by applying the ω to Ω conversion formula expressed as Eq. 7.2.

$$\omega = \frac{P}{2}\Omega \qquad \textbf{Eq. 7.2}$$

$$V(t) = \frac{p}{2}\Omega NAB \sin\left(\frac{p}{2}\Omega\right)t \qquad \textbf{Eq. 7.3}$$

Since Ω (in rads/sec) is related to the RPM of the DC generator as stated in Eq. 7.4, Eq. 7.3 can be expanded in form of Eq. 7.5.

$$\Omega = \frac{2\pi n}{60}$$ **Eq. 7.4**

$$V(t) = \pi n \frac{p}{60} NABsin\left(\frac{p}{2}\Omega\right)t$$ **Eq. 7.5**

Since V(t) represents the sinusoidal form of the voltage generated by the dynamo, to derive the magnitude of DC, work producing, effective or RMS portion of this voltage we can equate the coefficient of the sinusoidal term in Eq. 7.5 to V_{peak}, V_{max} or, simply, V_p or V_m, and apply Eq. 1.3. This yields Eq. 7.6 for computation of the DC voltage produced by a dynamo or DC generator.

$$V_{DC} = V_{RMS} = V_{eff} = \frac{V_p}{\sqrt{2}} = \left(\frac{\pi}{2}\right)\cdot\left(\frac{n}{30}\right)\cdot\frac{pNAB}{\left(\sqrt{2}\right)} = \frac{\pi npNAB}{(60)\cdot\left(\sqrt{2}\right)}$$ **Eq. 7.6**

Where,

$$V_m = V_p = \left(\frac{\pi}{2}\right)\cdot\left(\frac{n}{30}\right)\cdot pNAB = \frac{\pi npNAB}{60}$$ **Eq. 7.7**

And,

n = Rotational speed of the dynamo, in rpm
p = Number of poles in the design and construction of the armature. For instance, **one** coil or set of winding with N turns, consist of two (2) poles.
N = number of turns constituting and armature coil.
A = Cross-sectional Area portended by the coil in m².
B = Magnet field intensity, in Tesla, or T.

Graphical depiction of the DC voltage computed through Eq. 7.7 is shown in Figure 7.2.

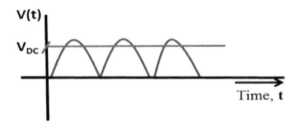

Figure 7.2: DC voltage output of a DC generator or dynamo

The relationship between the rotational speed of the generator, n_s, number of poles, p, the electrical frequency, f, and the angular speed, ω, corresponding to the electrical frequency, is given by Eq. 7.8 and Eq. 7.9.

$$n_s = \text{Synchronous speed} = \text{Rotational speed} = \frac{120f}{p} \qquad \textbf{Eq. 7.8}$$

$$\omega = 2\pi f \qquad \textbf{Eq. 7.9}$$

Example 7.1

The rotor of a single phase alternator is rotating at an actual mechanical rotational speed of 2400 rpm. The alternator consists of four pole construction. The effective diameter of the coil is 0.13m and the length of the coil loop is 0.22m. The coil consists of 20 turns. The magnetic flux density has been measured to be 1.15T. Calculate the RMS, effective, or V_{DC} produced by this generator.

Solution:

The RMS, effective or DC voltage produced through an alternator or generator can be computed by applying Eq. 7.6:

$$V_{DC} = V_{RMS} = V_{eff} = \frac{V_P}{\sqrt{2}} = \frac{\pi n p N A B}{(60).\left(\sqrt{2}\right)}$$

Given:

    n = 2400 rpm
    p = 4
    N = 20
    B = 1.15T
    A = (Eff. diameter of the coil conductor) x (Eff. length of the coil)
       = (0.22m) x (0.13m) = 0.0286m^2

$$V_{DC} = V_{RMS} = \frac{\pi n p N A B}{(60).\left(\sqrt{2}\right)}$$

$$= \frac{(3.14)(2400rpm)(4)(20)(0.0286m^2)(1.15T)}{(60).\left(\sqrt{2}\right)} = 233.7V$$

Example 7.2

A two-pole alternator/generator is producing electrical power at an electrical frequency of 60 Hz. (a) Determine the angular speed corresponding to the generated electrical frequency. (b) Determine the rotational (synchronous) speed of the armature/rotor. (c) Determine the angular velocity of the armature/rotor, assuming slip = 0.

Solution:

(a) Angular speed, ω, corresponding to the generated electrical frequency, f, can be calculated using Eq. 7.9:

$$\omega = 2\pi f = (2).(3.14).(60 Hz) = 377 \text{ rad/s}$$

(b) The rotational or synchronous speed of the armature/rotor is given by Eq. 7.8:

$$n_s = \frac{120f}{p} = \frac{(120)(60 \text{ Hz})}{2} = 3600 \text{ rpm}$$

(c) Angular velocity of the armature/rotor is simply the rotational speed, in rpm, converted into rad/s. Since there are 2π radians per revolution:

$$\omega_s = 3600 \text{ rev./min} = \left(\frac{3600 \text{ rev.}}{\text{min}}\right).\left(\frac{2\pi \text{ rad.}}{\text{rev.}(60\text{sec/min})}\right) = 377 \text{ rad/sec}$$

DC Motor

A DC motor can be perceived as a DC generator or dynamo operating in reverse. As in the case of a dynamo, a magnet—serving as a stator—provides the magnetic field (**B** and **H**) that interacts with the DC current flowing through the rotor. The DC current flowing through the rotor windings is supplied from an external DC voltage or current source, via the commutator rings that are stationary. Basic construction of a DC motor is illustrated in Figure 7.3.

In other words, a DC motor is a mechanically commutated electric motor powered from direct current (DC). The current in the rotor is switched by the commutator. The relative angle between the stator and rotor magnetic flux is maintained near 90 degrees, which generates the maximum torque.

In DC motors, different connections of the field and armature winding provide different inherent speed/torque regulation characteristics. In so far as the control of the *speed of a DC motor* is concerned, it can be controlled by changing either the *voltage applied* to the armature or by changing the *field*

Figure 7.3: DC motor

current. Since voltage is related to current by Ohm's law, V = I.R, speed control can be accomplished through introduction of variable resistance in the armature circuit or field circuit. However, modern DC motors are often controlled by power electronics systems called DC drives.

Historically, the introduction of DC motors to run machinery eliminated the need for steam or internal combustion engines. Case in point, the application of DC motors as the motive power in locomotives. An advantage of DC motors is that they can be operated ***directly*** from DC power supply, with variable voltage, without the need for an inverter. This positions DC motors well for electric vehicles application. Despite the dominance of AC induction motors in myriad applications, DC motors continue to play an important role in providing motive power in small toys and disk drives. Overall, DC motors continue to provide an alternative to inverter driven AC motors where motor speed variation is needed.

AC ALTERNATOR

The basic construction and premise of operation is similar to a dynamo with the exception of the fact that the commutator is ***unnecessary*** and therefore absent. Another salient difference between a dynamo and an AC alternator is that the roles and the properties of the stator and rotor are reversed. In an AC alternator, as shown in Figure 7.4 (a), the magnetic field is produced by the ***rotating rotor*** and the ***stator serves as an armature***. The key reason for the armature—the segment of the generator where the generated current and EMF (voltage) are harnessed—to serve as a stationary "exoskeleton" is that large induced currents require robust insulation of the armature windings. In addition, with large currents, larger magnetic forces and torques are in play, which makes it important to secure or anchor the windings in a rugged structure. As with a DC generator, the power in AC alternators is fundamentally produced in *sinusoidal form*. Since the output of an AC alternator does not need to be rectified to DC form, commutator function is not needed. Construction of a basic AC alternator or generator is shown in Figure 7.4 (a). The output of a typical AC alternator is depicted Figure 7.4 (b).

As observable in Figure 7.4 (a), the magnetized rotor is being rotated counter-clockwise by external means. As the magnetized rotor rotates the magnetic field emanating from the north pole and terminating into the south pole cuts through the three phase coils in the armature. Movement of the magnetic field through the coils initiates current flow and potentials in the three coils. As

(a) Generator/Alternator Construction

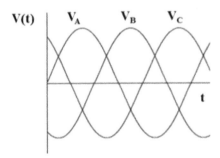

(b) Generator/Alternator Construction

Figure 7.4: AC alternator construction and output waveform.

noted in Figure 7.4 (a), the voltage thus developed across coil A is V_A, voltage developed across coil B is V_B, and the voltage developed across coil C is V_C.

The overall principle of operation and construction of a *single phase AC alternator/generator* is similar to a *three phase AC alternator/generator*, with the exception of the fact that the armature windings consist of one set of windings for the harnessing of single phase AC.

While the complete representation of the AC sinusoidal voltage generated by single phase AC generator is given by Eq. 7.5 (angle of the voltage is assumed to be "0"), the RMS and peak voltages would be stipulated by Eq. 7.6 and Eq. 7.7, respectively. For the sake of reader's convenience, these equations are restated below:

$$V(t) = \pi n \frac{p}{60} NAB \sin\left(\frac{p}{2}\Omega\right)t \qquad \textbf{Eq. 7.5}$$

$$V_{DC} = V_{RMS} = V_{eff} = \frac{V_P}{\sqrt{2}} = \left(\frac{\pi}{2}\right)\cdot\left(\frac{n}{30}\right)\cdot\frac{pNAB}{\left(\sqrt{2}\right)} = \frac{\pi npNAB}{(60)\cdot\left(\sqrt{2}\right)}$$

$$\textbf{Eq. 7.6}$$

$$V_m = V_P = \left(\frac{\pi}{2}\right)\cdot\left(\frac{n}{30}\right)\cdot pNAB = \frac{\pi npNAB}{60} \qquad \textbf{Eq. 7.7}$$

Example 7.3

A four pole single phase AC generator consists of windings constituting 90 series paths and is driven by a propane prime mover (engine). The effective or mean length of the armature is 20 cm and the cross-sectional radius of the armature is 5 cm. The armature is rotating at 2000 rpm. Each armature pole is exposed to a magnetic flux of 1.5 T. The efficiency of this generator is 92% and it is rated 1.5 kW. Determine the following:
(a) The maximum voltage generated. (b) The RMS voltage generated. (c) The horsepower rating of the generator. (d) The horse power output of the prime mover.

Solution:

The maximum voltage, V_m, generated by this alternator is given by Eq. 7.7.

$$V_m = V_P = \left(\frac{\pi}{2}\right)\cdot\left(\frac{n}{30}\right)\cdot pNAB = \frac{\pi npNAB}{60} \qquad \textbf{Eq. 7.7}$$

Given:

n = 2000 rpm

p = 4

N = Number of series paths = 90

B = 1.5 T

A = (Eff. diameter of the coil conductor) x (Eff. length of the coil)

$\quad$ = (2 x 5cm) x (20cm) = (0.1m) x (0.2m) = 0.02m²

(a) $\quad V_m = \dfrac{\pi npNAB}{60} = \dfrac{(3.14).(2000rpm)(4)(90)(0.02m^2)(1.5T)}{60}$

(b) The RMS voltage could be calculated by using Eq. 7.6, or simply by dividing V_m, from part (a), by the square root of 2 as follows:

$$V_{RMS} = V_{eff} = \frac{V_m}{\sqrt{2}} = \frac{1130}{\sqrt{2}} = 799 \text{ V}$$

(c) The horsepower rating of this generator is the power *output* rating specified in hp, premised on the stated *output capacity* of 1.5 kW. Therefore, application of the 0.746 kW/hp conversion factor yields:

$$P_{hp} = \frac{P_{kW}}{0.746kW/hp} = \frac{1.5kW}{0.746kW/hp} = 2.01hp \cong 2 \text{ hp}$$

(d) The horsepower rating of the prime mover—or the propane fired engine—would need to offset the inefficiency of the AC generator. Therefore, based on the given 92% efficiency rating of the generator:

$$P_{hp\text{ - Prime Mover}} = \frac{P_{hp\text{ - Gen.}}}{Eff._{Gen}} = \frac{2.01 \text{ hp}}{0.92} = 2.19 \text{ hp}$$

AC INDUCTION MOTOR

An induction motor is also referred to as an asynchronous motor; primarily, because a typical AC induction motor has a certain amount of "slip" (discussed in detail in the next section) and the shaft speed is less than the motor's synchronous speed. An AC induction motor can be considered as a special case of a transformer with a rotating secondary (rotor) and a stationary primary (stator). The electromagnetic field in the primary (stator) rotates at synchronous speed, n_s, as given by Eq. 7.9. The stator magnetic field, rotating at the synchronous speed, cuts through the rotor windings. This rotating or moving magnetic field—in accordance with the Faradays law—induces EMF and current through the rotor windings. Due to the fact that the rotor windings are inductive—and possess inductive reactance, X_L—the current induced in the rotor windings lags behind the induced voltage, resulting in a definite amount of slip.

$$n_s = \text{Synchronous speed} = \frac{120f}{p} = \frac{60\omega}{\pi p} \qquad \textbf{Eq. 7.9}$$

In essence induction motors allow the transfer of electrical energy from the primary (stator) windings to the secondary (rotor) windings. The primary or secondary windings—or the stator and the rotor—are separated by an air gap. In a wound rotor the wire wound construction is similar to the rotor construction of AC alternators. However, with AC motors, a more common alternative to a wound rotor is a *squirrel cage rotor*, which consists of copper or aluminum bars—in lieu of insulated wire—embedded in slots of the cylindrical iron core of the rotor. See Figure 7.5. Construction of a common, TEFC, totally enclosed fan cooled, motor is illustrated in Figure 7.5. Major components of an AC induction motor, i.e. squirrel cage rotor, motor shaft, armature windings, cooling fan and bearings, are labeled and highlighted in Figure 7.5.

Figure 7.5: AC induction motor construction.

Motor Slip

The rotational speed of rotor (secondary) of an induction motor—rotational speed of the rotor is the also the actual speed of the motor shaft—lags behind the synchronous speed, $\mathbf{n_s}$. This difference in rotational speeds can range between 2% to 5% and is called *slip*. Slip can also be defined as the difference between synchronous speed and operating speed, at the same frequency, expressed in rpm or in percent or ratio of synchronous speed. Slip can be defined mathematically as represented by Eq. 7.10 and Eq. 7.11.

$$\text{Slip} = \left(\frac{n_s - n}{n_s} \right) = \left(\frac{\Omega_s - \Omega}{\Omega_s} \right) \qquad \qquad \textbf{Eq. 7.10}$$

$$\% \text{ Slip} = \left(\frac{n_s - n}{n_s} \right).(100)\% = \left(\frac{\Omega_s - \Omega}{\Omega_s} \right).(100)\% \qquad \textbf{Eq. 7.11}$$

Where,

$\mathbf{n_s}$ = Synchronous speed of the AC induction motor, in rpm.
    = Rotational speed of the armature AC induction EMF, in rpm.
$\mathbf{n}$ = Actual speed of the AC induction motor shaft, in rpm.
    = Rotational speed of the EMF in the rotor, expressed in rpm.
$\mathbf{\Omega_s}$ = Rotational speed of the armature AC induction EMF, in rad/sec.
    = Synchronous speed of the AC induction motor, rad/sec.
Ω = Rotational speed of the rotor EMF, in rad/sec.
    = Actual speed of the rotor, in rad/sec.

Induction motors are made with slip ranging from less than 5% up to 20%. A motor with a slip of 5% or less is known as a ***normal-slip motor***. A normal-slip motor is sometimes referred to as a ***constant speed motor*** because the speed changes very little from no-load to full-load conditions. Higher slip characteristics of motors are not always "undesirable." Since low slip is often accompanied by instantaneous imparting of a large amount of torque—which can be detrimental to the material and mechanical integrity of the shaft—to the motor shaft, motors with slip over 5% are often used for hard to start applications.

Typically, at the rated full load, slip ranges from more than 5% for small or special purpose motors to less that 1% for large motors. Speed variations due to slip can cause load-sharing problems when differently sized motors are mechanically connected. Slip can be reduced through various means. Due to the progressive decline in the cost and continuous technological improvements in the VFD, variable frequency drive, technology, they offer the most optimal solution for mitigating undesirable effects associated with slip.

A common four-pole motor with a synchronous speed of 1,800 rpm may have a no-load speed of 1,795 rpm and a full-load speed of 1,750 rpm. The rate-of-change of slip is approximately linear from 10% to 110% load, when all other factors such as temperature and voltage are held constant.

MOTOR TORQUE AND POWER

Torque generated or produced by a motor is equivalent to the amount of *external work* performed *by* the motor. In the absence of transactions involving other forms of energy, such as heat, the amount of work performed, or torque applied by a motor, is equal to the energy stored in the external system in form of potential energy or kinetic energy. Therefore, torque, work and energy are equivalent entities. This is further corroborated by the fact that torque, work and energy can be quantified or measured in terms of equivalent units, i.e. ft-lbf, Btu, calories, therms, N-m, joules, kWh, etc.

On the other hand, power, as explained earlier in this text, is the rate of performance of work, rate of delivery of energy rate of application of torque, etc. In order to illustrate the difference between torque and power let's consider the example of a crank wheel that is being turned at a certain rate. Suppose the wheel has a one-foot long crank arm takes a force of one pound (lbf) to turn at steady rate. The torque required would be one pound times one foot or one foot-pound, 1 ft-lbf. If we were to apply this torque for two seconds, the power delivered would be ½ ft-lbf/sec. Now let's assume that we continue to apply the same amount of torque, i.e. 1 ft-lbf, but twice as fast, or for 1 second. Then we would have a case where the torque remains the same, or 1 ft-lbf, but the rate of delivery of torque would double to 1 ft-lbf/sec.

Application of motors in mechanical or electromechanical systems often challenges engineers to assess the *amount* of torque involved. However, unlike power output or input of a motor which can be assessed through simple and safe measurement of voltage and current—with the help of help of held instruments, such as the handheld voltmeter or clamp-on ammeter -measurement of torques produced by large motors requires large, stationary and rather unwieldy systems called dynamometers. See Figure 7.6. Therefore, if the motor power is known in, hp or kW, and if the actual shaft speed, **n** (rpm), is available, the torque can be calculated by using Eq. 7. 12 and Eq. 7.13.

$$\text{Torque}_{(ft-lbf)} = \frac{5250 \times P_{(horsepower)}}{n_{(rpm)}} \qquad\qquad \textbf{Eq. 7.12}$$

Figure 7.6: Motor dynamometer

$$\text{Torque}_{(N-m)} = \frac{9549 \times P_{(kW)}}{n_{(rpm)}}$$ **Eq. 7.13**

The torque computation equations stated above are derived from the basic torque, angular speed and power equation stated as Eq. 7.14.

$$P = T \cdot \omega \text{ or, } T = \frac{P}{\omega}$$ **Eq. 7.14**

Where,

P = Power delivered by the motor, in hp or watts.

T = Torque delivered by the motor, in ft-lbf or N-m or Joules.

ω = Actual rotational speed of the motor shaft, in rad/sec.

SINGLE PHASE AND THREE PHASE MOTOR LINE CURRENT COMPUTATION

While copious tables, charts, handbooks and pocket cards include full load current information for various (standard) motor sizes, at standard voltages, when accuracy is of the essence, one must *compute* correct current values pertaining to specific voltage, efficiency, power factor and actual load. Therein rests the value of line current calculation equations for single phase and three phase motors. Those equations and their derivation are summarized below.

Single Phase AC Induction Motor Current:

$S_{L,1\text{-}\phi}$ = Full load apparent power on the line side of single phase motor

$$- \frac{\text{Full Load Raging of Motor (in Watts)}}{\text{(Pf).(Eff.)}} \qquad \textbf{Eq. 7.15}$$

Or,

$|S_{L,I\text{-}\phi}|$ = Magnitude of apparent power (drawn) on the line side of a single

$$\text{phase motor} = \frac{P_{1\text{-}\phi} \text{ (in Watts)}}{\text{(Pf).(Eff.)}} \qquad \textbf{Eq. 7.16}$$

Since the magnitude of single phase apparent power, $|S_{L,I\text{-}\phi}| = |V_L|.|I_L|$,

$$\text{then } |L_L| = \frac{|S_{L,I\text{-}\phi}|}{|V_L|} \qquad \textbf{Eq. 7.17}$$

Therefore,

$$\text{Single Phase AC Line} \atop \text{Current drawn, } |I_L| = \frac{|S_{L,I\text{-}\phi}|}{|V_L|} = \frac{|P_{L,I\text{-}\phi}| \text{ (in Watts)}}{|V_L|(\text{Pf}).(\text{Eff.})} \qquad \textbf{Eq. 7.18}$$

THREE PHASE AC INDUCTION MOTOR CURRENT

$S_{L,3\text{-}\phi}$ = Full Load Apparent (3-ϕ) Power on the Line Side

$$= \frac{\text{Full Load (3-}\phi\text{) Rating of Motor (in Watts)}}{\text{(Pf).(Eff.)}} \qquad \textbf{Eq. 7.19}$$

Or,

$|S_{L,3\text{-}\phi}|$ = Magnitude of Full Load Apparent Power on the Line Side of

$$\text{a 3-phase motor} = \frac{P_{3\text{-}\phi} \text{ (in Watts)}}{\text{(Pf).(Eff.)}} \qquad \textbf{Eq. 7.20}$$

Since the magnitude of a (3-ϕ) (three phase) apparent power is given as:

$$|S_{L,3\text{-}\phi}| = \sqrt{3} \, (|V_L|).(|I_L|),$$

by rearranging this equation, the 3-ϕ phase line current, $\dfrac{|S_{L,3\text{-}\phi}|}{\sqrt{3}\,(|V_L|)}$ **Eq. 7.21**

Therefore,

The 3-ϕ phase line current, $|I_L| = \dfrac{|S_{L,3\text{-}\phi}|}{\sqrt{3}\,(|V_L|)} = \dfrac{P_{L,3\text{-}\phi}\ (\text{in Watts})}{\sqrt{3}\,(|V_L|)(\text{Pf}).(\text{Eff.})}$ **Eq. 7.22**

Where,

$|I_L|_{1\text{-}\phi}$ = Single phase RMS line current, measured in amps.
$|I_L|_{3\text{-}\phi}$ = Three phase RMS line current, measured in amps.
$|V_L|$ = RMS line to line, or phase to phase, voltage, measured in volts.
$|S_L|_{1\text{-}\phi}$ = Apparent power drawn by a single phase motor, in VA or kVA, etc.
$|S_L|_{3\text{-}\phi}$ = Apparent power drawn by three phase motor, in VA or kVA, etc.
$P_{1\text{-}\phi}$ = Real power demanded by a single phase motor load, in W, or kW, etc.
$P_{3\text{-}\phi}$ = Real power demanded by a three phase motor load, in W, or kW, etc.
Pf = Power factor of the motor, as specified on motor nameplate.
Eff. = Motor efficiency, as specified on motor nameplate.

Example 7.4

A three phase, four pole, AC induction motor, rated 150 hp, is operating at full load, 50 Hz, 480 V_{rms}, efficiency of 86%, power factor of 95%, and a slip of 2%. Determine the (a) motor shaft speed, in rpm, (b) torque developed, in ft-lbf. (c) line current drawn by the motor and (d) the amount of reactive power, Q, sequestered in the motor under the described operating conditions.

Solution
Given:

$P_{L,3\text{-}\phi}$ = Real power or rate of work performed by the motor = 150 hp
        = (150 hp).(746 W/hp) = 111,900 W
p = 4 poles
V_L = 480 V_{RMS}
Pf = 95% or 0.95
Eff. = 86% or 0.86
n_s = Synchronous speed, in rpm = ?
Slip, s = 2%
f = Frequency of operation = 50 Hz

(a)   **Shaft or motor speed:** Rearrange and apply Eq. 7.10:

$$\text{Slip} = s = \left(\frac{n_s - n}{n_s} \right)$$

And, by rearrangement: $n = n_s (1 - s)$

Next, we must determine the synchronous speed of the motor by applying Eq. 7.9:

$$n_s = \text{Synchronous speed} = \frac{120f}{p} = \frac{(120).(50)}{4} = 1500 \text{ rpm}$$

$\therefore n = (1500 \text{ rpm})(1-0.02) = 1470 \text{ rpm}$

(b) **Torque developed, in ft-lbf:** There are multiple methods at our disposal for determining the torque developed. Formulas associated with two common methods are represented by Eq. 7.12, Eq. 7.13 and Eq. 7.14. Since the power is available in hp and the rotational speed in rpm, apply Eq. 7.12:

$$\text{Torque}_{(\text{ft-lbf})} = \frac{5250 \times P_{(\text{horsepower})}}{n_{(\text{rpm})}} = \frac{5250 \times 150 \text{ hp}}{1470 \text{ rpm}} = 536 \text{ ft-lbf}$$

Note: The reader is encouraged to prove this result through application of Eq. 7.14.

(c) **Line current drawn by the motor:** Full load line current is the current the motor draws from the power source, or utility, in order to sustain the full load. Therefore, the determination of line current at full load requires that the full load output of 150 hp be escalated in accordance with motor efficiency and power factor to compute the line current. Application of Eq. 7.22 takes this into account in the computation of 3-phase motor line current, after the motor hp rating (or actual load) is converted into watts:

The 3-ϕ phase line current, $|I_L| = \dfrac{|S_{L,3\text{-}\phi}|}{\sqrt{3}\ (|V_L|)} = \dfrac{P_{L,3\text{-}\phi} \text{ (in Watts)}}{\sqrt{3}\ (|V_L|)(\text{Pf}).(\text{Eff.})}$ **Eq. 7.22**

Three phase (total) real power was converted into watts under "Given" as $P_{L,3\text{-}\phi} = 111,900$ W

Therefore,

The 3-ϕ phase line current, $|I_L| = \dfrac{111{,}900 \text{ Watts}}{\sqrt{3}\ (480\ V_{RMS})(0.95).(0.86)} = 165A$

(d) **Reactive power, Q:** There are multiple approaches available to us for determination of reactive power Q for this motor application. We will utilize the power triangle or apparent power component vector method:

$$S = P + jQ, \text{ and } S^2 = P^2 + Q^2, \text{ or, } Q^2 = S^2 - P^2 \text{ and } Q = \sqrt{S^2 - P^2}$$

Real power "P" was computed earlier as 111,900 W, and apparent power S can be assessed using Eq. 7.20

$$|S_{L,3\text{-}\phi}| = \frac{P_{L,3\text{-}\phi} \text{ (in Watts)}}{(\text{Pf}).(\text{Eff.})} = \frac{111{,}900 \text{ W}}{(0.95).(0.86)} = 136{,}965 \text{ VA}$$

Therefore,

$$Q = \sqrt{S^2 - P^2} = \sqrt{136{,}965^2 - 111{,}900^2} = \sqrt{6{,}237{,}665{,}433}$$
$$= 78{,}979 \text{ VAR}$$

Example 7.5

A four-pole induction motor operates on a three-phase, 240 V_{rms} line-to-line supply. The slip is 5%. The operating (shaft) speed is 1600 rpm. What is most nearly the operating frequency?

Solution:
Given

$$\begin{aligned}
p &= 4 \text{ poles} \\
n &= \text{Shaft or motor speed} = 1600 \text{ rpm} \\
n_s &= \text{Synchronous speed, in rpm} = ? \\
\text{Slip, } s &= 5\% \text{ or } 0.05 \\
f &= \text{Frequency of operation} = ?
\end{aligned}$$

Since this case involves synchronous speed and slip, it requires the application of Eq. 7.9 and 7.10. As apparent from examination of Eq. 7.9, the operating frequency can be determined by rearranging the equation if the synchronous speed n_s were known.

$$\text{Slip} = s = \left(\frac{n_s - n}{n_s} \right) \text{ or, } n_s = \left(\frac{n}{1-s} \right) = \left(\frac{1600}{1-0.05} \right) - 1684 \text{ rpm} \qquad \textbf{Eq. 7.10}$$

$$n_s = \text{Synchronous speed} = \frac{120f}{p} \text{ or, } f = \frac{(p).(n_s)}{120} = \frac{(4).(1684)}{120} = 56 \text{ Hz}$$

$$\textbf{Eq. 7.9}$$

SYNCHRONOUS MOTORS

Synchronous motors are AC induction motors that have no slip. In other words, in synchronous motors, the speed of rotor or motor shaft is the same as the rotational speed of the stator winding magnetic field. This is accomplished through the excitation of the rotor through the field windings such that the rotor is able to *"catch up"* to the rotational speed of the stator. This characteristic permits synchronous motors to induce *leading currents* in the branch circuit. Therefore, synchronous motors can serve as alternative means for power factor correction; albeit, application of power factor correcting capacitors is more economical and effective in most lagging power factor situations.

Example 7.6

A three phase induction motor delivers 550 kW at a power factor of 82%. Determine the size of synchronous motor—in kVA—that should be installed to carry a load of 250 hp load and, at the same time, raise the (combined) power factor to 95%.

Solution:
Given

P_1 =Real power delivered by the 3-ϕ induction motor = 550 kW
P_S =Real power contributed by the synchronous motors = 250 hp = (250 hp) x (0.746 kW/hp) = 186.5 kW
Pf_i =Initial Power Factor = 82% = 0.82
Pf_f =Final Power Factor = 95% = 0.95

To solve this problem, as in many others, we will begin with the final state and work our way upstream. With the final combined power factor of 95% in mind, we will work our way back to the required synchronous motor specifications. The apparent power rating, **S**, of the synchronous motor that must be installed to achieve the combined "system" power factor of 95% (0.95), while contributing 250 hp toward the system's real power, P_T, requirement, can be determined if we could assess the Q_T (kVAR) contributed by the synchronous motor in an effort to raise the combined power factor of the two motor system to 95% (0.95). This can be accomplished through assessment of final (combined) S_T and P_T values and application of the power triangle equation:

$$S_T = P_T + jQ_T, \text{ and } S_T{}^2 = P_T{}^2 + Q_T{}^2, \text{ or, } Q_T{}^2 = S_T{}^2 - P_T{}^2 \text{ and,}$$
$$Q_T = \sqrt{S_T{}^2 - P_T{}^2}$$

The combined real power of the induction motor and the synchronous motor:

$$P_T = 550 \text{ kW} + 186.5 \text{ kW} = 736.5 \text{ kW}$$

Since $S_T \text{Cos}\theta = P_T$, $S_T = \dfrac{P_T}{\text{Cos}\theta}$,

And since θ_R, the final power factor angle $= \text{Cos}^{-1}(0.95) = 18.19°$,

$$S_T = \frac{P_T}{\text{Cos}\theta_T} = \frac{736.5 \text{ kW}}{\text{Cos}(18.19°)} = 775.25 \text{ kV}$$

Therefore,

$$Q_T = \sqrt{S_T{}^2 - P_T{}^2} = \sqrt{775^2 - 736^2} = \sqrt{59,337} = 244 \text{ kVAR}$$

Now, in order to determine Q_S, the reactive power contributed by the synchronous motor, we must subtract the original reactive power, Q_O, from the final, total, reactive power, Q_T. However, Q_O is unknown and can be determined through the power triangle in Figure 7.7.

Since $\text{Tan}(\theta_O) = \dfrac{Q_O}{P_O}$, or, $Q_O = P_O \text{ Tan}(\theta_O) = 550 \text{ kW Tan}(\theta_O)$

S_O = Original, induction motor apparent power

Q_O, reactive component of original induction motor apparent power S_O

P_O, real power component of original induction motor apparent power S_O

Figure 7.7: Power triangle

And, $\theta_O = \text{Cos}^{-1}(\text{Pf}_O) = \text{Cos}^{-1}(0.82) = 34.92°$

Therefore,

$$Q_O = 550 \text{ kWTan}(34.92°) = 550 \text{ kW}(0.698) = 384 \text{ kVAR}$$

And, reactive power contributed by the synchronous motor would be:

$$Q_S = Q_O - Q_T = 384 \text{ kVAR} - 244 \text{ kVAR} = 140 \text{ kVAR}$$

Therefore,

$$S_S = P_S + jQ_S, \text{ and } S_S^2 = P_S^2 + Q_S^2, \text{ or, } S_S = \sqrt{P_S^2 + Q_S^2}$$

$$\text{Or, } S_S = \sqrt{(186.5 \text{ kW})^2 + (140 \text{ kW})^2} = 233 \text{ kVA}$$

MOTOR STARTING METHODS FOR INDUCTION MOTORS

Often, the starting method for an induction motor depends on the type of motor. The five basic types of small induction motors are as follows:

1)   Single-phase capacitor-start
2)   Capacitor-run
3)   Split-phase
4)   Shaded-pole
5)   Small polyphase induction motors

Because of the absence of a rotating EMF field in the armature, a single-phase induction motor requires separate starting circuitry to provide a rotating field to the motor. A starting circuit is needed to determine and "set" the direction of rotation due to the fact that the regular armature windings of a single-phase motor can cause the rotor to turn in either direction. Therefore, rotation direction is initiated in certain smaller single-phase motors by means of a "*shaded pole*" with a copper wire "*turn*" around a section of the pole. Then, the current induced in the shaded pole turn lags behind the supply current, creating a delayed magnetic field around the shaded part of the pole face. This imparts sufficient rotational field energy to start the motor in a specific direction. The shaded pole starting method is often employed in motors installed in applications such as small fans and pumps.

A common approach for staring larger single phase motors is to incorporate a second stator (armature) winding that is fed with an "out-of-phase" current. The out of phase current may be created by feeding the winding through a *capacitor* or through an *impedance* that is different from the main winding. Often, the second "starting" winding is disconnected once the motor has accelerated up to normal steady state speed. This is accomplished, commonly, through either by a *centrifugal switch* or a *thermistor*—the thermistor heats up and increases its resistance, thereby reducing the current through the starting winding to a negligible level. Some designs maintain the starting winding in the circuit to supplement the regular motor torque.

Since three phase, or polyphase, motors possess a rotating armature emf, the direction of rotation is determined by phase sequence, ABC versus ACB. Hence, during start-up or commission of three phase motors, if the motor rotation is incorrect, two of the three phase conductors are "swapped" to reverse the direction of rotation. This reversing starter design—in *FVR, Full Voltage Reversing starters*—is premised on the "phase-swapping" principle to reverse the direction of rotation of motors, such as in conveyor or fan applications where direction reversal is required. Self-starting polyphase induction motors produce torque even when stationary. Common starting methods for larger polyphase induction motors are as follows:

1)   Direct-on-line starting,
2)   Reduced-voltage reactor or auto-transformer starting,
3)   Star-delta starting
4)   Application of solid-state soft starting systems
5)   Use of VFDs, Variable Frequency Drives, for electronically controlled starting—and normal motor operation—through variation of frequency and voltage.

In squirrel cage polyphase motors, the rotor bars are designed and shaped according to the desired speed-torque characteristics. The current distribution within the rotor bars varies as a function of the frequency of the induced current. In locked rotor scenarios—or when the rotor is stationary—the rotor current has the same frequency as the stator current, and the current tends to travel at the outermost parts of the cage rotor bars due to the "*skin effect.*" The squirrel cage rotor bars are designed to meet the required speed-torque characteristics and to limit the inrush current.

DC Motor Speed Control

While DC motor speed control through armature or field windings was briefly discussed earlier in this chapter, specific approaches to DC motor shaft speed control can be categorized as follows:

1) **Armature Control:** The armature based speed control technique involves changing the voltage across the armature through variation of parallel or series resistance, while holding the field voltage constant.

2) **Field Control:** The field control approach involves variation of the field voltage while holding the armature voltage constant. The field voltage is varied through the variation of series or shunt resistance.

3) **Electronic Control:** This approach involves the use of electronic controls for the variation of armature and/or the field voltage and current. Due to the fact that features like programmability, closed loop control and automation are inherently available with electronic controls, this approach tends to provide better control over a wide range of speeds and torques.

4) **Combination of basic approaches:** In certain applications, combination of the three approaches described above is more suitable.

AC Motor Speed Control

A common and efficient means for controlling speed of AC motors is through the application of AC inverters, VFD (Variable Frequency Drives) or ASD (Adjustable Speed Drives). The premise of AC motor speed control is the ability to vary the frequency of AC power being supplied to the motor through a variable frequency drive. The shaft speed in such cases can be determined through the application of Eq. 7.9 and 7.23.

$$n_s = \text{Synchronous speed} = \frac{120f}{p} \text{ or, } f = \frac{(p).(n_s)}{120} = \frac{(4).(1684)}{120} = 56 \text{ Hz}$$

Eq. 7.9

$$\text{Slip} = s = \left(\frac{n_s - n}{n_s} \right) \text{ or, } n_s = \left(\frac{n}{1-s} \right) = \left(\frac{1600}{1-0.05} \right) = 1684 \text{ rpm}$$

Eq. 7.23

Where,

n_s = Synchronous speed of the AC induction motor, in rpm.
  = Rotational speed of the armature AC induction EMF, in rpm.
n = Actual speed of the AC induction motor shaft, in rpm.
  = Rotational speed of the EMF in the rotor, expressed in rpm.
s = Motor slip.
f = Frequency of the AC power source feeding the motor, in Hz.
p = Number of poles, as stated on the motor nameplate.

MOTOR CLASSIFICATIONS

There are numerous motor classifications and these classifications tend to change and evolve over time as new applications for AC and DC motors emerge. Discussion and listing of all possible classifications and categories of AC and DC motors is beyond the scope of this text. However, some of the more common categories and classifications are stated below for reference.

a. Self-commutated motors
b. AC Asynchronous motors (i.e. typical induction motor)
c. AC Synchronous motor
d. Constant speed motors
e. Adjustable speed motor
f. General duty
g. Special purpose

Classification based on the winding insulation type: Motor insulation class rating is determined by the ***maximum allowable operating temperature*** of the motor; which, subsequently, depends on the type/grade of insulation used in the motor. Seven mainstream insulation classifications and associated maximum allowable temperatures are listed below:

    I. Class A: 105° C
   II. Class B: 130° C
  III. Class F: 155° C
  IV. Class H: 180° C
   V. Class N: 200° C
  VI. Class R: 220° C
 VII. Class S: 240° C

Aside from exercising proper care in specifying the correct insulation type for a motor when installing new motors—based on the operating environment of a motor—one must ensure that if a motor is rewound or repaired, the windings are replaced with *original classification of insulation*. For example, if the windings of a 100 hp *Class H* motor are replaced with *Class B* insulation windings, the motor will lose its capacity to operate in higher temperature environment.

TYPICAL MOTOR NAMEPLATE INFORMATION FOR LARGE THREE PHASE INDUCTION MOTOR

Electric motors are typically labeled with a nameplate that displays pertinent specifications of that motor. While there are myriad possible specifications, ratings and characteristics of motors, some of the more common ones are described and illustrated through a motor nameplate example in Figure 7.8.

Model No.	B200	TYPE:	Marine Duty, IEEE 45				
HP:	200	kW	150	RPM:	1760		
Volt	460	AMP	233	PH.	3	MAX. AMB:	40°C
Hz	60	S.F.	1.15	P.F.	84.5	Code: G	
NEMA NOM EFF:	96.2			DUTY:	CONT.	WT: 1052 kg	2314 lbs
FRAME:	447T	ENCL:	TEFC	TYPE:	TKKH	NEMA: B	IP: 56

Figure: 7.8: Motor nameplate example,

Interpretation of this nameplate information would be as follows:

Model No: B200
Type: Marine Duty, IEEE-45

Power Rating: 200 hp

Standard Operating Voltage: 460 V_{RMS}.

Standard Frequency of operation: 60 Hz

NEMA based efficiency of the motor: 96.2%

Motor frame size: 447T. This NEMA designated frame classification often signifies the mechanical construction, mounting or installation of a motor.

Power rating of the motor, in kW = 150 kW

AMP or Full Load Current rating of the motor: 233 Amp

Service factor of the motor: 1.15 or 115%

ENCL. or NEMA rating of the motor enclosure: TEFC; or, Totally Enclosed Fan Cooled.

RPM: Full load shaft or rotor speed: 1760 rpm.

PH. Or number of phases: 3

PF, or Power Factor of the motor: 84.5%

Duty, or duration of operation: Continuous.

TYPE: TKKH, High Efficiency Motor

NEMA—B: Class B winding insulation

IP56: European enclosure rating, comparable to NEMA 4.

Maximum Allowable Ambient Temperature: 40°C

Weight of the motor: 1052 kG or 2413 lb.

Self-assessment Problems & Questions—Chapter 7

1. A gas powered prime mover is rotating the rotor of a single phase alternator at a speed of 1200 rpm. The alternator consists of six pole construction. The effective diameter of the coil is 0.15m and the length of the coil loop is 0.24m. The coil consists of 20 turns. The magnetic flux density has been measured to be 1.2T. Calculate the power delivered by this generator across a resistive load of 10Ω.

2. A four pole alternator/generator is producing electrical power at an electrical frequency of 50 Hz. (a) Determine the angular speed corresponding to the generated electrical frequency. (b) Determine the rotational (synchronous) speed of the armature/rotor. (c) Determine the angular velocity of the armature/rotor (rad/sec).

3. A four pole single phase AC generator consists of windings constituting 80 series paths and is driven by a diesel engine. The effective or mean

length of the armature is 18 cm and the cross-sectional radius of the armature is 5 cm. The armature is rotating at 1800 rpm. Each armature pole is exposed to a magnetic flux of 1.0 T. The efficiency of this generator is 90% and it is rated 2 kW. Determine the following:

(a) The maximum voltage generated. (b) The RMS voltage generated. (c) The horsepower rating of the generator. (d) The horse power output of the prime mover.

4.   A three phase, four pole, AC induction motor is rated 170 hp, is operating at full load, 60 Hz, 460 V_{rms}, efficiency of 90%, power factor of 80%, and a slip of 4%. Determine the (a) motor shaft speed, in rpm, (b) torque developed, in ft-lbf. (c) line current drawn by the motor and (d) the amount of reactive power, Q, sequestered in the motor under the described operating conditions.

5.   A three phase, four pole, AC induction motor is tested to deliver 200 hp, at 900 rpm. Determine the frequency at which this motor should be operated for the stated shaft speed. Assume the slip to be negligible.

6.   A three phase induction motor delivers 600 kW at a power factor of 80%. In lieu of installing power factor correction capacitors a synchronous motor is being considered as a power fact correction measure. Determine the apparent power size of the synchronous motor—in kVA—that should be installed to carry a load of 300 hp and raise the (combined) power factor to 93%. The source voltage is 230 V_{rms}.

Chapter 8

Power Distribution Equipment, Instrumentation and Electronic Safety Devices

Introduction

In this chapter will take an exploratory tour of power distribution equipment through the review of motor control center (MCC), disconnect switches, motor starters, breakers, power switchgear, variable frequency drives (VFD), etc. Pictures and diagrams are used in this discussion to give the reader a "hands- on" feel of common electrical and electronic equipment. Of course, as with the rest of this text, the reader will have an opportunity to test their knowledge through self-assessment problems end of the chapter.

VOLTAGE CATEGORIES IN POWER DISTRIBUTION SYSTEMS

Before we delve into the specific voltage categories of electrical systems, let's examine the flow of power from the point of generation to the point of consumption through Figure 8.1.

Figure 8.1: Electric power distribution and voltage levels. US DOE.

As shown in Figure 8.1, after AC power is generated at a typical power plant in the US, it is "stepped up" for transmission, to voltages ranging from 138 kV to 765 kV. As explained earlier in this text, the higher the transmission voltage, the lower the current required to deliver a certain amount of power to the consumers. As shown in Figure 8.1, once the power arrives within reasonable proximity of the consumers, it is stepped down to lower voltages suitable for use by the consumers. As we describe the major categories of voltage systems in the sections below, the reader is cautioned against interpreting these categories in an absolute fashion. As explained below, the boundaries between the various categories vary to some degree depending on the entity or organization performing the interpretation.

Low Voltage Systems

The low voltage category includes systems with voltages that **range 50 to 1000 V_{RMS} AC or 120 to 1500 Volts DC**. This category has the following subcategories:

> **Extra low voltage:** The voltage in this category is typically below $50V_{RMS}$ AC or below 120 Volts DC. The extra low voltage category is associated voltage, which typically can't harm humans, due to the low magnitude of potential difference. This category applies to equipment and wiring widely used in bathrooms, showers, swimming pools, toys and other electric devices, which might be in open contact with human.
>
> **Low voltage in power supplies for fluorescent lamps:** This low voltage category pertains to fluorescent lamp power supplies that use low DC voltage as source.
>
> **Low-voltage connectors:** This low voltage category is associated with low voltage connectors and low voltage plugs. Common examples include cigarette lighter 12V plugs, low voltage power adapters such as those used for charging rechargeable domestic and office cordless electronic, DC, equipment
>
> **Low-voltage overhead power lines:** This low voltage category pertains to power lines that bring low voltage, 110-240VAC power, to most homes and small commercial establishments.
>
> **LVDS voltage level:** This low voltage category pertains to LVDS systems. The term LVDS stands for low voltage differential signaling. LVDS

represents electrical digital signaling standard that pertains to high speed digital communications. It specifies the electrical-level details for interface between inputs and outputs on integrated circuit chips. Some common applications of low voltage LVDS standards are high-speed video, graphics, video camera data transfers, and general purpose computer buses.

Microphone preamp voltage level: This category pertains to microphone preamplifiers used to amplify a microphone's low output voltage to a higher, more usable level.

Medium Voltage Systems

Medium voltage is typically the voltage going from the substations to the service drops. The medium voltage **category ranges from 1 kV up to 15 kV, or, 35 kV,** by some standards. According to ANSI/IEEE 1585-2002 medium voltage ranges from 1000V to 35 kV_{RMS}. NEMA 600-2003, on the other hand, refers to medium voltage cables as "medium voltage cables rated from 600 volts to 69,000 volts AC."

High Voltage Systems

The high voltage category, by most standards, ranges from 15,000 V (35 kV, by some standards) up to 230 KV. High voltage is typically the voltage going from substation to substation or, as shown in Figure 8.1, the voltage level at which power is typically transmitted in the US from the point of generation to the point of consumption. Transmission-level voltages are usually considered to be 110 kV and above. Lower voltages such as 66 kV and 33 kV are usually considered sub-transmission voltages.

Extra High Voltage (EHV) Systems

There is a general consensus among power experts that voltage ranging between **230 kV to 600 kV** could be aptly termed as extra high voltage, or **EHV.** However, some standards refer to voltages ranging from 230 kV to 600 kV as simply "High Voltage."

Ultra-High Voltage Systems

There is a general consensus that **voltages exceeding 600 kV** could be referred to as ultra-high voltage, or **UHV.** Ultrahigh voltage (UHV) transmission lines are rare, however. There are a very few of these in the world. A 1.20 MV (1,200,000 V) Ultra High Voltage line, spanning from eastern to western part of India, when commissioned into operation is expected to be the highest voltage commercial power transmission line in the world.

Power Equipment Measurement Categories

The measurement category system pertains to the classification of live electric circuits used in measurement and testing of installations and equipment.

There are four measurement categories that take into account the total continuous energy available at the given point of circuit, and the occurrence of impulse voltages. Of course, this energy can be limited by circuit breakers or fuses, and the impulse voltages by the nominal level of voltage. Figure 8.2 shows the four categories in relation to the typical power distribution components.

Figure 8.2: Power measurement categories.

There are four categories, which are always stated with the designated voltage, for instance "CAT III, 150 V" or "CAT IV, 1000 V." This has important safety implications for impulse voltages and insulation clearances.

CAT I: This category is applicable to instruments and equipment, which are not intended to be connected to the main supply. Because the available energy is very limited, this category is typically not labeled on the equipment. Examples: Low voltage electronic circuits, load circuits of typical test bench power supplies, etc.

CAT II: This category defines circuits which are intended for direct connection into mains sockets. The energy in such installations is typically limited to below 100 A continuous. In CAT II systems, maximum available continuous power must be limited to 22k VA or less through fuses or circuit breakers.

CAT III: This category pertains to circuits which can be connected to the main feeders. The energy in CAT III systems is limited by circuit breakers to less than 110 kVA with the current not exceeding 11 kA.

CAT IV: CAT IV systems include circuits which are connected directly to the source of power, or utility, for a given building. The level of energy in CAT IV systems is high, limited to an extent by the power transformer impedance. CAT IV systems, typically, carry higher arc flash hazard during energized work.

MCCs or Motor Control Centers

The term motor control center can be somewhat misleading. The application of MCCs is not typically limited to the "control of motors." While majority of power controls housed in most MCCs are motor starters, often controls for lighting and other types of loads are included, as well. While our examination of MCCs will be limited to Allen-Bradley/Rockwell brand of MCCs and associated components, there are many other world class brands available, i.e. Square-D, ABB and Siemens.

As you tour an industrial or commercial facility, gray or blue cabinets such as the one depicted in Figure 8.3 can hardly escape notice. Those who are not engineers or technicians are likely to walk by one of these cabinets unaware of the electromechanical equipment performing critical functions behind the doors of those individual cubicles. Through examination of a series of MCC pictures and components, thereof, we will explore vital specifications of MCCs, the devices behind those MCC cubicle doors, how they function and what they consist of.

An Allen-Bradley CENTERLINE Motor Control Center, pictured in Figure 8.3, is made up of one or more vertical sections. A standard section is 90"

•20" (508mm) wide standard

•15" (381mm) or 20" (508mm) deep

•90" (2286mm) high standard,

   – 71" (1790mm) high available

•Front mounted or back-to-back

Figure 8.3: Physical size specifications of a typical MCC.

(2286mm) high, 20" (508mm) wide, and either 15" (381mm) or 20" (508mm) deep for front mounted configurations. Greater widths are sometimes supplied when larger equipment is required. Back-to-back configured MCCs are also available in 30" (762mm) and 40" (1016mm) designs.

NEMA Enclosure Types

Rockwell Automation offers a variety of NEMA type enclosures to meet specific requirements. The standard enclosure is NEMA Type 1. This compares

•Type 1

•Type 1 with gaskets

•Type 12 (also in stainless steel)

•Type 3R

•Type 4 stainless steel

Figure 8.4: MCC enclosure NEMA specifications.

to the IEC enclosure IP40. NEMA Type 1 with gaskets, which is unique to MCCs, provides gasketing for unit doors. This compares to the IEC enclosure IP41. NEMA Type 12 provides gasketing for unit doors, bottom plates, and all cover plates.

MCCs in stainless steel NEMA Type 12 enclosures are also available for corrosive environment applications. This is comparable to the IEC enclosure IP54. For outdoor use, manufacturers offer NEMA Type 3R enclosures. This enclosure is essentially a metal shell around a NEMA Type 1 inner enclosure. This is comparable to the IEC enclosure IP44. For indoor or outdoor use, the manufacturers offer a NEMA Type 4 enclosure. This enclosure is essentially a stainless steel shell around a NEMA Type 1 inner enclosure. This is comparable to the IEC enclosure IP65. All metallic parts (except stainless steel) are painted or plated before assembly, providing protection on all mating surfaces.

MCC bus bars and bus connections

The motor starters, contactors, VFDs and other devices housed in MCCs are mounted on draw out rack assemblies. These devices or systems can be "drawn out" and removed for replacement or service, after their disconnect switches have been turned off. This does not imply that the main power disconnect switch for power fed *into* the MCC must be disconnected or turned off while removing starters/cubicles for service.

If all cubicle assemblies were removed the view would be similar to what is depicted in Figure 8.5. As apparent in Figure 8.5, behind the Micarta insulating panel, or equivalent, an array of bus bars are visible. Three visible bus bars are shown magnified on the right side of Figure 8.5. The three bus bars represent phases A, B and C of three phase AC. These bus bars are, essentially, *silver plated copper conductors* forming a network of conductors carrying three phase 480 VAC (or other specified/rated voltage) to all MCC cubicles. Power bus connections are, typically, made with at least two bolts. The vertical-to-horizontal bus connection is made with two bolts, as pictured in Figure 8.5. The horizontal bus splice connections are made with at least two bolts on each side of the splice (depending on bus size). The two-bolt connection requires less maintenance and the extra bolt guards against the occurrence of "hot spots" (a result of loose connections), and arcing faults (a result of an open connection).

Horizontal Ground Bus

A 1/4" x 1" (6.4mm x 25.4mm) horizontal ground bus, rated 500A, is supplied, as standard, with each vertical section. A 1/4" x 2" (6.4mm x 50.8mm) horizontal ground bus, rated 900A, is optional. The bus can be unplated or supplied with optional tin plating. The horizontal ground bus can be mounted in the top or bottom horizontal wireway, or top and bottom horizontal wireways of the MCC. See Figure 8.6.

Vertical Ground Bus

Proper grounding is critical for safety of personnel and property. A substantial part of the NEC®, National Electrical Code, is dedicated safe grounding design, grounding equipment, and grounding methods. Hence, it is no

Figure 8.5: MCC bus bar network.

Figure 8.6: Vertical and horizontal ground bus bars in a typical MCC.

surprise that grounding of MCCs and the connected equipment is facilitated through a network of ground buses in various segments of MCCs. Figure 8.6 and 8.7, for instance, show vertical and horizontal ground bus bars located inside the MCC cabinet in.

Each standard vertical section is supplied with a steel vertical "plug-in ground bus" on the left side of the section. The vertical ground bus is bolted to the horizontal ground bus, providing positive grounding for all plug-in units. The load ground wires, from various field locations, can be brought directly to the MCC.

Incoming Lug Compartment

Power must be brought into the MCC. Therefore, the MCC needs to have some type of incoming compartment. There are 3 ways to accommodate the incoming power: incoming lug compartment, main fusible disconnect, and main circuit breaker.

Typically, an incoming lug compartment is used when the main disconnecting means for the MCC is located in switchgear near the MCC. See Figure 8.8. Incoming lug compartments are available for top or bottom entry of the power cables. The incoming lug compartment is designed so the cables are brought straight into the compartment. There is no power bus in the way and therefore, there is no need for sharp cable bends. This configuration is available sizes range from 600A to 3000A.

Main Fusible Disconnect Switch

Another way to accommodate the incoming cables is with a main fusible disconnect switch. See Figure 8.9. Main fusible disconnect switches are avail-

•1/4" x 1" (6.4mm x 25.4mm)
or
1/4" x 2" (6.4mm x 50.8mm)

•Unplated or
tin-plated copper

•Located in wireway
- Top or bottom
- Top and bottom

Figure 8.7: Horizontal ground bus bars in an MCC.

•Top or bottom

•Straight pull for cables

Figure 8.8: Incoming lug compartment on a typical MCC.

•Top or bottom

•Frame mounted

•600A-2000A utilize "Bolted Pressure Switch"

•Visible blade

Bolted
Pressure
Switch

Figure 8.9: Incoming power fusible disconnect switch in a typical MCC.

able for top or bottom entry of power cables. They are frame-mounted (non plug-in), and hard-wired to the horizontal bus. For 600A to 2000A applications, a "Bolted Pressure Switch" is used. The bolted pressure switch features a contact system that tightly holds the blades during closure to provide a reliable current path and high withstand. All main disconnect switches provide visible blade indication.

Main Circuit Breaker

A main circuit breaker is another way that the incoming power cables can be accommodated. See Figure 8.10. Main circuit breakers are available for top or bottom entry of power cables and, are frame-mounted (non plug-in)

•Top or bottom

•Frame mounted

•Ground fault protection available

Figure 8.10: Incoming power circuit breaker in a typical MCC.

and hard-wired to the horizontal bus. They are available in sizes up to 2000A. Ground fault protection is available for 600A to 2000A main circuit breakers.

There are pros and cons associate with fusible disconnect switches and breakers. When fuses clear (or "blow"), they must be replaced with exact equivalents. Incidents have been reported where improper fuse substitution has resulted in catastrophic failure of fuses, resulting in arc flash incidents. If breakers are applied as disconnecting and over current protection means in a power distribution system, they simply need to be reset when they trip in response to over-current conditions. One possible disadvantage associated with breakers is that they don't offers as much flexibility in terms of current limiting and fast action as compared with some of the current limiting and fast acting fuses, i.e. Ferraz Shawmut class J and RK.

Stab Assembly

The stab assembly housing isolates each phase at the rear of the unit. See Figure 8.11. Since the power wires are isolated within the stab assembly, a fault barrier is effectively formed between the units and the vertical bus.

The tin-plated stabs, rated 240 amperes, are directly crimped to the power wires, minimizing any chance for a loose connection. The steel spring that backs the stab ensures a reliable high-pressure four-point connection on the vertical bus. The stabs are also free-floating and self-aligning, meaning they will position themselves for easy unit insertion.

•Housing
  – Isolates incoming
    phases acting as a
    fault barrier

•Stabs
  – Tin-plated
  – Rated 240A
  – Directly crimped
  – Steel spring backed
  – Free-floating and
    self-aligning

Figure 8.11: Stab assembly behind a typical MCC cubicle.

Unit Grounding Provisions

A unit ground stab is provided on the back of each plug-in unit. See Figure 8.12. The ground stab engages the ground bus before the power stabs contact the vertical bus, and disengages after the power stabs are withdrawn from the vertical bus. The standard unit ground stab is made of a copper alloy. An unplated or tin-plated solid copper unit ground stab is also available.

•Unit ground stab
  – Used with unit plug-in
    ground bus
  – Copper alloy (standard)
  – Solid copper (optional)

•Unit load ground connector
  – Used with unit load
    ground bus
  – Solid copper
  – For easier termination of load
    ground wire

Standard
Plug-in
Ground
Stab

Optional
Plug-in
Ground
Stab

Unit Load
Ground
Connector

Figure 8.12: Ground system bus stabs in a typical MCC.

Unit Handle

The unit handle is flange mounted, and therefore stays in control of the disconnecting means at all times—whether the door is opened or closed.

The unit handle has positive status indication:
— Color-coded: **red** for **ON, green** for **OFF.**
— Labeled: ON and OFF (international symbols—**I** is used for **ON** and **O** for **OFF**).

Unit Handle

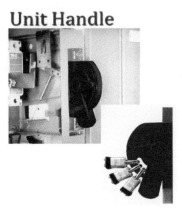

•Remains in control of disconnecting means

•Positive status identification

•Interlocked with door

•Padlockable

Figure 8.13: LOTO feature in the incoming power disconnect switch in a typical MCC.

The handle position is depicted in the ON and OFF positions (and TRIPPED position for circuit breakers). See Figure 8.13. The unit door is interlocked to the unit handle. The door cannot be opened when the handle is in the ON position unless the operator "defeats" the mechanism using a screwdriver. The unit handle can be locked in the OFF position with up to 3 padlocks for LOTO, Lock Out Tag Out, purposes.

Unit Interlock

Plug-in units are supplied with a "unit interlock" that prevents the unit from being inserted into or withdrawn from the section while the handle is in the ON position. See Figure 8.14. When the handle is in ON position, the interlock mechanism moves upward engaging the unit support pan above the unit. The unit interlock can also be used to secure the unit in a service position to guard against accidental unit insertion. This is shown in the picture on the left.

Lastly, the interlock can be padlocked during servicing to prevent unit insertion, shown in the photo on the right, even with the handle OFF.

MOTOR STARTER—A-B BULLETIN 2100

A motor starter is designed to control the flow of power to a motor, while,

•Prevents inserting or
withdrawing unit with
handle in ON position

•Padlockable

Figure 8.14: Interlock on MCC cubicle.

simultaneously, protecting the motor branch circuit against over-current conditions. Motor starters assemblies, such as the A-B Bulletin 500, are designed for up to 10 million operations. See Figure 8.15. Additional information on the design, specifications and operation of some of the motor starter components will be discussed in greater detail in Chapter 10.

We will use the picture of a Bulletin 500 A-B starter, as captured in Figure 8.15, to examine the design and operation of a typical FVNR, Full Voltage Non-Reversing motor starter. Beginning at about 3'o clock, in the motor starter picture below, we notice the handle of disconnect switch. The handle of the dis-

Figure 8.15: Motor starter details.

connect switch is mechanically linked to the mechanism labeled as the "Fusible Disconnect or Circuit Breaker." The three knife switches visible in the picture, at about 12'o clock, allow the three phase AC voltage to be applied to the fuse holder segment below. The three over current protection fuses are not shown installed in the starter assembly. The three terminals, with three emerging black wires, provide the continuity of the three phase voltage to the top of the contactor; labeled Bulletin 500 NEMA

Contactor

A contactor, similar to a typical electrical relay, is simply an electro-mechanically controlled device that closes or opens contacts, to apply or disengage AC voltage, respectively, from the load. The load, in this case, is an electric motor.

The two wires visible directly in front of the contactor are *control wires* that energize the solenoid of the contactor. When the solenoid coil is energized, the core rod or plunger pushes the contact of the contactor shut and allows the flow of current and power downstream toward the motor, via the solid state overload protection device shown at about 8'o clock. The solid state overload device is labeled as "Bulletin 592 Melting Alloy or Solid State Overload Protection" in Figure 8.15.

Notice the white terminal strips or terminal blocks visible along the bottom of the starter unit depicted in Figure 8.15. These terminal strips are called the "pull-apart terminal blocks." These pull apart terminal blocks are, typically, used in plug-in units and they represent a significant improvement in wire termination method. Prior to the advent of pull-apart terminal blocks, replacement, removal and re-installation of electrical required meticulous and arduous effort on the part of technicians to correctly terminate wires. The tedious task of disconnecting and connecting wires often resulted in miss-wiring and workmanship defects. The pull-apart terminal blocks have a front half and a rear half that detach for easy unit removal. The back half of the terminal block is factory wired, and the front half is where the customer terminates field wires.

Pilot Devices

Pilot devices are shown mounted on the left side of the starter chassis as depicted in Figure 8.15 and shown again in the top left corner of Figure 8.16, labeled as "Bulletin 800T." Typical set of pilot devices include indicating lamps and control switches. The pilot device set shown in Figure 8.16 consists of one motor/load status indicating light at the top (typically, red or green). The two push buttons located below the indicating light are momentary (spring loaded) push button type switches. The middle switch, typically green in color, is the "START" switch and the bottom switch, typically red in color, functions as a "STOP switch. The control sequence and algorithm associated with the pilot devices is explained in the wiring diagram discussion, in chapter 10.

Figure 8.16: Motor starter pilot devices.

FVR, Full Voltage Reversing Starter and Variety of Other Power Control Devices

As discussed earlier, when "on the fly," instantaneous, reversal of a three phase AC motor is desired, a second contactor is incorporated into the motor starter cubicle to reverse the three phase AC sequence from ABC to ACB, and vice and versa. Of course, the two starters in such cases are interlocked to ensure that simultaneous connection of the two opposite phase sequences is avoided.

An FVR starter is pictured on the top right side of Figure 8.17. See labels "Contactor No. 1" and "Contactor No. 2" in Figure 8.17. An FVNR starter—with a single contactor—is depicted to the left of the FVR starter in Figure 8.17 for contrast. The bottom half of the picture in Figure 8.17 shows two other power control devices: (1) A soft start or two-speed starter unit, and (2) A full voltage lighting contactor unit.

VARIABLE FREQUENCY DRIVES—UP TO 250HP

In addition to the power control devices discussed above, engineers can design or specify the integration of VFDs, or variable frequency drives, into

Full Voltage Non-Reversing Starters

Full Voltage Reversing Starters

Contactor No. 1

Contactor No. 2

Feeders

Lighting Transformers

Plus: Full Voltage Contactors, Two-Speed Starters, Reduced Voltage Autotransformer Starters

Figure 8.17: Motor starter pilot devices.

MCC systems. When VFDs are housed into an MCC cabinet, it is advisable to maintain *electrical shielding* and adequate physical separation between power control components and more sensitive electronic systems like the VFDs and PLCs, Programmable Logic Controllers. In the absence of such caution, *electromagnetic interference*, or *EMI*, can adversely affect the operation of sensitive electronic systems.

Smaller variable frequency drives are often, specifically, designed to fit into plug-in units, or "**rack-out assemblies**." The VFD pictured on the left side of Figure 8.18, is a smaller, Bulletin 1305 drive. The smaller drives range from 3 hp at 240V to 5 hp at 480V. The VFD pictured on the right side of Figure 8.18, is a larger, Bulletin 1336 PLUS, drive. Some of the larger VFDs, such as the A-B 1336, fall in the following size categories:

- 1/2—30HP at 480V, plug-in design
- 40—125HP at 480V, frame-mounted design
- 150—250HP at 480V, roll-out design

SMART MOTOR CONTROLLERS (SMCS)—UP TO 500A

The Allen-Bradley, like other power and controls manufacturers, offers various smart motor controllers that allow motor loads to be operated at multiple, *discrete*, speeds to take advantage of energy savings in accordance with the *fan laws* or *affinity laws*. This essentially means that if the motor is oper-

Figure 8.18: VFD, Variable Frequency Drives.

ated at 50% of the normal full load speed (rpm), the power or energy consumption of the motor drops to 12.5%.

The smart motor controller featured in Figure 8.19 is an SMC-50 smart motor controller by Rockwell Automation. It is a solid-state motor controller that is premised on three-phase, solid-state, silicon-controlled rectifiers (SCRs, or thyristors).

Motor Soft Starts

Soft starters are typically employed in power control systems to limit excessive initial inrush of current associated with motor start up and provide a gentler ramp up to full speed. Contrary

Figure 8.19: SMC 50, Rockwell-Automation Smart Motor Controller.

to VFDs, soft starters are only used during startup of motors. See Allen-Bradley soft start pictured in Figure 8.20.

In pump applications, a soft start can be installed to avoid pressure surges. Soft starts can facilitate smoother starting, prevent jerking and stressing of mechanical drive components, e.g. in conveyor belt systems that are loaded with bulk materials, intermittently. Using soft starts, fans or other systems with belt drives can be started slowly to avoid belt slipping. In essence, a soft start *limits the inrush current*, *improves stability* of the power supply and *reduces transient voltage drops* that may affect other loads and the overall power

Figure 8.20: Rockwell-Automation Soft Start.

Among "non-electronic" methods for "soft-starting" are means such as the installation of a *series reactor* (coil or inductance, in general). Presence of a series reactance limits motor starting current.

COMPARISON BETWEEN SMCS, VFDS AND SOFT STARTS

The key differences between the application of VFDs versus SMCs are as follows:

1) VFDs have the ability to vary the armature frequency—and the motor speed—in practically *continuous*, gradual or *analog* fashion. Smart motor controllers, on the other hand, allow variation of the motor at multiple *discrete*, preset, levels or steps.

2) While VFDs tend to cost more than the SMCs, the continuous adjustability of the VFDs can provide "finer," higher resolution, control of AC motors. This could provide the end user better process control, in addition to the energy savings opportunity.

3) SMCs and Soft Start systems tend to be smaller, in physical size, than VFDs.

4) VFDs offer the greatest energy savings for fans and pumps.

5) Soft-starts tend to present the smallest footprint as compared to SMCs and VFDs.

PLC AND I/O CHASSIS

Even though the focus of discussion—through the exploration of Allen-Bradley MCCs—has been attainment of a better understanding of power or motor control devices, we will probe the hardware aspects of PLCs, Programmable Logic Controllers, in this section; primarily, because the SLC-500 and Bulletin 1771 PLCs can be integrated into a typical Allen-Bradley MCC. See Figure 8.21. Of course, smaller PLCs, i.e. the Micro-Logix series could be specified to be included in a section of a PLC, just as well.

Figure 8.21: PLCs incorporated into an Allen-Bradley MCC.

However, once again, care should be exercised to protect the PLC, and associated I/O, Inputs and Outputs, from potential EMI emitted from power components in close proximity. The leftmost picture of a PLC shown in Figure 8.21 (a) depicts a SLC 500 PLC and associated I/O blocks. In Figure 8.21 (a), adjacent to the CPU, Central Processing Unit—the PLC power supply is shown mounted to the "*back plane*" of the PLC. A backplane, often, consists of a circuit board with rigid connectors that the I/O modules, CPU module and the power supply can be plugged into. A PLC, not unlike a PC (Personal Computer), operates off of low DC voltage, 5-10 V. This low DC voltage is generated by the PLC power supply. The power supply, in turn, is powered by low AC voltage. The PLC *control transformer*, shown in Figure 8.21 (a), transforms the 480 V_{AC}—available at the 480 V_{AC} bus stabs—into a lower AC voltage ($\approx$14 V$_{AC}$). This lower AC voltage is converted from AC to DC by the rectifier in the power supply.

Larger Bulletin 1771 Allen-Bradley PLCs are pictured in Figures 8.21 (b) and (c). The PLC depicted in Figure 8.21 (b) is a Bulletin 1771 PLC with combined power supply, CPU and I/O on one panel. The PLC depicted in Figure 8.21 (c) is a Full Section Bulletin 1771 PLC, with power supply and the CPU on one panel and the I/O rack on a separate panel. The split panel configuration of the Full Section 1771 PLC system accommodates a larger number of inputs and outputs.

Figure 8.22: Complete control system in single control cabinet.

Often, PLCs are integrated into a single cabinet with peripheral control equipment such as variable frequency drives for automatic speed control and the control of higher current field devices through relays and contactors. Figure 8.22 depicts one such scenario.

Metering Units

Analog Metering Unit   Powermonitor II Metering Unit

Figure 8.23: MCC compatible power monitoring equipment

It is common for engineers to specify and incorporate power monitoring equipment into an MCC. Examples of typical power monitors are pictured in Figure 8.23. The power monitoring unit shown on the left reads the monitored power parameters in the traditional analog format, while the unit on the right is a more contemporary digital power monitor. The digital power monitors tend to be more versatile; they are often designed to capture and display electrical parameters like: three phase voltage, current, kW (or real power **P**), kVA (or apparent power **S**), kVAR (or reactive power **Q**), power factor, frequency, etc.

The rear of a multifunctional, digital, power monitor is displayed in Figure 8.24 below. The wires bringing the current signal from line current measuring CTs (current transformers) would be terminated at terminals I_1, I_2 and I_3. The voltage signal coming from the voltage or potential transformers would be terminated at points labeled V_1, V_2 and V_3. Also, note the Ethernet connection port in the top right corner of Figure 8.24. This Ethernet port allows the power monitor to transmit all data to a central power monitoring work station through an ***Ethernet LAN***, Local Area Network. Such network connectivity allows multiple monitors to feed data to a central location for display and control purposes. The digital communication attribute of digital power monitors—and the connectivity to LAN—also permits users to broadcast the data for off-site monitoring through ***routers*** and ***modems***.

MAIN SWITCH YARD AND MEDIUM VOLTAGE SWITCHGEAR

During the course of our discussion of power distribution equipment in this text we made mention of "Main Switch Yard." For illustration purposes,

Figure 8.24: Rear termination points on an Allen-Bradley Power Monitor

a picture of the main switch yard of a facility is shown in Figure 8.25. Main switch yards are compounds, on site at the consumer's facility, where the utility (power company) "lands" the power into the customer's facility and transforms it from high voltage to medium voltage, i.e. 100 kV to 13 kV.

MEDIUM VOLTAGE SWITCHGEAR—LOOP SWITCH

Switching within facilities, at the medium voltage level, is accomplished through loop switches and breakers. A typical, three phase, 13 kV loop switch is pictured in Figure 8.26. Some of the key functional components of the loop switch, i.e. the operating mechanism, viewing ports and switch status (Open or Closed) are annotated on Figure 8.26.

Figure 8.25: Main Switch Yard; 100 kV.

Power Factor Correcting Capacitors and
Step Down Transformer, Main Switch Yard

Power factor correcting capacitors, installed on the power company sided of the main switch yard are pointed out in Figure 8.27. Also annotated are the large 100 kV to 13 kV step down transformers, on the power company side of the switch yard.

CIRCUIT BREAKERS

A circuit breaker is an over current protection device that is an automatically operated. A circuit breaker, basically, is an electrical switch designed to protect an electrical circuit from damage caused by overload or short circuit.

Figure 8.26: 13 kV Loop Switch.

Figure 8.27: Power factor correcting capacitors and main step-down transformer in a main switch yard.

Its core function is to detect a fault condition and interrupt current flow. Unlike a fuse, which operates once and then must be replaced, a circuit breaker can be reset (either manually or automatically) to resume normal operation. The most common and basic type of circuit breaker is the type that is often applied in residential dwelling breaker panels. Circuit breakers are categorized based on the following features or functions:

a) Voltage
b) Current
c) Principle of operation

Circuit breakers for large currents or high voltages are equipped with sensing and transducing devices that detect fault current and operate the trip or circuit opening mechanism. The term "trip," essentially means "opening" of an electrical circuit such that the current ceases to flow. The trip function is typically executed by a solenoid that releases a latch that keeps the breaker closed (ON). In order to maintain the fail safe properties—typically required in control systems—the tripping solenoid is often energized by a separate battery. High-voltage circuit breakers are self-contained with current transformers, protection relays, and an internal control power source.

Upon detection of a fault, or fault current, contacts within the circuit breaker open to interrupt the circuit. Often, mechanically stored energy—in springs or compressed air—is used to separate and force open the contacts, thus interrupting the flow of current.

Low Voltage Circuit Breakers

As the name implies, low voltage breakers are applied in domestic, commercial and industrial application, operated at less than 1000 V_{AC}. The design of low voltage breakers is premised on thermal or thermal-magnetic operation. Low voltage power circuit breakers can be mounted in multi-tiers in low-voltage switchboards or switchgear cabinets. Low-voltage circuit breakers are also made for DC applications, i.e. electrical protection of subway lines.

A sectional diagram of a common 10 ampere DIN rail-mounted thermal-magnetic miniature circuit breaker is depicted in Figure 8.28.

Where,
1. Actuator lever—used to manually trip and reset the circuit breaker. The legend on the breaker typically indicates the status of the circuit breaker (ON or OFF).
2. Actuator mechanism—forces the contacts to "close" or "open."
3. Contacts—Allow the load current to flow when closed interrupt the flow when open.
4. Terminals.
5. Bimetallic strip for thermal operation or tripping of the breaker under prolonged overload conditions.
6. Calibration screw—allows the manufacturer to precisely adjust the trip

Figure 8.28: Sectional diagram of a common rail-mounted thermal-magnetic miniature circuit breaker. Dorman Smith circuit breaker, annotated by: Ali, UK.

      current of the device after assembly.

7.   Solenoid.

8.   Arc divider or extinguisher.

MEDIUM-VOLTAGE CIRCUIT BREAKERS

      Medium-voltage circuit breakers are classified based on the medium (or substance) used, within, to extinguish the arc. Some of the medium-voltage circuit breakers types are discussed below:

      Vacuum circuit breakers: The vacuum circuit breakers—rated up to 3000 A—interrupt the current by creating and extinguishing the arc in a near vacuum environment, within a sealed container. The vacuum circuit breakers are generally applied for voltages up to about 35,000 V; they tend to have longer life expectancies, and they offer longer service spans between overhauls than do air circuit breakers.

      Air circuit breakers: Air circuit breakers are rated, in current, up to 10,000 A. Their trip characteristics are often fully adjustable including configurable trip thresholds and delays. They are often electronically controlled.

SF$_6$ circuit breakers: The SF6 medium voltage circuit breakers extinguish the arc in a chamber that is filled with ***sulfur hexafluoride gas***.

HIGH-VOLTAGE CIRCUIT BREAKERS

High-voltage circuit breakers are often applied in electrical power transmission networks. High-voltage breakers are almost always solenoid-operated, with current sensing protective relays operated through current transformers. A high voltage circuit breaker, rated 110 kV, is depicted in Figure 8.29. This breaker is designed for 50 Hz operation and is equipped with dielectric oil based arc quenching system.

High-voltage breakers are broadly classified based on the medium used to extinguish the arc. Technically, any of the following can be used for arc quenching:

— Bulk oil
— Minimum oil
— Air blast
— Vacuum
— SF6, sulfur hexafluoride gas

However, due to environmental and cost concerns over insulating oil spills, most new breakers use SF$_6$ gas to quench the arc.

Figure 8.29: Oil circuit breaker MKP-110 (110 kV) on traction substation 110 kV 50 Hz/3.3 kV DC of railway, Toliatti City, Russia. By: Vivan.

Self-assessment Questions—Chapter 8

1. A substation in a manufacturing facility is being fed from a 13 kV transformer secondary. This switchgear in this substation would be categorized as:
   A. Medium voltage
   B. Low voltage
   C. High voltage
   D. None of the above

2. Power transmission lines would be categorized as:
   A. Medium voltage
   B. Low voltage
   C. High voltage
   D. Medium voltage
   E. None of the above

3. The breakers installed in residential breaker panels are:
   A. OCBs
   B. Thermal magnetic circuit breakers
   C. Low voltage thermal magnetic circuit breakers
   D. None of the above
   E. Both B and C

4. The vacuum circuit breakers tend to offer longer service spans between overhauls than do air circuit breakers.
   A. True
   B. False

5. The SF_6 type high voltage circuit breakers are not preferred due to environmental concerns.
   A. True
   B. False

6. MCCs are not designed to accommodate PLCs and VFDs.
   A. True
   B. False

7. The bus bars in MCCs are commonly constructed out of:
    A. Aluminum
    B. Silver plated copper
    C. Silver
    D. Iron

8. Pilot devices on MCCs:
    A. Control circuit breakers
    B. Indicate the status of MCC
    C. Indicate the status of motor/load
    D. Include "Start" and "Stop" controls
    E. Both (C) and (D).

9. Power control cubicles in MCCs are fixed and cannot be removed while the main fusible disconnect switch of the MCC is ON.
    A. True
    B. False

10. A control transformer, in a given MCC compartment:
    A. Steps down the voltage for control circuit operation.
    B. Provides power for MCC cabinet lighting.
    C. Is seldom needed.
    D. Serves as an isolation transformer.
    E. Both (C) and (D).

Chapter 9

National Electric Code, NFPA® 70 E and Electrical Standards

Introduction

The purpose of this chapter is to merely introduce the reader to NEC ®, National Electrical Code. The coverage of the NEC® in this chapter is not intended to provide the reader intermediate, advanced or expert level knowledge that is required and expected of a NEC® trained and practicing electrical Professional Engineer, specializing in power system design. Notwithstanding the foregoing clarification regarding the limited depth of NEC® knowledge provided in this chapter, the nonelectrical engineer or the non-practicing electrical engineering reader will find that this chapter on NEC® and electrical safety opens the door and allows the reader to appreciate the complexity and depth of the NEC®. And, most of all, if the nonelectrical reader of this chapter finds themselves leading a group of electrical engineers and electricians in a "2:00 AM triage situation," trying to troubleshoot and reinstate an important piece of equipment back into operation, the knowledge and familiarity they will gain in this chapter should prepare them better to comprehend the code related jargon used by EEs and electrical technicians.

The NEC® is revised every three years. At the time this text was authored, the 2011 revision was in effect. The reader is reminded that the objective of this text is not to provide precise code content and references at the time this text is read. Instead, the this text aims to give the reader *general* references of NEC® articles that have, *traditionally*, addressed the minimum requirements associated with equipment and appurtenances, i.e. conductors, conduits, raceways, fuses, breakers, grounding systems, etc.

Non-electrical engineering professionals often wonder if they should invest in the NEC® book. If so, what format and version would be most suitable? While the most appropriate answer to this question lays in the depth and extent of study intended, most non-electrical engineers might find the *NEC® handbook* more beneficial. The handbook version of the NEC® is replete with copious pictures, diagrams and illustrative example problems that aptly facilitate quicker comprehension of code and associated concepts.

In keeping with the approach utilized in the rest of this text, as we get introduced to certain commonly applied codes, we will take our knowledge and comprehension to the next level through example problems, end of chapter problems and the solutions at the back of the text.

NEC ® Articles

Listed below are major NEC® articles that are introduced in this chapter, followed by a few details that explain the significance of the articles. Later, in this chapter, we will illustrate the significance of some of these NEC ® articles through examples.

- Art. 90—Introduction to NEC ® & Outline.
- Art. 100—Definitions, Including Enclosure Ratings
- Art. 110.6.—Conductor Sizes, AWG and Circular Mils.
- Article 110.16—Arc Flash Regulations:
- Art. 110.26—Clearances and Working Space Requirements.
- Art. 210—Load Configurations and Voltages in Branch Circuits.
- Art. 210.9—Autotransformers.
- Art 210.20—Branch Circuit Ampacity Determination & Over current Protection.
- Art. 240—Over current Protection
- Art. 240.50—240.101—Circuit Breaker and Fuse Types
- Art. 250—Grounding
- Art. 310—Conductor Insulation Rating
- Art. 310.15—Conductor Ampacity
- Art. 358—392—Conduit and Cable Trays
- Art. 408.13—408.35—Panel Boards

Art. 90—Introduction to NEC® and Outline

Article 90 of the NEC® is a basic introduction to the intention of the codes. This article states the purpose of the code provision of uniform and practical means to safeguard people and equipment from electrical hazards. These safety guidelines are not meant to describe the most convenient or efficient installations and don't guarantee good service or allow for future expansion. The NEC ® articles of code are designed to provide a standard for safety that protects against electrical shock and thermal effects, as well as dangerous over-currents, fault currents, overvoltage, etc. The NEC® comports, for the most part, with the principles for safety covered in Section 131 of the International Electrotechnical Commission Standard for electrical installations. International Electrotechnical Commission, or IEC, is the world's leading organization that prepares and publishes *International Standards* for all electrical, electronic and related technologies.

Art. 100—Definitions

Professionals who are not electrical engineers or electricians are likely to find this article on of the most useful articles. In that, this article contains definitions of terms that are essential in the understanding and interpretation of the code. As an introduction, a few of the terms described in Article 100 are listed and explained below.

Ampacity: Ampacity is defined as "The *maximum* current, in amperes, that a conductor can carry continuously under the conditions of use without exceeding its temperature rating."

Note: Ampacity does vary depending on several factors. Appropriate NEC ® Tables and derating rules must be applied to determine the correct ampacity. See Article 310 for additional explanation.

Bonded (Bonding): Bonding is defined as equipment or objects "Connected to establish electrical continuity and conductivity." In other words, bonding constitutes permanent joining of metallic parts to form an electrically conductive path that ensures electrical continuity and the capacity to conduct electrical current safely.

Note: This is not the same as grounding, but bonding jumpers are essential components of the bonding system, which is an essential component of the grounding system. Furthermore, note that the NEC does not authorize the use of the earth as a bonding jumper because the resistance of the earth is more than 100,000 times greater than that of a typical bonding jumper.

Branch Circuit: Branch circuit is defined as "The circuit conductors between the final overcurrent device protecting the circuit and the outlet(s)." As we will discuss and illustrate later, a branch circuit, as defined by the NEC ®, can be the entire circuit that lays between the overcurrent fuses and a motor load, in a typical motor branch circuit.

Continuous Load: A continuous load is defined by the NEC ® as "A load where the maximum current is expected to continue for 3 hours or more." Note:

The maximum running current referred to in this definition is exclusive of the *starting current.*

Feeder: A feeder is defined by the NEC ® as "All circuit conductors between the service equipment, the source of a separately derived system, or other power supply source and the final branch-circuit overcurrent device."

Ground: A ground is defined by the NEC ® as simply "The earth." On the other hand, the term **Grounded (Grounding)** is defined as "Connected (Connecting) to ground or to a conductive body that extends the ground connection."

Note: Simply driving an electrode into the earth does not constitute grounding a circuit. The ground must be made with the source or supply in mind, as the flow of electrons—or current—always tries to *return* to the source.

In Sight From (Within Sight From, Within Sight): The NEC ® defines this as "Where this Code specifies that one equipment shall be 'in sight from,' 'within sight from,' or within sight of,' and so forth, another equipment, the specified equipment is to be visible and not more than 15 m (50 ft) distant from the other." In other words, lockable disconnecting means must be provided in sight of the load, unless the specific installation meets the criteria stated under article 430.102 (B) (2) (a) or (b).

Labeled: The NEC ® defines this as "Labeled. Equipment or materials to which has been attached a label...acceptable to the authority having jurisdiction and concerned with product evaluation,...."

Note: The reader is advised to read the entire original definition in order to attain full understanding of this term.

Neutral Conductor: The NEC ® defines this as "The conductor connected to the neutral point of the system that is intended to carry current under normal circumstances.

Overcurrent: The NEC ® defines overcurrent as "Any current in excess of the rated current of equipment or the ampacity of a conductor. It may result from overload, short circuit or ground fault."

Overload: The NEC ® defines overload as "Operation of equipment in excess of normal, full-load rating, or of a conductor in excess of rated ampacity that, when it persists for a sufficient length of time, would cause damage or dangerous overheating. A fault, such as short circuit or ground fault, is not an overload."

Raceway: The NEC ® defines raceway as "An enclosed channel of metal or nonmetallic materials designed expressly for holding wires, cables, or bus-bars.

Note: The reader is advised to read the entire original definition in order to attain full understanding of this term.

Short-Circuit Current Rating: The NEC ® defines this term as "The prospective symmetrical fault current at a nominal voltage to which an apparatus or system is able to be connected without sustaining damage exceeding defined acceptable criteria."

Voltage, Nominal: The NEC ® defines nominal voltage as "A nominal value assigned to a circuit or system for the purpose of conveniently designating its voltage class (e.g., 120/240 volts, 480/277 volts, 600 volts). The actual voltage at which a circuit operates can vary from the nominal within a range that permits satisfactory operation of equipment."

Art. 110.6.—Conductor Sizes, AWG and Circular Mils.

Overall Article 110 lays out the requirements for electrical installations. Article 110.6, specifically, addresses conductor sizes. This article clarifies the fact that the conductor sizes are expressed in AWG, American Wire Gage or in circular mils. For copper, aluminum or copper clad aluminum conductors up to size 4/0 AWG, the code identifies the conductor sizes in AWG. Conductors larger than 4/0 AWG are sized in circular mils, beginning with 250,000 mils; formerly referred to as 250 MCM. Where first "M" stands for 1000. The unit *kcmil* was adopted 1990. Therefore, the more contemporary identification of a 250,000 mil conductor would be 250 *kcmil.* Note: 1 mil is equal to 1/1000th of an inch.

Article 110.16—Arc-Flash Hazard Warning

This article addresses stipulation associated with potential hazard of arc-flash incidents initiated by electrical faults. The essence of the *Code*, in this article is that switchboards, panel-boards, industrial control panels, meter socket enclosures, and motor control centers in other than dwelling units, which are likely to require examination, adjustment, servicing, or maintenance while energized, shall be field marked to warn qualified persons of potential electric arc flash hazards. The marking or label shall be located so as to be clearly visible to qualified persons before examination, adjustment, servicing, or maintenance of the equipment. An example of a label generated as a result of arc flash hazard analysis is shown in Figure 9.1. Look for additional discussion on the topic of arc flash later in this chapter.

⚠ WARNING	
Arc Flash and Shock Hazard	
Appropriate PPE Required	
120 inch	**Flash Hazard Boundary**
468	**cal/cm^2 Flash Hazard at 18 inches**
DANGER!	**Dangerous!!! No FR Class Found!! DO NOT WORK ON EQUIPMENT WHILE ENERGIZED!!**
480 VAC	**Shock Hazard when cover is removed**
42 inch	**Limited Approach**
12 inch	**Restricted Approach**
1 inch	**Prohibited Approach**
Bus Name: Water panel 1, Prot Device: CB XXXX	

Figure 9.1: Arc-flash hazard label, produced through arc-flash hazard analysis.

Art. 110.26—Clearances and Working Space Requirements

This article of the Code states: "Access and working space shall be provided and maintained about all electrical equipment to permit ready and safe operation and maintenance of such equipment." The essence of this article is that one cannot install electrical equipment in just *any* available physical space. Equipment must be installed such that it can be accessed, safely, at all times by maintenance personal.

Art. 210—Branch Circuits

This article applies to branch circuits that supply power to motor and non-motor loads. Article 430 must be consulted for situations where branch circuits feed motors only. The NEC ® handbook contains myriad illustrations of different types of branch circuits and associated code requirements.

Art. 210.9—Circuits Derived from Autotransformers

This article stipulates: "Branch circuits shall not be derived from autotransformers unless the circuit supplied has a grounded conductor that is electrically connected to a grounded conductor of the system supplying the autotransformer." In addition to explanation of this code, the handbook shows circuit diagrams of autotransformers connected in various configurations.

Art. 210.20—Overcurrent Protection

According to 210.20, an overcurrent device that supplies continuous and non-continuous loads must have a rating that is not less than the sum of 100 percent of the non-continuous load plus 125 percent of the continuous load, calculated in accordance with Article 210. Because grounded/neutral conductors are generally not connected to the terminals of an overcurrent protective device, this requirement for sizing conductors subject to continuous loads does not apply.

Art. 240—Overcurrent Protection Device

This article of the *Code* provides the requirements for selecting and installing overcurrent protection devices (OCPDs). On one hand an OCPD protects a circuit by opening the circuit when current reaches a value that would cause an excessive temperature rise in the conductors. If one were to apply the rising water analogy, current rises like water in a tank, and at a certain level, the OCPD shuts off the faucet. On the other hand, an OCPD protects equipment by opening the circuit when it detects a short circuit or ground fault. All electrical equipment must have a short-circuit current rating that permits the OCPDs to clear short circuits or ground faults without a catastrophic failure.

Art. 240.50—240.101—Circuit Breaker and Fuse Types

Articles, ranging from Art. 240.50-240.101, address requirements associated with selection, specification and installation of various overcurrent protection devices such as circuit breakers and fuses. Once again, in addition to explanation of the code, the handbook shows diagrams and pictures of various types of fuses and a breaker tripping unit.

Art. 250—Grounding and Bonding

Overall, this article addresses general requirements for grounding and bonding of electrical installations, types and sizes of grounding and bonding equipment, methods of grounding and bonding, and situations when guards, isolation, or insulation may be substituted for grounding.

Art. 310—Conductors and General Wiring

This NEC ® article addresses general requirements for conductors, their types, insulations, markings, mechanical strengths and ampacity ratings. Table 310.15 lists ampacities of various conductors and is, possibly, the most frequently used page in the *Code*. See additional discussion below on conductor ampacity.

Art. 358—Electrical Metallic Tubing: Type EMT

This article of the *Code* addresses the application, installation and construction specifications for electrical metallic tubing (conduit) and associated tubing.

Art. 408—Switchboards and Panelboards

This article of the *Code* covers switchboards and panelboards associated with equipment operating at 600 volts or less.

AMPACITY OF CONDUCTORS—TABLE 310-15

In order to provide a measure of familiarity with the most frequently used section of the NFPA, NEC, National Electric Code, an older version of NEC Table 310-15 has been included under Tables 9.1 and 9.2. Closer examination of this table shows that the ampacities of various conductors are listed under two separate sections. The left section represents copper conductors and the right section represents the aluminum or copper clad aluminum conductors. These two sections are divided further into three columns each. These three columns list the ampacities of various commercially available conductors under three temperature categories: 60°C(140°F), 75°C(167°F) and 90°C(194°F). Note that these three separate temperature columns lump different types of commercially available insulations. The temperature and insulation columns are selected in accordance with specific application, environment and ambient temperature. The bottom section of Table 9.2 represents the conductor *ampacity correction factors* based on the ambient temperatures above or below 30°C.

The methods associated with the application of NEC® articles are illustrated through example problems listed below. However, the NEC® articles cited are subject to change, without notice, due to periodic code revisions and were meant to be valid at the time of development of these problems. Therefore, these NEC® article references may not be current and should not be used for actual practice of electrical engineering design. Instead, reader should refer to the most current NEC® articles in their practice of engineering.

Example 9.1
Over Current Protection:

Applicable Code/Codes: Articles 210.19 (A) (1), 210.20 (A) and 310.15. Tables 310.15(B)(2)(a) and 310.15(B)(16).

The branch circuit in the exhibit below consists of two (2) 8 amp continuous loads. Over current protection in the branch circuit is provided through a 20 amp circuit breaker. (a) Verify the size/specifications of the circuit breaker

Table 9.1: NEC® Ampacity table superseded by current NEC Table 310.15(B) (16). Included for general illustration purposes, only. Courtesy, NEC, NFPA.

Table 310.15(B)(2)(a) Adjustment Factors for More Than Three Current-Carrying Conductors in a Raceway or Cable

Size AWG or kcmil	COPPER			ALUMINUM OR COPPER-CLAD ALUMINUM			Size AWG or kcmil
	60°C (140°F) — Types TW, UF	75°C (167°F) — Types RHW, THHW, THW, THWN, XHHW, USE, ZW	90°C (194°F) — Types TBS, SA, SIS, FEP, FEPB, MI, RHH, RHW-2, THHN, THHW, THW-2, THWN-2, USE-2, XHH, XHHW, XHHW-2, ZW-2	60°C (140°F) — Types TW, UF	75°C (167°F) — Types RHW, THHW, THW, THWN, XHHW, USE	90°C (194°F) — Types TBS, SA, SIS, THHN, THHW, THW-2, THWN-2, RHH, RHW-2, USE-2, XHH, XHHW, XHHW-2, ZW-2	
18	–	–	14	–	–	–	–
16	–	–	18	–	–	–	–
14*	20	–	25	–	–	–	–
12*	25	20	30	20	20	25	12*
10*	30	25	40	25	30	35	10*
8	40	35	55	30	40	45	8
6	55	50	75	40	50	60	6
4	70	65	95	55	65	75	4
3	85	85	110	65	75	85	3
2	95	100	130	75	90	100	2
1	110	115	150	85	100	115	1
1/0	125	130	170	100	120	135	1/0
2/0	145	150	195	115	135	150	2/0
3/0	165	175	225	130	155	175	3/0
4/0	195	200	260	150	180	205	4/0
250	215	230	290	170	205	230	250
300	240	255	320	190	230	255	300
350	260	285	350	210	250	280	350

Table 9.2: NEC® Ampacity table superseded by current NEC Table 310.15(B)(16). Included for general illustration purposes, only. Courtesy, NEC, NFPA.

Table 310.15(B)(2)(a) Adjustment Factors for More Than Three Current-Carrying Conductors in a Raceway or Cable

Size AWG or kcmil	Temperature Rating of Conductor						Size AWG or kcmil
	COPPER			ALUMINUM OR COPPER-CLAD ALUMINUM			
	60°C (140°F)	75°C (167°F)	90°C (194°F)	60°C (140°F)	75°C (167°F)	90°C (194°F)	
	Types TW, UF	Types RHW, THHW, THW, THWN, XHHW, USE, ZW	Types TBS, SA, SIS, FEP, FEPB, MI, RHH, RHW-2, THHN, THHW, THW-2, THWN-2, USE-2, XHH, XHHW, XHHW-2, ZW-2	Types TW, UF	Types RHW, THHW, THW, THWN, XHHW, USE	Types TBS, SA, SIS, THHN, THHW, THW-2, RHH, RHW-2, USE-2, XHH, XHHW, XHHW-2, ZW-2	

CORRECTION FACTORS

For ambient temperatures other than 30°C (86°F), multiply the allowable ampacities shown above by the appropriate factor shown below.

Ambient Temp. (°C)	60°C	75°C	90°C	60°C	75°C	90°C	Ambient Temp. (°F)
21-25	1.08	1.05	1.04	1.08	1.05	1.04	70-77
26-30	1	1	1	1	1	1	78-86
31-35	0.91	0.94	0.96	0.91	0.94	0.96	87-95
36-40	0.82	0.88	0.91	0.82	0.88	0.91	96-104
41-45	0.71	0.82	0.87	0.71	0.82	0.87	105-113
46-50	0.58	0.75	0.82	0.58	0.75	0.82	114-122
51-55	0.41	0.67	0.76	0.41	0.67	0.76	123-131
56-60	–	0.53	0.71	–	0.58	0.71	132-140
61-70	–	0.33	0.58	–	0.33	0.58	141-158
71-80	–	–	0.41	–	–	0.41	159-176

* Small Conductors. Unless specifically permitted in 240.4(E) through (G), the overcurrent protection shall not exceed 15 amperes for 14 AWG, 20 amperes for 12 AWG, and copper-clad aluminum after any correction factors for ambient temperature and number of conductors have been applied.

and the 12 AWG conductor, assuming conductor temperature at 60°C (140°F) or less. (b) If the ambient temperature were to rise to 50°C, how would the conductor size be impacted?

Conductor and OCPD, Over Current Protection Device, verification:

Solution:

In accordance with article 210.20 (A), which stipulates:

"Branch-circuit conductors and equipment shall be protected by overcurrent protective devices that have a rating or setting that complies with 210.20(A) through (D).

(A) Continuous and Non-continuous Loads. Where a branch circuit supplies continuous loads or any combination of continuous and non-continuous loads, the rating of the overcurrent device shall not be less than the non-continuous load plus 125 percent of the continuous load." NEC® 2011.

The over current protection should be rated = 1.25 x Continuous Load
= 1.25 x (8 x 2A) = 20 amps.

∴ The 20 amp circuit breaker as an over current protection device is adequate for the given branch circuit.

In accordance with article 210.19 (A) (1): "Branch circuit conductors shall have an ampacity not less than maximum load to be served... (and) shall have an allowable ampacity not less than the non-continuous load plus 125% of the continuous load.

∴ The conductor ampacity for the given branch circuit
    = 1.25 x Continuous Load + 1.00 x Non-Continuous Load
    = 1.25 x (16A) + 1.00 x (0) = 20 amps.

According to Table 310.15(B)(16), for 60°C operation, with Type TW or UF insulation, AWG 12 conductor carries an allowable ampacity of 20 amps for conductors that are insulated, rated for 0-2000 volt operation, in situations with no more than three (3) current carrying conductors in raceway, cable earth (directly buried); under ambient temperature (not exceeding) 30°C (85°F); with no required/applicable derating.

∴ Selection of AWG 12 would be adequate for this scenario.

(B) Ambient Temperature Rise and Conductor size: According to Article 310.15, Tables 310.15(B)(2)(A) and 310.15(B)(16), when ambient deviates from 30°C to 50°C, a derating multiplier of 0.58 must be applied to adjust the ampacity of the conductor.

∴ The adjusted or derated ampacity of AWG #12 conductor, in this case, would be:
    = 0.58 x 20 amps = 11.6 amps

Since the derated ampacity of AWG #12, for this application, falls below the 20 amp capacity mandated by 210.19 (A) (1), AWG #12 would no longer be adequate. Therefore, AWG #10, which is the next size above AWG#12, must be considered. According to Table 310.15(B)(16), for 60°C operation, with Type TW or UF insulation, AWG#10 conductor carries an allowable ampacity of 30 amps. Then, if the 50°C adjustment rating of 0.58 is applied to 30 amp ampacity of AWG #10, the derated ampacity would be:

    = 0.58 x 30 amps = 17.4 amps

Since the 17.4 amp derated ampacity of AWG #10 still fall short of the 20 amp requirement, it would *not* meet the code. So the next larger size conductor,

AWG #8, with an ampacity of 40 amps should be considered.

If the 50°C adjustment rating of 0.58 is applied to the 40 amp ampacity of an AWG # 8 conductor, the derated ampacity would be:

= 0.58 x 40 amps = 23.2 amps

The 23.2 amp derated ampacity of AWG #10 exceeds the 20 amp requirement, therefore, it would meet the code.

Example 9.2
Over current Protection and Minimum Conductor Size
Applicable Code/Codes: 210.20 (A), 240.6(A), 210.19 (A) (1), 310.15 and Table 310.15 (B) (16).

Determine the size of over current protection device and the minimum conductor size for the following scenario assuming that no derating applies:
— Four (4) current carrying copper conductors in a raceway.
— Operating temperature and OCPD, Over current Protective Device, terminal rating: 60°C
— Insulation: THWN
— Load: (calculated) 26 amps, continuous.

Solution
Size of the OCPD:
In accordance with article 210.20 (A), over current protection should be rated = 1.25 x Continuous Load = 1.25 x (26A) = 32.5 amps.

According to article 240.6(A), standard ampere rating above 30 amp is 35 amp.

∴ The minimum standard size or rating of the **OCPD device should be 35 amps.**

Minimum Conductor Size
In accordance with article 210.19 (A) (1), the branch circuit conductors shall have an ampacity not less than maximum load to be served… (and) shall have an allowable ampacity not less than the non-continuous load plus 125% of the continuous load.

∴ The conductor ampacity for the given branch circuit = 1.25 x Continuous

Load + 1.00 x Non-Continuous Load = 1.25 x (26A) + 1.00 x (0) = 32.5 amps. **Select AWG 8 which has an ampacity of 40 amps.**

Example 9.3
Appliance Load—Dwelling Unit(s):
Applicable Code/Codes: 220.53

Determine the feeder capacity needed for a 120/240 VAC, fastened-in-place, appliance load in a dwelling unit for the following appliances:

— Water Heater, Rated: 4000 W, 240 V; Load: 4000 VA (PF = 1)

— Kitchen Disposal, Rated: 0.5 hp, 120 V; Load: 1176 VA (PF & Eff. <<100%)

— Dishwasher, Rated: 1200 W, 120 V; Load: 1200 VA (PF = 1)

— Furnace Motor, Rated: 0.25 hp, 120 V; Load: 696 VA (PF & Eff. <<100%)

— Attic Fan Motor, Rated: 0.25 hp, 120 V; Load: 696 VA (PF & Eff. <<100%)

— Water Heater, Rated: 0.5 hp, 120 V; Load: 1176 VA (PF = 1)

Solution
Total Load = 4000 VA + 1176 VA + 1200 VA + 696 VA + 696 VA + 1176 VA
= 8944 VA

Since the total load consists of more than four (4) appliances, according to article 220.53, a demand factor of 75 % is permissible.

∴ The size of the service and feeder conductors may be based on net load:
= 0.75 x 8,944 VA
= **6,708 VA**

Example 9.4
Outlets in Dwelling Unit(s):
A 120 V dwelling branch circuit supplies four outlets, one of which has four receptacles. What is the total volt-ampere load?

Solution

According to article 220.14(I), a single outlet is counted as a 180 VA load. So the three outlets, out of the four would constitute a load of:

= (3)(180 VA) = 540 VA

Each of the receptacles, in the fourth outlet—with four receptacles—according to article 220.14(I), would constitute a load of 90 VA. Therefore, the total load contribution from the four receptacles would be:

= (numbers of receptacles) (90 VA/receptacle)

= (4)(90 VA) = 360 VA

∴ The total load is = 540 VA + 360 VA = **900 VA**

Arc Flash

While some background information on the subject of arc flash is presented under Article 110.16 of the NEC®, the National Electric Code, NEC®, is not a core source for information on arc flash regulations. The subject of arc flash is comprehensively addressed by **NFPA 70 E.** The latest version of NFPA 70 E in effect at the time of publication of this book was NFPA 70E 2015. In the recent years, NFPA 70 E, similar to the NEC ®, has been revised every three years. Like the NEC®, NFPA 70 E is available in print, on-line and in other electronic format, i.e. CD. Arc flash is being introduced in this text primarily due to the importance and gravity of arc flash hazard in the electrical work environment. The introduction of arc flash in this text is at the basic level and *does not* enable the reader to adequately practice arc flash safety.

Basic facts related to arc flash are listed below:

• Arc flash is the result of a rapid release of energy due to an arcing fault between a **phase bus bar and another phase bus bar, neutral or a ground.**

• Arc faults are typically limited to systems with the bus voltage is in *excess of 50 volts.*

• An arc fault is similar to the arc obtained during *electric welding.*

• The massive energy discharged during and arc fault phenomenon has the *capacity to burn bus bars, vaporize the copper and cause an explosive volumetric increase.*

- An arc blast is estimated to cause explosive expansion of gas or air **67,000** times the nominal volume.

- The essence of relative magnitude or intensity of arc flash can be understood and appreciated through Eq. 9.1:

Arc Flash Energy, or Incident Energy = (V).(I).(t)   **Eq. 9.1**

In Eq. 9.1, the arc flash energy, or incident energy, is the energy released by an arc flash fault. This energy can be measured in kWh, Watt—sec, Joules, calories, Btus, therms, etc. In arc flash analysis, arc flash classifications and arc flash PPE (Personal Protective Equipment) ratings, arc flash energy "**intensity**" term "**Calories/cm^2**" is commonly used. In Eq. 9.1, "**V**" is the rms (root mean square) voltage, "**i**" is the rms (root mean square) fault current and "**t**" is the duration of the arc fault, in seconds. The arc fault current magnitude, typically, ranges in 1,000s of amps and the fault duration is typically in milliseconds. As obvious, this equation stipulates that arc fault energy is directly proportional to the voltage, current and fault duration. Note that, over the time span or duration of the fault, the voltage change is negligible in comparison with the order of magnitude rise in the current level. Therefore, mostly, the magnitude of the fault current is responsible for the enormity of the arc flash energy.

- Temperature in ARC Plasma is approx. 5,000°F.

- There are approximately 5-10 arc flash incidents (explosions) recorded per day, in the US.

- Average medical cost associated with remediation is estimated to exceed $1.5 Million per incident. Total cost, including litigation, is estimated to be $8-10 million per incident.

- OSHA, Occupational Safety and Health Administration, carries the Arc Flash regulation enforcement responsibility.

Compliance with OSHA involves adherence to a six-point plan:

- A facility must provide, and be able to demonstrate, a safety program with defined responsibilities.

- Calculations for the degree of arc flash hazard, at electrical equipment rated or operating at 120V, or greater.

- Warning labels on equipment. Note that the labels are provided by the equipment owners, not the manufacturers.

- Provision of proper personal protective equipment (PPE) for workers, as prescribed by proper arc flash risk analysis.

- Initial training, and subsequent refresher training, for workers on:
  — The hazards of arc flash
  — The proper interpretation of arc flash risk and PPE labels
  — Proper use of arc flash PPE and voltage rated tools and gloves.

Physical and Thermal Background of Arc Flash

This section is devoted to the exploration of the physical, chemical and thermal aspects of arc flash phenomena. This discussion is intended to enhance technicians' and engineers' appreciation of the reasoning behind arc flash hazard analysis requirements and the need for arc flash PPE. Pictures from two different arc flash simulations are depicted in Figures 9.2 and 9.3.

The picture in Figure 9.2 depicts a simulated arc-flash incident, conducted at 250V, 13.1 KA (13,100 amps). The energy intensity for this simulated 13,100 amp fault is estimated to be 1.48 Cal/cm^2. The picture in Figure 9.3 depicts a simulated arc-flash incident, conducted at 250V, 44 KA (44,000 amps). The energy intensity for this simulated 44,000 amp fault is estimated to be 8.48 Cal/cm^2, almost *six times the fault energy* released by the lower current fault. Note that the voltage in both simulations was 250 V. This observation supports the directly proportional relationship between fault current and the arc flash fault energy as stipulated in the fault energy equation, Eq. 9.1.

Figure 9.2: Simulated arc-flash incident, conducted at 250V, 13.1 KA (13,100 amps). Energy intensity: 1.48 Cal/cm^2.

Some of the hazards associated with an arc flash incident are evident from pictures in Figures 9.2 and 9.3. These hazards are as follows:

- **Extreme Heat**—Energy contained in explosive arc flash events raises the temperature thousands of degrees; to the extent that it not only melts copper components but vaporizes them. Immense heat energy contained in vaporized copper can burn through or ignite regular work clothes. In addition, human skin coming in contact with vapors at such high temperatures would result in third degree burns. Temperature in ARC plasma is approx. 5,000°F. Proper PPE, including hood, balaclava, arc rated coveralls and jacket can provide a measure of protection against the radiated heat and the heat contained in the vapors. See the PPE section below.

- **Brilliant Flash**—Without adequate tinting and shielding, the intense light that accompanies the release of immense explosive energy can result in retina damage and can cause blindness.

- **High UV Emission**—The ultra-violet light that accompanies arc flash events can damage human epithelial (skin) layer. Once again, proper arc rated clothing can provide a measure of protection.

- **Shock Wave**—Vaporization of copper accompanies 67,000 fold expansion of air and gases. The intense instantaneous expansion of gases is tantamount to explosion of ordinance and results in a shock wave. The shock wave can launch workers off their feet resulting in broken limbs and bones. The explosive energy laden shock waves have the capacity to subject anterior of a human body to immense pressure, potentially, fracturing ribs, puncturing and collapsing lungs. Best practices associated with safe posture can provide a measure of protection against effect of a shock wave.

- **Concussion (Noise)**—Due to the fact that expanding vaporized copper is an explosive event, it generates a loud report. This is the reason why hearing protection is required by NFPA 70 E.

- **Projectiles**—As visible in Figures 9.2 and 9.3, energy contained in arc flash events projects components outward in solid or molten form. Note the bright streaks in the picture of the simulated arc faults. The projectiles launched in arc flash events can result in shrapnel injuries.

- **Electrical Shock**—The explosion associated with arc flash events dislodges components, insulation and conductors; thus, exposing personnel, in close proximity, to live energized components. This exposure can result in electrical shock hazard.

Figure 9.3: Simulated arc-flash incident, conducted at 250V, 44 KA (44,000 amps). Energy intensity: 8.48 Cal/cm².

Arc Flash PPE

Having gained some appreciation of the hazards, and causes thereof, associated with arc flash events, let's consider some of the PPE required by NFPA 70E. The sets of PPE introduced here pertain to situations when (1) incident energy is 1.2 – 12 cal/cm2 and (2) when incident energy is > 12 cal/cm2. Arc flash PPE for these two categories as listed below:

PPE, for Incident Energy of 1.2 – 12 cal/cm2:

Cal/cm²	Classification	PPE
1.2 - 12	CL-1	Arc-rated long-sleeve shirt and are-rated pants or arc-rated coverall or arc flash suit (SR) (See Note 3) Arc-rated face shield and arc-rated balaclava or arc flash suit hood (SR) (See Note 1) Arc-rated jacket, parka, rainwear (AN) **Other PPE:** Hard hat Arc-rated hard hat liner (AN) Safety glasses or safety goggles (SR) Hearing protection Heavy duty leather gloves or rubber insulating gloves with leather protectors (SR) (see Note 4) Leather footwear

PPE, for Incident Energy > 12 cal/cm2:

Cal/cm²	Classification	PPE
> 12	N/A	Arc-rated long-sleeve shirt and are-rated pants or arc-rated coverall or arc flash suit (SR) (See Note 3) Arc-rated arc flash suit hood (SR) (See Note 1) Arc-rated gloves Arc-rated jacket, parka, rainwear (AN) **Other PPE:** Hard hat Arc-rated hard hat liner (AN) Safety glasses or safety goggles (SR) Hearing protection Arc rated gloves or rubber insulating gloves with leather protectors (SR) (see Note 4) Leather footwear

The notes referenced above pertain to NFPA 70 E 2015, Table H.3 (b).

An example of a PPE system, similar to what is required for a typical situation pertaining to > 12 cal/cm2, is depicted in Figure 9.4:

Hard hat, safety glasses and ear plugs

Class 4 hood with tinted lens/face shield

Two component voltage rated glove system

Class 4 Suit, rated 8 to 40 cal/cm²

Figure 9.4: Example of NFPA 70 E Class 4 PPE

PPE requirement can be assessed on the basis of NFPA 70 E "category" method or through NFPA 70E, Annex D, based incident energy analysis. The label depicted in Figures 9.5 shows the format of the Arc Flash labels and type of information displayed as a result of arc flash risk analysis. Especially noteworthy is the calculated "incident energy" and the "level of PPE" required when transacting with the pertaining electrical equipment.

Figure 9.5:
Label based on arc flash risk analysis, NFPA 70 E, 2015.

On the other hand, the label shown in Figure 9.6 pertains to electrical equipment that has the potential for drawing enough fault current to exceed the energy intensity level of 40 cal/cm2. When fault energy intensity of 40 cal/cm2 is exceeded, as stipulated on the label in Figure 9.6, work on energized equipment is *not permitted*.

Figure 9.6: Example of a label pertaining to higher than 40 cal/cm2 arc flash hazard, based on NFPA 70 E.

ELECTRICAL SAFETY CERTIFICATIONS

Certifications and certification labels on electrical equipment are intended to provide the end user, installer or integrator a measure of assurance that the labeled equipment is safe when applied or used as prescribed by the manufacturer. The certification labels can be found at the bottom, rear or sides of non-custom, off-the-shelf, standard electrical equipment. See Figure 9.7.

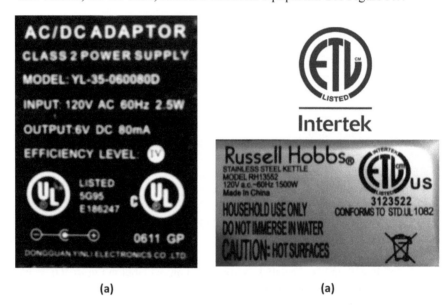

(a)   (a)

Figure 9.7: Examples of a safety certification labels—ETL and UL Listings.

Various certifications exist in different parts of the world. A few, more prominent ones, are mentioned in this text as matter of introduction. These certifications are listed below:

- UL®, Underwriters Laboratories, United States. UL®, tests equipment to be certified for safety, either at laboratories owned and operated by UL or at laboratories owned and operated by its sub-contractors, such as ETL^{CM}.

- ETL^{CM}, Intertek Listing. The ETL Listed Mark is proof of product compliance (electrical, gas and other safety standards) to North American safety standards. Following authorities having jurisdiction in 50 states and Canada accept the ETL Listed Mark as proof of product safety: UL, ANSI (American National Standards Institute), CSA (CSA Group),

ASTM (formerly known as: American Society for Testing and Materials), NFPA (National Fire Protection Agency), and NOM (Norma Official Mexicana).

• **NOM Mark**: (Norma Official Mexicana) NOM is a mark of product safety approval for virtually any type of product exported into Mexico.

• **ULC**, Underwriters Laboratories of Canada.

• **IEC**, International Electrotechnical Commission. The IEC plays an important role in developing and distributing electrical standards. IEC was instrumental in developing and distributing standards for units of measurement, particularly the gauss, hertz and weber. IEC first proposed a system of standards, the Giorgi System, which ultimately became the SI, or Système International d'unités (in English, the International System of Units).

• **IP Rating**: IP rating stands for *International Protection* rating. Sometimes interpreted as *Ingress Protection* rating, the IP rating consists of the letters *IP* followed by two digits and an optional letter. As defined in IEC 60529, it classifies the degrees of protection provided against the intrusion of solid objects (including body parts like hands and fingers), dust, accidental contact, and water in electrical enclosures. See **Table 9.2** for correspondence between the American **NEMA® ratings** and their counterparts in the **IP** realm. Also see Example 9.5 for illustration of enclosure ratings application.

• **CE Certification**: The CE Mark stands for Conformité Européenne, a French term that can be literally translated into English as European Conformity.

• **Safety Compliance Statement** from the manufacturer, engineering firm, general contractor, turn-key installer or system integrator is often required on custom engineered systems or equipment. This requirement must be clarified and agreed upon—at the quotation/bidding phase of the project—between the project manager, end-user/customer, plant safety manager, engineering firm, general contractor, turn-key installer and the system integrator (if applicable). Absence of such documented agreement on Safety Compliance Statement requirement, at the very outset of a custom engineered project, can result in system commissioning and start-up delays, in addition to penalties, unforeseen costs, potential breach of contract, future business prospects, etc.

Additional information on NEMA and IP ratings

The digits or numeral in the NEMA or IP code indicate conformity with the conditions summarized in the Table 9.3. The first digit indicates the level of protection that the enclosure provides against the ingress of solid foreign objects. For example, an electrical socket rated IP22 is protected against insertion of fingers and will not be damaged or become unsafe during a specified test in which it is exposed to vertically or nearly vertically dripping water. IP22 or IP2X are common minimum requirements for the design of electrical accessories for indoor use.

NEMA VERSUS IP ENCLOSURE RATINGS

Example 9.5

Electrical specifications for a factory call for a junction box that must be submersed into a water tank. (a) Determine the NEMA rating of junction box for the US installations. (b) Determine the IP rating of the junction box for the European installations.

Solution:

(a) Examination of the NEMA—IP rating table in this chapter shows that NEMA 6 enclosure is rated as:

"Submersible. Design depends on specified conditions of pressure and time; submersible in water; used in quarries, mines, and manholes."

Therefore, a **NEMA 6 enclosure should be specified** for the US installation.
(b) Since the European installation would be exposed to the same worst case conditions, US NEMA 6's European counterpart, **IP 67 should be specified**.

Common Acronyms Associated with Electrical Standards Organizations

A few organizational acronyms, commonly used in the field of electrical engineering, are listed below with respective background information and website addresses to facilitate further exploration of the roles of these organizations by the reader:

Table 9.3: NEMA and IP electrical enclosure ratings descriptions, comparison and correspondence.

NEMA Type	Definition	IEC Equivalent
1	General-purpose. Protects against dust, light, and indirect splashing but is not dust-tight; primarily prevents contact with live parts; used indoors and under normal atmospheric conditions.	IP10
2	Drip-tight. Similar to Type 1 but with addition of drip shields; used where condensation may be severe (as in cooling and laundry rooms).	IP11
3 and 3S	Weather-resistant. Protects against weather hazards such as rain and sleet; used outdoors on ship docks, in construction work, and in tunnels and subways.	IP54
3R	Intended for outdoor use. Provides a degree of protection against falling rain and ice formation. Meets rod entry, rain, external icing, and rust-resistance design tests.	IP14
4 and 4X	Watertight (weatherproof). Must exclude at least 65 GPM of water from 1-in. nozzle delivered from a distance not less than 10 ft for 5 min. Used outdoors on ship docks, in dairies, and in breweries.	IP66
5	Dust-tight. Provided with gaskets or equivalent to exclude dust; used in steel mills and cement plants.	IP52
6 and 6P	Submersible. Design depends on specified conditions of pressure and time; submersible in water; used in quarries, mines, and manholes.	IP67
7	Hazardous. For indoor use in Class I, Groups A, B, C, and D environments as defined in the NEC ®.	N/A

Table 9.3 (*Continued*)

8	Hazardous. For indoor and outdoor use in locations classified as Class I, Groups A, B, C, and D as defined in the NEC ®.	N/A
9	Hazardous. For indoor and outdoor use in locations classified as Class II, Groups E, F, or G as defined in the NEC ®.	N/A
10	MSHA. Meets the requirements of the Mine Safety and Health Administration, 30 CFR Part 18 (1978).	N/A
11	General-purpose. Protects against the corrosive effects of liquids and gases. Meets drip and corrosion-resistance tests.	N/A
12 and 12K	General-purpose. Intended for indoor use, provides some protection against dust, falling dirt, and dripping noncorrosive liquids. Meets drip, dust, and rust resistance tests.	IP52
13	General-purpose. Primarily used to provide protection against dust, spraying of water, oil, and noncorrosive coolants. Meets oil exclusion and rust resistance design tests.	IP54

NEMA: National Electrical Manufacturers Association; www.nema.org

NEMA, created in the fall of 1926 by the merger of the Electric Power Club and the Associated Manufacturers of Electrical Supplies, provides a forum for the standardization of electrical equipment, enabling consumers to select from a range of safe, effective, and compatible electrical products.

ANSI: American National Standards Institute; www.ansi.org

The American National Standards Institute (ANSI) is a private, non-profit organization that administers and coordinates the U.S. voluntary standardization and conformity assessment system.

IEC: International Electrotechnical Commission.

IEC is the authoritative worldwide body responsible for developing consensus global standards in the electrotechnical field. IEC is the European counterpart to NEMA and ANSI.

IEEE: Institute of Electrical and Electronic Engineers: www.ieee.org

The IEEE is a non-profit, technical professional association for Electrical and Electronics Engineers. IEEE makes vital contributions in the electrical engineering realm, in many ways. Some of IEEE's more prominent contributions are as follows:

- Publishing of texts and reference material that promote education and training on various electrical and electronic technologies and subjects.
- Development of Arc Flash Hazard Calculation formulas through IEEE 1584 committee.
- Facilitation of the development of protocols that promote communication between various electronic devices.

International Society of Automation (ISA): www.isa.org

Founded in 1945, the International Society of Automation is a leading, global, nonprofit organization that is setting the standard for automation by helping members and other professionals solve difficult technical problems. ISA is based in Research Triangle Park, North Carolina. ISA develops standards, certifies industry professionals, provides education and training, publishes books and technical articles, and hosts conferences and exhibitions for automation professionals.

RIA—Robotics Industries Association: www.robotics.org.

Robotic Industries Association publishes information to help engineers, managers and executives apply and justify robotics and flexible automation. The RIA website includes a proprietary search engine algorithm that makes it easy to find and compare leading companies, products and services. Robotics Online is dedicated to news, articles and information specifically for the robotics industry.

COMMON ELECTRICAL/ELECTRONIC SAFETY DEVICES

This section introduces the reader to electrical and electronic devices commonly employed in automated control systems and process controls, in general, for safety related actions and events. We will first familiarize the reader to each of the devices, individually. Then, we will discuss integration of the devices in an automated manufacturing system example.

Similar to the pictorial tour approach utilized with the MCC and power distribution system discussion, we will explore Rockwell Automation® safety control devices to gain familiarization with what these devices look like and gain a better understanding about their functions.

Safety E-stop Devices

E-stop switches are, functionally, passive mechanical devices, similar to the light switches in offices and homes. Rockwell/A-B® offer Series 800T/800E Push-Buttons and Self-Monitoring Contact Block based E-stop switches. The emergency stopping function is implemented through a combination of two components: (1) Push pull operator and (2) Contact block.

The description and specification of these two components are as follows:

- E-Stops Operator
  — Available in 30mm & 22mm sizes
  — Metal and plastic construction
  — Meet EN418 and IEC 60947-5-5 standards
  — Push-pull, push-pull/twist release, illuminated, or key-operated devices

- Self-Monitoring Contact Blocks
  — For use with 800T & 800E E-Stops
  — If contact block becomes separated from E-stop, monitoring circuit automatically opens and shuts down the controlled process. This feature essentially eliminates contact separation concerns from improper installation, damage or high-vibration applications.

Safety Light Curtain System

Even though safety light curtains often perform emergency stopping function, similar to the emergency stop switches introduced above, they are

Contact Blocks   Operators

Figure 9.8: Rockwell® Safety E-stop switch operators and contact blocks.

powered, active, somewhat automated, and substantially more sophisticated than the e-stop switches. See Figures 9.9 and 9.10.

●Safety Light Curtain
System Consists of:
  – Transmitter/Receiver
  – Control Reliable Controller
  – Interface Cables
  – Optional Power Supply
  – Optional Muting Blanking
    control

Figure 9.9: Rockwell® Safety Light Curtain.

Some of the key components of a safety light curtain system are listed in Figure 9.9. As depicted in Figure 9.9, the transmitting "column" transmits an array of invisible (infra-red) light beams. When the path between the transmitter and the receiver is clear and unobstructed, the transmitted beams are received by the receiver. This, in most applications of light curtains, constitutes the norm. If, however, the light beams are interrupted by equipment or personnel, typically, an emergency stop command is generated, thus, shutting down the protected system. The interface cable pictured in Figure 9.9, as simple and unsophisticated as it appears, it constitutes a reliable and robust approach to electrical/control connections. Electrical connections, a decade or so ago, had to be made one wire and one terminal at a time. With hundreds of connections required in mid to large size control systems, the old wire to connector method often resulted in miss-wiring, loose connections, unreliability and delayed system start-ups.

Other important features and components associated with the light curtains are included in Figure 9.10. Rockwell's® DNet module (hardware), is an interface module and is pictured in Figure 9.10. The software or protocol that permits the safety devices, safety PLCs and other Rockwell® control devices to communicate with each other is referred to as DeviceNet®.

The laptop PC shown in Figure 9.10 allows control engineers and technicians to configure or "program" the light curtain to respond to safety events in a desired fashion, In addition, the Rockwell® application software loaded on the

- DNet module can be interfaced to any SafeShield product at any time
- Diagnostic information can be transferred via DeviceNet
  (Output status / EDM active / Restart required / Device faulted / Weak signal / Blanking selected)
- DNet Communication via DNet Module (Black)
- SCD Software for L.C. Configuration only, not for DNet

Figure 9.10: SafeShield DeviceNet: Rockwell® Safety Light Curtain and Device-Net interface.

laptop allows the controls engineer to ***configure "diagnostics"*** such that safety incidents and other associated events can be troubleshot promptly.

AAC (Area Access Control)

Area access control system depicted in Figure 9.11 offers a simpler, less costly, less complex alternative to light curtains. Typical area access control system consists of a small single beam transmitter that transmits an infra-red beam across the protected opening, to a pair of prism shaped reflectors. The reflectors return the beam across the opening to a receiver. When the beam is interrupted by personnel or equipment, an emergency stop is triggered by the Area Access Control System. As obvious from a comparison between Figures 9.10 and 9.11, the protected opening area of cross-section covered by a light curtain is greater than the two dimensional area covered by the AAC.

Cable Pull Switches

Cable pull safety switches are designed to trip an emergency stop circuit when the cord, cable or chain connected to the pull switch is pulled. Rockwell® cable pull switches incorporate a reset button on the front of the switch.

Some of the cable pull safety switches are equipped with a tension view-

- Long Scan Range up to 70M (230Ft.)
- Die Cast Aluminum Housing
- Two Ranges available
-    - .5M to 18M (19.5" to 59")
-    - 15M to 70M (4' to 230')
- Easy Installation
- Heated Front Screen, I.e. can be used in outdoor applications
- Fast response time 22ms
- 24vd/115vac standard / 230vac (special order)
- Built in monitored safety relays
-    - 2 NO/1NC / 2A Max switching Current
- IP 65 enclosure rating
- Operating temp. -25°C to 50°C
- PG connector IP 67

Direction of Entry
into Hazardous Area

Figure 9.11: Rockwell® Area Access Control System

ing window to facilitate set-up of the cable tension. Among other important features incorporated in the Rockwell® cable pull switches is a function that latches out the contacts electrically and mechanically in accordance to EN 418. The reason is that if a person were to pull an ordinary cable pull switch as they were being dragged into a machine, the system would stop, but it would not prevent the system from being restarted at the other end by an operator, who did not see the switch being pulled. Thus the operator could start the machine again, dragging the other subject back into the machine. In order to reset the lifeline 4 switch, an operator has to physically go up to the switch and reset the device by moving the designated lever into the run position. This allows for inspection of the area, before the machine is restarted. The switch is yellow in accordance to EN 60204-1 which stipulates that all e-stops have a red button and yellow background.

Key Interlock Solenoid Switches

The next family of safety switches offered by Rockwell® are solenoid locking switches. See switches depicted in Figure 9.12. These switches are used to prevent access to a hazardous area until the hazardous motion has been contained. An example would be to have a gate locked until a set of cutting shears have come to rest, after which time a voltage is applied to the coil of the solenoid, releasing the key, allowing access to the area.

TLS GD2   **Atlas 5 w/Trap Key**

Atlas 4

Atlas 5 w/Key

Spartan

Figure 9.12: Rockwell® Key interlock solenoid safety switch system.

GuardMat™ Safety Mats

Guardmaster offers a full range of standard mats and controllers for all applications. The GuardMat™ system is a tripping device. When a person steps on the mat, or equipment rolls onto it, their presence is detected and the safety output opens, shutting down the machine. See Figure 9.13.

The design of the mat includes two sheets of hardened rolled steel which are separated by small insulators. The separated sheets of metal have an approximate 24 volt potential difference across them. When pressure exceeding 70 psi is applied to the top sheet, the sheets make contact, creating a short circuit condition, changing the resistance, which is detected by the GaurdMat controller. The controller, in turn, executes emergency shut down through system interlocks and safety control circuits.

GuardEdge™ Safety Edges

The GuardEdge system, is a trip device which lends itself to several applications, as depicted in Figure 9.14. The principle behind this system, is that when the protruding rubber strip—referred to as profile—is depressed, emergency stop signal is issued and the machine shuts down. The rubber profile is impregnated with a conductive carbon powder, which creates a conductive rub-

Mats

Controllers

Figure 9.13: Rockwell® GaurdMat Switch System

ber. The system reacts to change in resistance. When the profile is pressed the resistance changes; this change is detected by the controller. The controller, in turn, executes emergency shut down through safety control circuits.

As apparent in Figure 9.14, a variety of profiles are available to match specific applications. These profiles can be made up to 50 meter in length. The profiles can also be bent to a radius of 200mm to accommodate the most non-linear applications. The safety edges can be wrapped around corners, incorporating active corner connections. The profiles can be wired in series or parallel, depending on system configurations.

Safety PLCs

Traditionally, National Fire Protection Association's "Electrical Standard for Industrial Machinery" (NFPA79) has required hard wiring and the use of electro-mechanical components for safety circuits. With the advent, subsequent development, and enhanced reliability of safety PLCs, electromechanical safety circuits continue to be replaced by safety PLCs. Some of the more significant advantages offered by safety PLCs are as follows:

- Safety PLC based control circuits require fewer wires and terminations as compared to electromechanical relay and contactor based safety control circuits.

Profiles/Rails and Controllers

Figure 9.14: Rockwell® GaurdEdge System

•   Safety PLC based control circuits require fewer "moving parts," fewer contactors, timers and relays, if any, as compared to electromechanical relay and contactor based safety control circuits.

•   Safety PLC based control circuits are programmable. Therefore, their logic and functionality can be modified through modification of application program or code. While, with electromechanical relay and contactor based safety control circuits, significant hardware and wiring modifications are necessary for accomplishment of functional modification.

Functionally and logically, safety PLCs are similar to regular PLCs (Programmable Logic Controllers). Notwithstanding the similarities, there are some notable differences between regular PLCs and Safety PLCs. These differences are as follows:

•   Safety PLCs are color-coded red to signify the fact that they are "control reliable" and safety rated.

•   Safety PLCs employ ***robust diagnostics and operational verification***. Standard PLC inputs provide no internal means for testing the functionality of the input circuitry. By contrast, Safety PLCs have an internal 'output' circuit associated with each input for the purpose of periodic testing and verification of the input circuitry. Simulated Highs (1s) and Lows (0s) are presented, automatically, to the inputs, to verify their functionality.

- A safety PLC has *redundant microprocessors*, Flash and RAM memory; that are continuously monitored by a *"watchdog" circuit* and a synchronous detection circuit. Regular PLCs, are typically equipped with *one microprocessor*.

- A regular PLC has one output switching device, whereas a safety PLCs digital output logic circuit contains a test point after each of two safety switches located behind the output driver and a third test point downstream of the output driver. Each of the two safety switches is controlled by a unique microprocessor. If a failure is detected at either of the two safety switches, the operating system of a safety PLC will *automatically acknowledge the anomaly* and will default to a known state; thus, facilitating an orderly equipment shutdown.

Typical specifications of safety PLCs are listed in Figure 9.15.

- GuardPLC 2000 and 1200 shipping since August
  - TÜV Certified (Entire System) - IEC 61508 SIL 3, DIN VDE 19250 AK6, EN 954-1 Category 4,
  - UL Listed
  - Primary Target Market - Machinery Safety
- GuardPLC 2000 - 6 I/O slot, modular design
  - 24 Input / 16 Output digital
  - 8 Channel Analog Input & Output (12 bit resolution)
  - 2 Channel HSC (100kHz, 24 bit)
- Guard 1200 - packaged design
  - 20 Inputs / 8 Outputs + 2 HSC Inputs (100khz, 24 bit)
- Communications
  - Proprietary GuardPLC Ethernet + ASCII
  - Peer to Peer Safety Communications
- RSLogix Guard Software (2 versions)
  - Lite and Professional Versions
  - Windows NT/2000
  - RSLogix "Look & Feel"
  - IEC 1131 Function Block Programming
  - User Defined Function Block Capabilities (1755-PCS)

Figure 9.15: Rockwell® Safety PLC (Programmable Logic Control) System

Safety Relays

A regular relay consists of an electromagnetic coil or solenoid and associated contacts. The solenoid functions in the same manner as described earlier in this text; it operates or controls multiple normally open or normally closed contacts. When a relay is off, its coil is de-energized, its normally open contacts are OPEN, and its normally closed contacts are CLOSED. Conversely, when a relay is turned on, the coil is energized, the normally open contacts are CLOSED and the normally closed contacts are OPEN. A regular relay is shown applied in an alarm circuit in Figure 9.16.

In the schematic diagram of the regular relay shown in Figure 9.16, a relay labeled CR1 (Control Relay 1) is being used to annunciate the status of an **alarm switch**, to the left. The coil of CR1 is connected to neutral "N" on the right side. The other side of CR1 is connected to 110 V only when the alarm switch is operated. As shown in the circuit diagram, the control relay CR1 has two normally open contacts, depicted as two parallel lines and two normally closed contacts shown with a diagonal bar across the parallel lines. When the alarm switch is operated, or closed, the CR1 coil is energized. This closes the normally open CR1 contacts and opens the normally closed contacts. The two

Figure 9.16: Regular relay applied in an alarm circuit.

normally open contacts are being used to turn on the ***alarm horn*** and the ***red alarm light***. One of the normally closed contacts is being used to maintain the ***green light*** on, as long as the safety switch is ***not*** closed. The second normally closed contact is left unconnected, as a spare contact for possible future use.

Safety relays operate somewhat similar to regular relays, with the following exceptions:

1)   Safety relays are equipped with redundant coils and contacts.

2)   Safety relays are equipped with diagnostics features.

Manufacturers, such as Rockwell®, also offer solid-state relays for applications involving high cycle rates. Safety relays offered by Rockwell® are shown in Figure 9.17, along with respective features and functions.

Now that we have gained a measure of familiarity with electrical control and safety devices, let's explore typical application of some of these devices through a tour of an automated manufacturing cell depicted in Figure 9.18. If we begin at approximately 11:00 o'clock and move clock wise, we

- Emergency Stop Relays
  - Monitors the E-stop Circuit
  - Monitors Safety Gate Limit Switches
  - Monitors Light Curtains
  - Monitors Rope Pull Switches

- 2-Hand Control & Safety Gate Monitors
  - 2-Hand Anti-tiedown & Anti-repeat relay
  - Controls machine from safety gate limit switches

- Provides additional safety contacts

Figure 9.17: Rockwell® Safety Relays and Contactors

Figure 9.18: Rockwell® safety equipment applied to an automated manufacturing cell.

notice the yellow cabinet housing the *safety PLC* and other safety devices such as safety relays, safety contactors, controllers for various safety devices, i.e. safety mat, safety edge, safety laser scanners, etc. The control cabinet housing the *safety* control devices is painted yellow to distinguish it from typical, gray, control cabinets for regular, *non-safety*, control systems.

At about 1:00 o'clock, the common *E-stop switch* is shown mounted on the right side of the processing machine opening. In the diagram, the E-stop switch is labeled as "Safety Button."

The *safety limit switch* pointed out in the middle of the automated cell diagram limits inadvertent rotation of the robot about its major axis, beyond a safe point. Such application of a safety limit switch, typically, serves as a back-up to a software based rotational limit.

Trapped key safety switch is shown installed on one of the man doors at about 3:00 o'clock. A *safety cable pull switch* is shown spanning the length of the conveyor. A *safety guard* is shown applied just outside a short conveyor section, at about 6:00 o'clock, for the purpose of preventing *authorized* person-nel in the area from coming too close to moving parts, i.e. the conveyor rollers.

A safety laser scanner is shown mounted on the frame that sustains the two gantry robots. This laser scanner is not clearly visible in Figure 9.18. There-fore, a picture of a Rockwell/Allen-Bradley, multi-zone, laser scanner is shown in Figure 9.19. Safety laser scanner represents a sophisticated and relatively new approach to three dimensional protection that is not available through two dimensional safety light curtains, and other less sophisticated safety devices. However, not unlike light curtains, safety laser scanners must be programmed and configured for desired function.

Figure 9.19: Rockwell® safety laser scanner.

Self-assessment Problems and Questions—Chapter 9

1. A given circuit is meant to carry a continuous lighting load of 16 A. In addition, four loads designed for permanent display stands are fastened in place and require 2 A each when operating. What is the rating of the over current protective device (OCPD) on the branch circuit?

2. A three-phase, four-wire feeder with a full-sized neutral carries 14 A continuous and 40 A non-continuous loads. The feeder uses an over current device with a terminal or conductor rating of 60°C. What is the minimum copper conductor size? Assume no derating applies. Use Tables 9.1 and 9.2.

3. Electrical specifications for a brewery company call for a fusible disconnect switch enclosure that must be able withstand occasional splashing of water during periodic wash downs required by the local health codes. This design will be applied in breweries in the US as well as Europe. The water flow from the 1-in wash down nozzles is expected to less than 60 GPM from a distance of 11 ft for less than 4 min. (a) Determine the NEMA rating of enclosure for the US installations. (b) Determine the IP rating of enclosure for the European installations.

4. Over current Protection & Conductor Ampacity:
    Applicable Code/Codes: Articles 210.19 (A) (1), 210.20 (A) and 310.15, and Tables 9.1 and 9.2 of this text (Note: This is not a current table and is only reproduced for exercise and illustration purposes).

    The branch circuit in the exhibit below consists of three continuous loads. Over current protection in the branch circuit is provided through a 20 amp circuit breaker. (a) Determine the size of copper conductor based on the

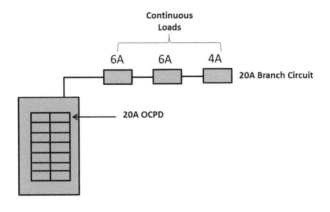

ampacities given in Tables 9.1 and 9.2 assuming conductor temperature is at 75°C or less. Assume 75°C operation and selection is allowed. (b) Verify the size/specifications of the circuit breaker. Assume no derating applies. (c) If the ambient temperature were to rise to 50°C, how would the conductor size be impacted?

5. A US appliance manufacturer is planning to market a new appliance in Mexico. The most appropriate safety certification for this appliance would be:
    A. UL
    B. ULC
    C. ETL
    D. NOM

6. Assume that the alarm switch in the control circuit depicted below is opened after being closed for a prolonged period of time. Which of the following conditions would best describe the status of the annunciating lights and the horn when the switch is opened?

A. Alarm horn will turn off
B. Red light will turn off
C. Green light will turn on
D. All of the above

Chapter 10

Electrical and Controls Drawings

Introduction

The purpose of this chapter is to merely ***introduce*** the reader to established electrical power distribution drawings, controls drawings, associated best practices and standards. In this chapter, the coverage of electrical design drawings, and the application of NEC® in the associated design work, is ***not*** intended to provide the reader expert level knowledge that is required and expected of an NEC® trained and practicing electrical Professional Engineer or electrical power system designer.

However, it is our hope that the nonelectrical engineering professional or non-engineer reader will find this chapter helpful in equipping them with enough knowledge to be able to stay abreast of discussion at hand when those electrical and controls drawings are spread across the table in the process of troubleshooting and reinstating an important piece of equipment back into operation. In addition, through a brief introduction to PLC relay ladder logic, we will introduce the reader to the programming technique utilized by most control engineers to control electrical and mechanical systems with PLC based control systems. Of course, as before, we will illustrate the concepts and practices discussed, through examples, end of chapter problems, and the solutions at the back of the text.

ELECTRICAL DRAWINGS

In so far as electrical drawing standards are concerned, electrical engineers, electrical designers and electrical CAD (Computer Aided Design) operators can draw pertinent, applicable, references from the Electrical Engineering Drawing Standards listed in Table 10.1. Of course, the litany of standards listed in Table 10.1 are but a few to draw from. There are copious other standards applicable to the formulation of electrical drawings in the US and abroad. In addition, CAD drawing software, i.e. AutoCAD Electrical, come incorporated with universally recognized electrical drawing symbols and standards.

Three common types of electrical drawings are discussed. These include a one-line power distribution schematic, a wiring diagram and electrical control drawings. The objective of this chapter is to inculcate basic understanding of electrical symbols, electrical drawing conventions and electrical design. In ad-

Table 10.1: Electrical Drawing Standards

Electrical Engineering Drawing Standards	
Name	**Description**
ISO 01.100.25	Electrical and Electronics Engineering Drawings
NASA	NASA Engineering Drawing Standards
NEMA ICS 19-2002 (R2007)	NEMA Diagram, Device Designations and Symbols for Industrial Controls and Systems
STS 904	Hunter Water Corporation Standard Technical Specifications for Preparation of Electrical Engineering Drawing
IEEE Std 100-1996	IEEE Standard Dictionary of Electrical and Electronic Terms, Sixth Edition
IEEE Std 315™	IEEE Std 315™, IEEE Standard Graphic Symbols for Electrical and Electronics Diagrams (Including Reference Designation Letters).
IEEE Std 280-1985	American National Standard for Mathematmatical Signs and Symbols for Use in Physical Sciences and Technology
IEEE Std 280-1985	IEEE Standard Letter Symbols for (R1997) Quantities Used in Electrical Science and Electrical Engineering
ANSI Y32.9	American National Standard Graphic Symbols for Electrical Wiring and Layout Diagrams Used in Architecture and Building Construction.

dition, the reader will be shown how NEC and other methods for application of the code are employed in the electrical power distribution system design.

ONE-LINE SCHEMATIC DIAGRAM

The one-line schematic diagram is also, simply, referred to as a "one-line drawing." This drawing is called a "one-line" drawing because it depicts electrical circuits and design through representation of just one of the three phases. This is predicated on a reasonable assumption that all three phases on the three phase loads and sources are substantially identical in current, voltage, impedance and other important considerations. As obvious, one important benefit derived from one-line representation of electrical circuits is that a greater number of circuits can be captured on one drawing. In other words, one can examine a large portion of the overall electrical system at one glance. This facilitates a quicker and more effective comprehension of a large segment of the electrical system being examined without flipping from one drawing to another. This is of considerable value in a triage situation when trouble shooting a system that is down. A simplified one-line schematic is shown in Figure 10.1. This

Figure 10.1: One line schematic for a power distribution system

drawing pertains to an MCC (Motor Control Center) based power distribution system. Due to the extensity of information captured in this diagram, certain segments and annotation are somewhat illegible. Therefore, those segments of the schematic that are examined in greater detail in this chapter are excerpted and duplicated in Figure 10.2.

As we examine the one-line schematic, and other drawings, we will note the more conventional symbols and nomenclature, as well as, certain practices adopted by the electrical engineer/designer of these drawings that deviate from the more universally accepted methods. As we examine the top portion of the one-line schematic in Figure 10.1, the first piece of information we notice is the rating of the MCC power distribution system:

480 V Bus, 3 φ, 4W, 600A, 60 Hz

This notation represents the rating of the MCC and the specification for the MCC bus bars. This notation encircles one of the three phase bus bars—represented by the long solid line—and it stipulates the following:

— This MCC is rated 480 V
— It is a three phase system
— The MCC is designed to accommodate four wire loads
— It is rated for a maximum of 600A, for 60 Hz application.

The long dashed line represents the outer chassis or MCC cabinet. The solid horizontal line, representing one of the three energized 480 V phases, has multiple vertical lines "hanging" below nodes. These vertical "drops represent the branch circuits pertaining to specific loads, catered to by this MCC. The dot at the junction of the vertical branch circuit and the horizontal bus represents a "node." The first branch circuit is identified by a unique identifier "D2," at the bus bar, on the line side. The other branch circuits are identified as D4, D5, and so on. Note the chevron symbols directly below each branch circuit drop. These symbols represent the MCC cubicle bus stabs, pointed out earlier in the MCC pictures. Focusing on branch circuit D2, and following this circuit down to the load, the next component we notice is the fusible disconnect switch. Refer to Figure 10.2 for a detailed excerpt of branch circuit D2. The three phase, or three pole, "switch" component of the fusible disconnect switch is followed by the associated overcurrent protection fuse. The branch circuit bifurcates after the overcurrent protection fuse into the motor pilot device control circuit and the remaining branch circuit leading to the 3-hp motor in the field. The function and operational logic of the motor pilot device control circuit will be described

Figure 10.2: Branch circuit D2 excerpt, from one line schematic for MCC.

in the wiring diagram section. As we follow the remaining branch circuit, leading to the motor in the field, the next entity we see is the solid state overload device. The branch circuit exits the MCC chassis on the load side of the overload device, via the terminal strip in the cubicle.

As labeled in Figure 10.2, the power in branch circuit D2, flows from the MCC to the 3-hp motor, through a bundle of three energized conductors, accompanied by one ground conductor. This bundle of conductors and the conduit are specified on the one-line schematic, and interpreted, as follows:

- **¾" Conduit**: The electrical designer of this power distribution system selected a ¾" (ID) conduit, or pipe, to house the four conductors specified for this branch circuit.

• **3—1/C #12**: The electrical designer has selected or specified a three, single (1/C), AWG #12 conductors to supply three phase AC power to the 3-hp motor.

• **1/C #12 Gnd**. The electrical designer has specified one (1/C), AWG #12 conductor to serve as ground for the grounded three phase AC service to the 3-hp motor.

The description of branch circuit symbols and notations provided above should be sufficient for understanding the symbols, nomenclature and notations used for other branch circuits in the one-line schematic depicted in Figure 10.1 with the exception of branch circuit D4. Branch circuit D4 differs from all other branch circuits in that it represents a *lighting load*. Note the square block labeled "LC-5" used to represent the lighting load, instead of the circles used for the motor branch circuits.

Examination of Branch Circuit D2

Having introduced the symbols, nomenclature and labels employed in the representation of power distribution branch circuits, we are better poised to explore and examine the design criteria and specifications incorporated in this one-line schematic. We will verify the following design specifications— pertaining to **branch circuit D2** (Figure 10.1 and 10.2)—in order to give the reader an appreciation of the design process an electrical engineer might go through in designing electrical power distribution systems:

1) Current carrying conductor size
2) Ground conductor size
3) Conduit size
4) Overload setting
5) Starter size
6) Overcurrent protection fuse size
7) Fusible disconnect switch size

CURRENT CARRYING CONDUCTOR SIZE

As discussed in the coverage of NEC®, in chapter 9 of this text, a conductor expected to support a certain continuous load must me rated 125% of the full load amps. So, the next step would be to determine the full load amps (FLA) expected to be drawn by the 3-hp motor. As discussed in chapter 9, the code requires that the full load amp information must be based on the "nameplate" of the motor. Since, actual motor data is not available in this text, for illustration purposes we will base our discussion and analysis on the information available in the Buss® table, shown in Figure 10.3; with some verification through NEC®.

As we focus on the circled/highlighted, section of the Buss® table in Figure 10.3, we can spot most of the code compliant design parameters associated with the 3-hp load in branch circuit D2. According to this table, a fully loaded 3-phase motor, operating at 480V, will demand 4.8 amps. In addition,

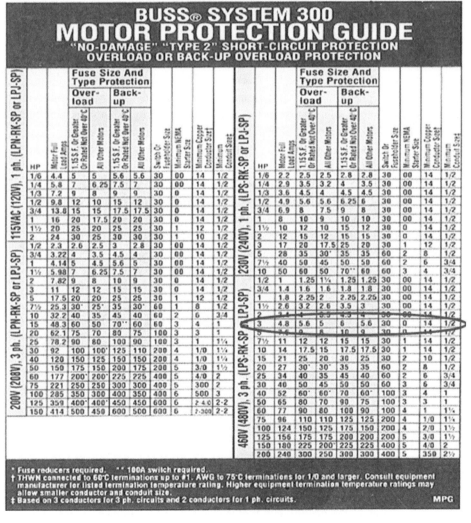

Figure 10.3 – Bus ® Table, used for electrical design verification.

according to this table, in order to comply with the code, the three current carrying conductors must be a minimum of AWG #14. If we follow the NEC® requirement that the conductor ampacity must be a minimum of 125% (1.25) times the motor FLA (Full Load Amps), each of the three current carrying conductors must be capable of carrying a minimum of:

(1.25) x (4.8 A) = 6A

Examining Table 10.2 (Superseded by NEC Tables 310.15(B)(2)(a)), we see that an AWG #14 has an ampacity of 20 A, for THHW and THWN insulation (and other types of insulation) at 60°F terminal or conductor temperature. Note, however, that the 20 A ampacity is predicated on other conditions and stipulations, i.e. overcurrent protection restrictions. Further discussion on additional stipulations, exceptions and code implications is outside the scope and context of this text. For simplicity, acknowledging the fact that AWG #14 has the ampacity to carry 20 A (amps), we establish that AWG #14—as stated in the BUS® table (Figure 10.3)—is adequate for this motor branch circuit.

Nevertheless, as we refer back to the schematics in Figures 10.1 and 10.2, we notice the electrical engineer/designer in this case, selected size AWG #12 copper conductors for this circuit; which is rated 25 A, and it clearly exceeds the minimum requirement. This decision by electrical engineer/designer *could* serve as an example of how some engineer/designers and engineering firms subscribe to a **"hard deck"** on the low end of conductor sizes. In other words, in some cases, the design engineer and the firm, as result of *"in-house best practice,"* don't design power circuits with conductors smaller than AWG #12. However, in this case, the more compelling reason for the designer selection of an AWG #12 conductor is, perhaps, the fact that the load—3-hp Roof Exhaust Fan Motor—is located at a distant location, on the roof. The long conductor run to the roof – due to higher resistance – would result in excessive voltage drop. See NEC 2014, Article 215 (2) (A) (1)(a) and (b) & Informational Note 2. So, prudently, the designer selected a larger conductor (AWG #12) to prevent excessive voltage drop.

Ground Conductor Size

Next, let's examine the ground conductor selected for branch circuit D2. As apparent from examination of Bus table in Figure 10.3, this table does not specify the **ground conductor** size for various branch circuits. One could specify the ground conductor size to be the same as the **current carry-**

Table 10.2: NEC ® Ampacity table superseded by current NEC Table 310.15(B) (16). Included for general illustration purposes, only. Courtesy, NEC, NFPA.

Table 310.15(B)(2)(a) Adjustment Factors for More Than Three Current-Carrying Conductors in a Raceway or Cable							
	Temperature Rating of Conductor						
	60°C	75°C	90°C	60°C	75°C	90°C	
	(140°F)	(167°F)	(194°F)	(140°F)	(167°F)	(194°F)	
Size AWG or kcmil	Types TW, UF	Types RHW, THHW, THW, THWN, XHHW, USE, ZW	Types TBS, SA, SIS, FEP, FEPB, MI, RHH, RHW-2, THHN, THHW, THW-2, THWN-2, USE-2, XHH, XHHW, XHHW-2, ZW-2	Types TW, UF	Types RHW, THHW, THW, THWN, XHHW, USE	Types TBS, SA, SIS, THHN, THHW, THW-2, THWN-2, RHH, RHW-2, USE-2, XHH, XHHW, XHHW-2, ZW-2	Size AWG or kcmil
	COPPER			ALUMINUM OR COPPER-CLAD ALUMINUM			
18	–	–	14	–	–	–	–
16	–	–	18	–	–	–	–
14*	20	20	25	–	–	–	–
12*	25	25	30	20	20	25	12*
10*	30	35	40	25	30	35	10*
8	40	50	55	30	40	45	8
6	55	65	75	40	50	60	6
4	70	85	95	55	65	75	4
3	5	100	110	65	75	85	3
2	95	115	130	75	90	100	2
1	110	130	150	85	100	115	1
Jan-00	125	150	170	100	120	135	Jan-00
Feb-00	145	175	195	115	135	150	Feb-00
Mar-00	165	200	225	130	155	175	Mar-00
Apr-00	195	230	260	150	180	205	Apr-00
250	215	255	290	170	205	230	250
300	240	285	320	190	230	255	300
350	260	310	350	210	250	280	350

Figure 10.4: Excerpt, one line schematic for a power distribution system

ing conductors, as done by the designer of branch circuit D2, in schematics
on Figures 10.1 and 10.2; where the designer specified AWG #12 conductors
for the three, 3-hp motor, current carrying conductors, as well as the ground
conductor. However, NEC 2014 prescribes the use of Table 25.122 for establishing the size of the ground conductor, based, primarily, on the overcurrent
protection device rating. According to Table 25.122, selection of an AWG #14
would have been sufficient in branch circuit D2. Note that the purpose of the
ground conductor is to carry fault current and to serve as a "mechanism" for
safely de-energizing circuits with electrical faults. For example, if the insulation of one of the energized phases in branch circuit D2 were to atrophy, or
break down due to other reasons, the 460 V potential could come in contact
with the motor chassis and raise its potential to 460 V. With the motor ground-

ed properly, such an electrically unsafe condition would be rectified through a *phase to ground fault* clearing the fuse. Once again the designer could be credited with the more appropriate decision of specifying a larger, AWG#12, ground conductor due to the long run of the all four wires to the remotely located motor. The discourse presented in this paragraph shows the complexities and intricacies associated with correct interpretation of the code, with the paramount interest of the safety of public and electrical workers in mind.

CONDUIT AND CONDUIT SIZE

As stated on schematics in Figures 10.1 and 10.2, the conduit specified by the engineer is ¾" ID. Before we determine if the ¾" conduit is adequate for this branch circuit, we need to recount the purpose and general characteristics of an electrical conduit. Electrical conduits can be rigid or flexible. Electrical conduits are similar to pipes used to convey fluids. Conduits can be constructed out of metal or PVC (polyvinyl chloride). The key function of an electrical conduit is to protect the electrical conductors against mechanical damage, exposure to adverse environmental ambient factors, i.e. moisture, corrosive liquids, solids or gases. Unlike pipes applied in fluid system, conduits used in electrical systems cannot be filled to completely. As apparent from Chapter 9 in this text, and pertinent tables therein, electrical conduits are permitted to be filled only partially. Two ostensible reasons for this constraint are: (1) It is, physically, difficult to pull too many wires through a conduit and associated elbows and fittings (2) a certain minimum, cross-sectional, open space is necessary in an electrical conduit to facilitate ventilation of any heat dissipated in the conductors due to I^2R losses.

The ¾" conduit specified for branch circuit D2, on schematics in Figures 10.1 and 10.2, is oversized based on the Bus® table in Figure 10.3. As noted earlier, the Bus® table identifies a ½ "conduit for the 3-hp branch circuit D2, in combination with the AWG #14 conductors. The ¾" conduit specified on the drawings is sized based on fact that four AWG #12 conductors were selected by the electrical designer. In order to understand the process of selection of conduits, using the NEC®, we will compare both alternatives: a ½" conduit and the ¾" conduit.

According to NEC®, Chapter 9, Table 4, a **½" Rigid Metallic Conduit (RMC),** with greater than two conductors, can only be filled up to 40% of its internal area of cross-section. The available, internal area of cross-section, according to NEC®, Chapter 9, Table 4 is 81 mm².

According to NEC® Chap. 9, Table 5, the total area of cross-section three AWG #14 current carrying conductors would occupy, would be:

= 3 x 8.968 = 26.9 mm²

Based on NEC®, Chapter 9, Table 4, 40% fill would require that the minimum internal area of the cross-section of the selected conduit must be:

$$\text{Minimum Area of Cross-section} = \frac{26/9}{0.4} = 67.26 \text{ mm}^2$$

Since this computed minimum area of cross-section of 67.26 mm² is well within the available internal area of cross-section of 81 mm² offered by a ½" RMC, *a ½" conduit for D2 branch circuit would have been adequate, had AWG #14 conductors been used.*

However, since the designer chose AWG #12 for branch circuit D2, according to NEC Chap. 9, Table 5, the total area of cross-section that the three AWG #12 current carrying conductors would occupy, would be:

= 3 x 11.68 = 35 mm²

Based on NEC®, Chapter 9, Table 4, 40% fill would require that the minimum internal area of the cross-section of the selected conduit must be:

$$\text{Minimum Area of Cross-section} = \frac{35}{0.4} = 87.6 \text{ mm}^2$$

Since this computed minimum area of cross-section of 87.6 mm² is well within the available internal area of cross-section of 141 mm² offered by a ¾" RMC conduit, a ¾" conduit for D2 branch circuit would be adequate with the AWG #12 conductors specified.

OVERLOAD PROTECTION SETTING

Overload protection devices are included in motor and branch circuits to protect motors, motor control apparatus, and motor branch circuit conductors against overheating due to overloads. A solid state motor overload device offers flexibility in setting of overload set-points. And motor overload device setting can determined using the Bus® Table or NEC® Article 430.32. In the case of branch circuit D2, according to NEC 430.32 (1)(For separate overload device, for motors other than those with Service Factor of 1.15 and marked operating temperature of 40°C), the over load device should be set at:

= 1.15 x Full Load Amps
= 1.15 x 4.8 = 5.52 A, or simply 5 amps.

According to the Bus® Table, overload current setting, for this specific case, should be 5 A. So, once again we see the Bus Table and the NEC® yielding, practically, the same results for design specification.

STARTER SIZE

The motor branch circuit starter size is, generally, based on NEMA standards. According to the Bus® Table, in Figure 10.4, the starter should be "**Size 0.**" However, once again, we see that the designer of the branch circuit selected a size that is greater than the minimum requirement. The designer of the 3-hp motor branch circuit incorporated a NEMA "Size 1" starter in the schematics shown on Figures 10.1 and 10.2.

OVERCURRENT PROTECTION FUSE SIZE

As with the motor overload protection device, the selection of motor overcurrent device can be accomplished by using the Bus® Table or NEC®. In the case of branch circuit D2, according to NEC 430.52 and Table 430.52 (For a polyphase motor protected by a time-delay fuse) the short circuit fuse must be rated as:

$$= 1.75 \text{ x Full Load Amps}$$
$$= 1.75 \text{ x } 4.8 = 8.4 \text{ A}$$

According to NEC® Article 240.6 (A), when overcurrent calculation yields an ampere specification that is non-standard, the next higher size may be used. Therefore, a standard 10A fuse could have been selected. Instead, the designer decided to exceed the requirements once again, and chose the standard 8A fuse. As long as the 8 amp fuse does not cause nuisance trips or fuse clearing, it should be acceptable.

FUSIBLE DISCONNECT SWITCH SIZE

The fusible disconnect switch specification can be determined through the Bus Table in Figure 10.4. According to the Bus® Table, at the point of intersection of the row representing 3-hp motor and the column titled: "Switch or Fuse Holder Size," we see that a 30 A disconnect switch is needed to meet the code. As shown on the one-line diagram in Figures 10.1 and 10.2, in compliance with the code, a 30 A disconnect switch is specified by the designer. This is also in compliance with NEC® Article 430.101 and NEC®Table 430.251 (B).

So, all in all, it appears that design of branch circuit D2 is in compliance with the code, and in some aspects, it exceeds the code. However, if one compares the design of branch circuit D2 with the rest of the branch circuits, it becomes apparent that, unlike the other branch circuits, D2 lacks a safety disconnect switch at the motor. And, as obvious from the label on the motor load, this motor pertains to a roof exhaust fan and is, possibly, located on the roof; out of line of sight from the MCC disconnect switch. According to NEC® Article 430.102 (B) (2), the disconnecting means should be located in sight of the motor, baring qualification for exceptions stated under this article. It is unknown if this particular installation qualified for the exceptions. Nevertheless, it is always a desirable, in the interest of safety, to have a lockable disconnect switch at each motor load.

Example 10.1

Answer the following questions pertaining to the branch circuit shown in the schematic diagram below:

(a) What is the maximum current the power distribution system for this branch circuit rated for?

(b) What is the turns ratio of the control transformer shown in the motor branch circuit?

(c) What is the full load current in this circuit?

(d) What should the solid state overload device be set for at commissioning of this system?

Solution:

(a) The maximum current the power distribution system for this branch circuit rated for is stated in the specification stated at the very top of the schematic diagram, as 600 Amps, within the caption:

480 V Bus, 3 φ, 4W, 600 A, 60 Hz,

480V

120V

O/L
Setting?

5

"480 V Bus, 3 φ, 4W, 600 A, 60 Hz."

Therefore, the answer is: **600 Amps.**

(b)   The turns ratio of the control transformer shown in the motor branch circuit would be based on the voltage transformation stated on the control transformer. As obvious from the schematic diagram, the control transformer has a 480V primary and a 120 V secondary. Apply Eq. 10.1, introduced earlier in this text:

$$\text{Turns Ratio} = \frac{N_P}{N_S} = \frac{V_P}{V_S}$$

Eq. 10.1

$$\text{Turns Ratio} = \frac{N_P}{N_S} = \frac{V_P}{V_S} = \frac{480 \text{ V}}{120 \text{ V}} = \frac{4}{1} = 4{:}1$$

(c)   What is the full load current in this circuit? This question can be answered through multiple approaches, including the calculation method based on the equation 10.2, provided the motor efficiency ($\eta_{\text{motor efficiency}}$) and the power factor are known.

$$|S| = \frac{P(\text{watts})}{(\text{pf}).(\eta_{motor\ efficiency})} = |V|.|I|, \text{ or}$$

Eq. 10.2

$$|I| = \frac{P(\text{watts})}{(\text{pf}).(\eta_{motor\ efficiency}).(|V|)}, \text{ or,}$$

$$|I| = \frac{P(\text{hp}).(746 \text{ W/hp})}{(\text{pf}).(\eta_{motor\ efficiency}).(|V|)}$$

Since the efficiency and the power factor are not given, as introduced earlier in this text, we will the Buss® table introduced earlier in this chapter. As stated in schematic diagram—in the circled motor symbol—the motor's full load rating is **5 hp**. As specified at the top of schematic diagram, the motor is being powered by a **480 V, 3-phase** source (**480 V Bus, 3 φ,** 4W, 600 A, 60 Hz). Therefore, according to the Buss® table, and as highlighted (circled)—under the "460 V(480V), 3-ph, section—**the motor full load current would be 7.6 A.**

(d)   Using the Buss® in Figure 10.5a, for **3—phase, 5 hp** motor, operating at a **480 V,** the overload device should be set at **8.0 A.**

Note: If NEC tables were used here, the 115% multiplier by NEC® would result in an overload setting of 8.7 A.

Examination of Branch Circuit D5

While we will not examine branch circuit D5 to the degree we examined D2, it is important to point out one apparent oversight on the part of the designer. This oversight pertains to the overcurrent protection fuse specification

BUSS® SYSTEM 300
MOTOR PROTECTION GUIDE
"NO-DAMAGE" "TYPE 2" SHORT-CIRCUIT PROTECTION
OVERLOAD OR BACK-UP OVERLOAD PROTECTION

Left section — 200V (208V), 3 ph. (LPN-RK-SP or LPJ-SP) / 115VAC (120V), 1 ph. (LPN-RK-SP or LPJ-SP)

HP	Motor Full Load Amps	Over-load 1.15 S.F. Or Greater Or Rated Not Over 40°C	Over-load All Other Motors	Back-up 1.15 S.F. Or Greater Or Rated Not Over 40°C	Back-up All Other Motors	Switch Or Fuseholder Size	Minimum NEMA Starter Size	Minimum Copper Conductor Size†	Minimum Conduit Size‡
1/6	4.4	5	5	5.6	5.6	30	00	14	1/2
1/4	5.8	7	6.25	7.5	7	30	00	14	1/2
1/3	7.2	8	8	9	9	30	0	14	1/2
1/2	9.8	12	10	15	12	30	0	14	1/2
3/4	13.8	15	15	17.5	17.5	30	0	14	1/2
1	16	20	17.5	20	20	30	0	14	1/2
1½	20	25	20	25	25	30	1	12	1/2
2	24	30	25	30	30	30	1	10	1/2
1/2	2.3	2.6	2.5	3	2.8	30	00	14	1/2
3/4	3.22	4	3.5	4.5	4	30	00	14	1/2
1	4.14	5	4.5	5.6	5	30	00	14	1/2
1½	5.98	7	6.25	7.5	7	30	00	14	1/2
2	7.82	9	8	10	9	30	0	14	1/2
3	11	12	12	15	15	30	0	14	1/2
5	17.5	20	20	25	25	30	1	12	1/2
7½	25.3	30*	25*	35	30*	60	1	8	1/2
10	32.2	40	35	45	40	60	2	6	3/4
15	48.3	60	50	70**	60	60	3	4	1
20	62.1	75	70	80	75	100	3	3	1
25	78.2	90	80	100	90	100	3	1	1¼
30	92	100	100*	125	110	200	4	1/0	1¼
40	120	150	125	150	150	200	4	1/0	1¼
50	150	175	150	200	175	200	5	3/0	1½
60	177	200*	200*	225	225	400	5	4/0	2
75	221	250	250	300	300	400	5	300	2
100	285	350	300	400	350	400	6	500	3
125	359	400*	400*	450	450	600	6	2 4/0	2-2
150	414	500	450	600	500	600	6	2-300	2-2

Right section — 230V (240V), 1 ph. (LPS-RK-SP or LPJ-SP) / 460V (480V), 3 ph. (LPS-RK-SP or LPJ-SP)

HP	Motor Full Load Amps	Over-load 1.15 S.F. Or Greater Or Rated Not Over 40°C	Over-load All Other Motors	Back-up 1.15 S.F. Or Greater Or Rated Not Over 40°C	Back-up All Other Motors	Switch Or Fuseholder Size	Minimum NEMA Starter Size	Minimum Copper Conductor Size†	Minimum Conduit Size‡
1/6	2.2	2.5	2.5	2.8	2.8	30	00	14	1/2
1/4	2.9	3.5	3.2	4	3.5	30	00	14	1/2
1/3	3.6	4.5	4	4.5	4.5	30	00	14	1/2
1/2	4.9	5.6	5.6	6.25	6	30	00	14	1/2
3/4	6.9	8	7.5	9	8	30	00	14	1/2
1	8	10	9	10	10	30	00	14	1/2
1½	10	12	10	15	12	30	0	14	1/2
2	12	15	12	15	15	30	0	14	1/2
3	17	20	17.5	25	20	30	1	12	1/2
5	28	35	30*	35	35	60	2	8	1/2
7½	40	50	45	50	50	60	2	6	3/4
10	50	60	50	70**	60	60	3	4	3/4
1/2	1	1.25	1¼	1.25	1.25	30	00	14	1/2
3/4	1.4	1.6	1.6	1.8	1.8	30	00	14	1/2
1	1.8	2.25	2	2.25	2.25	30	00	14	1/2
1½	2.6	3.2	3.5	3.5	3	30	00	14	1/2
2	3.4	4	3.5	4.5	4	30	00	14	1/2
3									
5	7.6	9	8	10	9	30	0	14	1/2
7½									
10	14	17.5	15	17.5	17.5	30	1	14	1/2
15	21	25	20	30	25	30	2	10	1/2
20	27	30*	30*	35	35	60	2	8	1/2
25	34	40	35	45	40	60	2	6	3/4
30	40	50	45	50	50	60	3	6	3/4
40	52	60*	60*	70	60*	100	3	4	1
50	65	80	70	90	75	100	3	3	1
60	77	90	80	100	90	100	4	1	1¼
75	96	110	110	125	125	200	4	1/0	1¼
100	124	150	125	175	150	200	4	2/0	1½
125	156	175	175	200	200	200	5	3/0	1½
150	180	225	200*	225	225	400	5	4/0	2
200	240	300	250	300	300	400	5	350	2½

* Fuse reducers required.   ** 100A switch required.
† THWN connected to 60°C terminations up to #1. AWG to 75°C terminations for 1/0 and larger. Consult equipment manufacturer for listed termination temperature rating. Higher equipment termination temperature ratings may allow smaller conductor and conduit size.
‡ Based on 3 conductors for 3 ph. circuits and 2 conductors for 1 ph. circuits.   MPG

Figure 10.5a: Bussman ® Motor Protection Guide.

of 125 A, as it appears on the one-line drawing in Figures 10.1 and 10.2.

See the circled data, in Figure 10.5b Bus® table, for full load amps associated with a 40-hp motor. In the case of branch circuit D5, which supports a 40-hp (52 FLA, Full Load Amps, as stated under the column adjacent to the HP column) motor, according to NEC 430.52 and Table 430.52 (For a polyphase motor protected by a time-delay fuse) the short circuit fuse must be rated as:

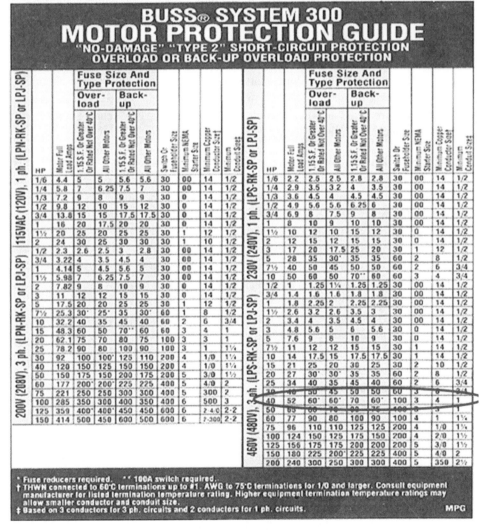

Figure 10.5b: Bussman ® Motor Protection Guide

$$= 1.75 \text{ x Full Load Amps}$$
$$= 1.75 \text{ x } 52 = 91 \text{ A}$$

According to NEC® Article 240.6 (A), when overcurrent calculation yields an ampere specification that is non-standard, the next higher size may be used. Therefore, standard 100 A fuses should have been selected. Instead, the designer chose a 125 A overcurrent protection fuse. In the absence of possible qualifying exceptions, this branch circuit should be equipped with a combination of a 100 A fusible disconnect switch (see 4th column from the right) and 100 A short circuit protection fuses.

Examination of Branch Circuit D4

As apparent from the load identifying symbol, LC-5, this branch circuit supplies power to lighting load of 5-hp. The 60 A disconnect switches and the 60 A fuses are based on NEC® Article 430. Note that the Bus® chart cannot be used in the case of lighting loads.

Wiring Diagram

Unlike a one-line schematic, a wiring diagram displays comprehensive information on all three phases of a three phase AC system, along with the associated grounds. See Figure 10.6. The wiring diagram in Figure 10.6 pertains to three motor loads: 40-hp supply air fan motor, 75-hp return air fan motor, and a 15-hp pump motor. We will limit our focus the 75-hp return air motor circuit, on the left. Notice the vertical lines, spanning from top to bottom, on the left side of the wiring diagram. These lines, as annotated on the drawing, represent the voltage bearing phase busses A, B and C; as well as the ground bus (G).

As we work our way from the left side of this 75-hp motor circuit to the right, we notice the three energized phase busses are "tapped" by the 75-hp branch circuit via the bus stabs, represented by the "chevron" symbols. The MCC ground is shown routed from the MCC to the three phase AC 75-hp return air motor. All three phases are shown bridged from the MCC bus stabs to the 200 A disconnect switch, labeled as "3 Pole, 200A." Each of the three phases is then routed to respective motor starter contactor contacts via the short circuit protection fuses. Note the strict adherence of unique labeling assigned to each continuous conductor, between consecutive current interrupting points. For instance, the wire label "1L1" is reserved for the conductor between the load side of phase A 150 A fuse and the motor starter contact, labeled "Size 4 M;" where "M" represents the motor contactor for this circuit. Unique numbering of conductors is crucial for the following few reasons:

a)   Troubleshooting and tracing of circuits, using a drawing.

b)   Ensuring that wires are not "cross-wired." Crossed wires can result in phase to phase and phase to ground faults, and control circuit malfunctions.

c)   Unique wiring facilitate accurate wiring or assembly of power distribution and control circuits.

Application of wiring numbers is not feasible when wire type conductor are not used to connect one current interrupting point to another, such as the connection between a motor starter contact and an overload contact, as evident on Figure 10.6 wiring diagram.

Wires T1, T2 and T3 connect phases A, B and C to respective line side terminals of the safety disconnect switch poles at the motor, in the field. Note that the identification of wires, completing the final connection of phases A, B and C to the motor terminals, changes to 1T1, 1T2 and 1T3, respectively.

The pilot device and control power for this 75-hp motor is shown "tapped" or branched off phases A and B. This phase to phase power, as expected, would be 480 V. However, the control circuit must be 120 V or less. Therefore, the tapped 480 V power, as shown on the wiring diagram, is "stepped down" to 120V through the 480/120 V control transformer. Note that the secondary of this control transformer is grounded on one leg. This leg, serves as a neutral, ground or low potential point. The other terminal, then, serves as the higher potential, 120 V, point, labeled as X1. This 120 V and neutral pair sustains and operates all of the pilot devices, switches, interlock contacts and motor contactor coil.

The 120 V control power is protected through a fuse and arrives at the "STOP" switch's normally open contacts via conductor numbered as "1." The spring loaded, momentary, STOP push button contact is normally closed, as shown in the wiring diagram. So, the 120 V potential crosses over to point 1A and three safety interlock contacts. These contacts are normally open. Such interlock contacts are often associated with system components such as doors— which must be closed before the motor is allowed to be energized. Let's assume that all safety conditions are met in this case. This would allow us to assume that all three of these interlock contacts are closed. Then 120 control voltage would bridge over to point 26 on wiring diagram. Point 26 and the terminal labeled "2" are at the same potential; there is an electrical short between these two points. Therefore, terminal 2 retains 120 V potential as long as the STOP button is *not* depressed. Then, if the START switch is depressed, its normally open contacts close allowing the 120 control voltage to be applied to the coil

Figure 10.6: Wiring diagram for power distribution system

of the motor starter contactor "M." Since neutral or ground side of the coil is always grounded through the normally closed solid state overload contacts (unless the overload protection device has tripped under an overload condition), the motor starter coil energizes when the START switch is depressed. The three motor starter contacts, shown in the motor branch circuit, close; thus, releasing three phase voltage and power to the 75-hp motor. Since the ***pilot light*** for the motor is connected in parallel with the motor starter coil, the motor "ON" light turns on, as well. One of the motor starter contacts, referred to as the "latching contact "M" is connected in parallel with the START switch contacts. This contact seals the coil in energized mode, such that when the START button is released, the motor starter coil stays latched and the motor continues to operate. The motor continues to operate as long as all of the safety interlocks stay closed, the overload protection device does not sense an overload, and the STOP switch is not depressed.

When the motor STOP switch is depressed, its normally closed contacts open and the continuity of the 120 V circuit is broken. This results in de-energization of the motor starter coil, thus ***unlatching*** the sealing contacts of the motor starter coil. The motor ceases to operate.

The control sequence and logic described above represents the essence of approaches applied in most motor starter circuits, with some application specific variations.

Control Diagram/Drawing

Our discussion in this section will be based on the control diagrams depicted in Figures 10.7 and 10.8. Both of these drawings are based on a control system premised on a PLC, Programmable Logic Controller. In other words, the control algorithm is being implemented through a control architecture where the PLC CPU, Central Processing Unit, makes all control decisions in accordance with the PLC control program. Of course, the CPU makes the control decisions in ***response*** to the current state or change of state of all pertinent inputs.

The control drawing, or control schematic, in Figure 10.7 shows some of the inputs interfaced to, or monitored by, the PLC. The inputs are, essentially, signals coming from sensors and switches in the field. These signals can be "discrete" or "digital;" meaning, they are in form of "ONs," "OFFs," "1s," "0s," switch ***closed*** or ***open***, etc. The column to the right of the drawing, labeled with input numbers, i.e. Input 1, Input 2, and so on, represents a PLC input module. Note that this module is labeled as 115 VAC Input, Modicon, Cat. No: B805-016. The role of this input module is to receive 115 VAC signals from various sensors and switches in the field and convert them into low voltage DC signals. A 115 or 120 VAC signal is transduced by the input module circuitry into a $5V_{DC}$ or 10 V_{DC} signal. An absence of AC voltage at a given input is converted into 0 V_{DC} and interpreted as a "0." When a 115 V signal is received at an input point, it is interpreted by the PLC as a "**1**." In other words, the PLC expects discrete **1s** and **0s** from its input modules.

The output module shown in Figure 10.8 serves as an interface between the PLC and the control devices/equipment in the field in a manner similar to the input module. However, there are a few distinct differences. A discrete 115 VAC output module, *receives* discrete signals, 1s and 0s, *from* the PLC CPU and *converts them* into 115 VAC or 0 VAC outputs or commands—that are conveyed to the control devices in the field, such as, motor starter coils, solenoids, lamps, horns, etc.—as shown on the right side of the control drawing in Figure 10.8. Notice that contiguous electrical connections or wires are identified by unique numbers or alphanumeric identifications on the input and the output control drawings.

INPUT CONTROL DIAGRAM

In order to get a better understanding of how the inputs from various field sensors and switches interface with the PLC, let's focus on the input control drawing in Figure 10.7 and follow a hypothetical, yet plausible, scenario. We will then extend the possible outcome of this hypothetical scenario to the output control diagram on Figure 10.8.

Let's assume that pressure switch, PSxxx56, shown in Figure 10.7, is monitoring the pressure in a vessel that must be maintained above a certain critical, level. This pressure switch is *normally open*, but *closed* when the vessel is pressurized, as shown in the control drawing. In this safe state, with the pressure switch closed, 120 V control voltage is passed on to Input #1 on the PLC input module. As described earlier, the 120 V potential at Input #1 gets transduced into a logic level "1" for interpretation by the CPU (microprocessor). As long as the pressure switch remains closed, Input #1 maintains a logic value of "1." If the pressure switch opens, 120 V at Input #1 would disappear and logic level "1" would be replaced by a logic level "0." Often switches— such as the pressure switch in this circuit—are applied such that they *open* under "*unsafe*" conditions. When electrical control devices are applied in this manner, they are said to be in "*fail safe*" mode. If the vessel in this scenario were to lose pressure to a level below a preset critical level, the pressure switch will open. The set point, in such a case could be set at the pressure switch. The temperature of a downstream process in this scenario, is being monitored by a temperature switch TSxxx60. See Figure 10.7. The process, downstream,

Figure 10.7: Input Control Drawing

Figure 10.8: PLC output module based control drawing.

relies on pressure in the vessel being maintained at an adequate level. This temperature switch is open under normal circumstances and would close only if safe operating temperature is exceeded. Note that this temperature switch is *not* applied in *fail safe* mode. Under normal operation, the temperature switch is open, and the 120 V control voltage stops at the left side of the temperature switch. This results in 0 V at Input #3. No voltage at terminal #3 of the PLC input module translates into logic level "0" at Input #3. The PLC CPU, or microprocessor, completes execution of one cycle of the control program, on average, in approximately, 30 milliseconds. So, as the CPU examines Input #3, every 30 milliseconds, it sees a "0" and interprets that as a normal situation, requiring no specific action by the PLC system.

Now, let's assume that pressurized vessel develops a leak and loses pressure, to the extent that the pressure switch opens. As the pressure switch opens, the 120 V control voltage at Input #1 disappears, resulting in the change of input #1 logic state from a "1" to a "0." At the same time, as the pressure in the vessel drops below critical level, the temperature of the process rises to critical level, and the temperature switch closes. The 120 V control voltage is bridged across to PLC input module input #3, changing its logic state from a "0" to a "1." At the very next scan, the PLC senses the change of state at inputs #1 and #3. The PLC logic and algorithm recognizes these two changes of state as an anomaly and develops a logical solution that it must execute through its outputs. This juncture serves as an appropriate segue into the output section of the control system.

OUTPUT CONTROL DIAGRAM

As described earlier, the 115 VAC Modicon output module, shown in Figure 10.8, serves as an interface between the PLC and the control devices in the field. Let's continue our analysis of the hypothetical scenario surrounding the loss of pressure in a vessel; with the focus, this time, on specific outputs associated with the proper response to the anomaly described above. Assume that the PLC's response, to the anomaly described above, involves some actions that yield instantaneous results and some that will require a finite amount of time to implement. Actions that the PLC based control system must initiate and execute, immediately, are as follows:

1)   Energize the reserve tank solenoid valve to boost the vessel pressure to a certain level.
2)   Turn ON the alarm horn.
3)   Turn ON the "Critical/Low Pressure Alarm light.

4)   Turn ON the pump motor starter.

In order to energize the reserve tank solenoid valve to boost the vessel pressure, the PLC CPU must change the logic level at Output #5 from a "0" to a "1." The output module transduces the "1" to a 120 V output at the terminal for output #1. This 120 V control voltage propagates to the solenoid through the 1.5A short circuit protection fuse FUxxx14 and turns on the solenoid. The solenoid opens the reserve tank valve and begins to pressurize the vessel.

In order to implement full restoration of pressure in the vessel, the PLC CPU must change the logic level at Output #1 from a "0" to a "1." As a result, the 120 V control voltage propagates to the Pump Motor Starter coil through the 1.5A short circuit protection fuse FUxxx06 and a normally open safety interlock contact. The safety interlock is assumed to be met, so, this safety interlock contact should be closed. The 120 VAC, therefore, arrives at the pump motor starter, "M," energizes it, and starts the 3-hp motor, shown on Figure 10.8.

At the same instant, the CPU changes the state of Output #7 from "0" to a "1," thus sending 120 V to the alarm horn through another 1.5A fuse. The horn continues to annunciate the critical alarm condition, audibly, until the CPU has received verification of re-pressurization of the vessel via Input #1 or the alarm is acknowledged and silenced through a horn silence circuit on the input side of the control system. Note that the horn-acknowledge and silence button is not shown on input control drawing in Figure 10.7.

In addition to turning ON Output #7, the CPU changes the logic state of Output #6 from a "0" to a "1," thus sending 120 V to an alarm light via 1.5A fuse FUxxx16. The alarm light turns on and stays on until the CPU has received verification of re-pressurization of the vessel via Input #1.

Relay Ladder Logic

An introduction to control circuits and control systems would be incomplete without some explanation an illustration of PLC programming method. Even though the brain of a PLC, like a PC (Personal Computers), consists of single or multiple microprocessors, unlike a PC it does not operate off a "Windows®" or Microsoft Office based software systems. PLC operation is based on proprietary software, provided by the PLC manufacturer. On the other hand, the PLC *programming* software, nowadays, is "Windows®" based. The programming software allows control engineers and technicians to program the operation of PLCs in form of a diagrammatic programming language called *Relay Ladder Logic*. This programming language emulates the "pre-PLC" era, electromechanical relay, counter and timer based control logic, in symbolic form. Of course, the symbols and associated functions are represented as "virtual" symbols and logical functions that are used to implement functional specifications of the overall control system. In the pre-PLC era, these electromechanical relay, timer and counter based functions had limited reliability, lacked speed, resolution, accuracy, flexibility, versatility and occupied a great deal of control cabinet or control room space. PLC manufacturers, such as Rockwell® provide training classes for control engineers and technicians, on PLC programming and installation, at local or regional offices.

A basic, relay ladder logic based, program example is shown in Figure 10.9. Brief excerpts of a comprehensive relay ladder logic program for a control system are shown in Figures 10.9 and 10.10. These excerpts are, actually, screen captures or "print screens" of the PLC program as viewed by a PLC programmer for program creation, program modification or troubleshooting purposes. These excerpts represent a specific example of an RSLogix5000 program for an Allen Bradley Control Logix Programmable Controller.

Fan Start and Stop Program Segment

Note that, at first glance, *this* page of the overall program for the control system resembles a typical Windows® based application program. However, a closer inspection reveals that—unlike computer programming languages like COBOL, Basic, C, C+, etc.—this is a diagrammatic logic code, wherein, program or algorithm is premised on symbols that appear as coils and contacts of relays, timers, counters, etc. As visible on the left side of the relay ladder program, there are seven tiers, lines or "rungs" depicted on this page of the program, ranging from rung #7 to rung #14. The logic symbols in each rung represent *inputs* and *outputs*. See chapter 8 for descriptions and illustration of typical PLC inputs and outputs. These inputs and outputs, in a typical PLC program, can be *"internal"* or *"external"* inputs. External inputs (**discrete**) are inputs that represent "tangible" or physical switches and contacts in the field. In other words, these inputs emanate from devices and equipment that are "peripheral" to the PLC. On program segment depicted in Figure 10.9, the logic elements, input and output symbols can be categorized as follows:

Batch MCR

Batch control system master control relay contact. This contact is "normally open," but shown closed—at the time the program was viewed—through high lighting (darker shading) of the contact.

Start Fan 3

This symbol or element represents the normally open contact of the

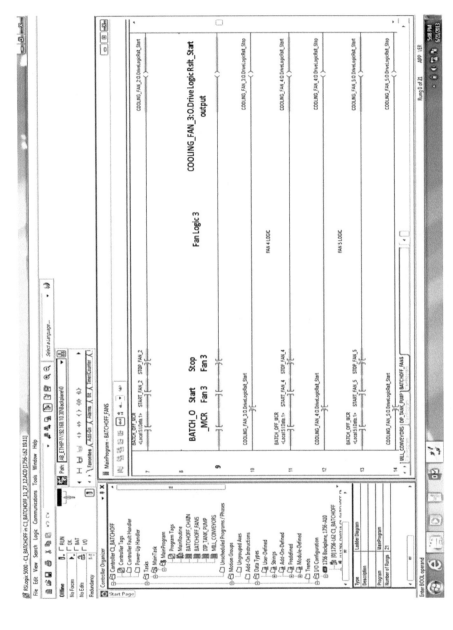

Figure 10.9: PLC Relay Ladder Logic Programming Example—Cooling Fan Control Logic.

START switch for Fan #3. Even though this is a "normally open" contact "--|
|--," it is shown *closed*—at the time the program was viewed—through high
lighting (darker shading) of the contact.

Stop Fan 3

This symbol or element represents the normally closed contact belonging
to the STOP switch for Fan #3. This is a "normally closed" contact "--|/|--"(look
for the "/" within the parallel lines) and it is shown *closed* (True)—at the time
the program was viewed—through high lighting (darker shading) of the contact. In other words, the STOP switch is not depressed at this time.

Cooling Fan # 3 Dr Start

The coil symbol "—( )—," typically, represents an output. When this
output is energized or "**True**," it actually turns ON an output, such as those
outputs we considered in the review of PLC Output Controls Drawing in Figure
10.8. In this case, this output is connected to a VFD, variable frequency drive.

Note that all elements in Rung #9 are highlighted, meaning, they were
conducting current, ON, or True, at the time the snap shot of the program was
taken. Note that when elements are *closed* or passing current, there status is
said to be "*True*." When elements are not energized, not activated, their status
is said to be "*False*." The current status of Rung #9 can be interpreted as follows:

At the time this snapshot of the logic was taken, **BATCH_OFF_MCR**
input was **True,** closed or High, the **START_FAN_3** bit (logical representation
of input emanating from **Fan #3** start switch) is **True,** and the **STOP_FAN_3**
bit (logical representation of input emanating from Fan #3 stop switch) is
false then the **COOLING_FAN_3:O. DriveLogicRslt Start** output (same as
Cooling Fan # 3 Dr Start output) will go high or *True*. This output is directly
connected via the software to the Allen Bradley® (PowerFlex 70) Variable Frequency Drive (VFD). This is accomplished through an Ethernet connection and
will cause the motor connected to the VFD to start. In this case there are seven
fans that are used to cool freshly made rubber so that it can be used in the next
process.

Example 10.2.

Consider the PLC relay ladder logic program in Figure 10.9 and answer
the following questions:

a) What would happen to the "run" signal going to the VFD is MCR, Master

Control Relay, drops out or de-energizes?

b)   How does Rung #10 respond to the energization (becoming True) of **COOLING_FAN_3:O. DriveLogicRslt Start** output?

(a) Answer: If the MCR, master control relay, for the system drops out or de-energizes, the continuity of rung #9 would break. This will result in de-energization of **COOLING_FAN_3:O.DriveLogicRslt Start** output to the VFD, and Cooling Fan #3 will **stop**.

(b) Answer: Upon inspection of Rung #10, we see that a normally closed contact belonging to output "**COOLING_FAN_3:O. DriveLogicRslt Start**" is in series in Rung #10. This normally closed contact opens when **COOLING_ FAN_3:O. DriveLogicRslt Start** output is energized. That is the reason why both elements in this rung are shown de-energized (no shading or highlighting). Note that the output of this rung is labeled as **COOLING_FAN_3:O. DriveLogicRslt Stop.** This implies that when **COOLING_FAN_3:O. DriveLogicRslt Start** output is turned ON, the **STOP** signal to the Cooling Fan VFD is disabled through the normally closed contact of **COOLING_FAN_3:O. DriveLogicRslt Start** "coil" in Rung #10. In essence, Rung #10 serves as an interlock to ensure that the STOP command to Cooling Fan #3 is disabled when Cooling Fan #3 START bit is ON.

Timer Based Starting Control Logic

The PLC control program examined in this section is shown in Figure 10.10. The logic elements, input and output symbols, employed in this section of the overall PLC program are shown on Rung 11, in Figure 10.10. The logic represented in this rung is more complex than the logic associated with Rung #9 on Figure 10.8. The essence of this program rung is as follows:

If all of the requirements, conditions, or bits for the Examine if ON (labeled as XIC, Examine if Closed) and Examine if Off (XIO), in Rung 11, are **True** then Timer P105.TD_CASCADE_START will start timing. Notice the rectangular block on the right side of this rung. This block represents the timer. The timer setting is annotated as 2500. The set-point of 2500, in this case, implies 2500 milliseconds. In other words, the timer will time for 2.5 seconds before energizing the output. When 2.5 seconds expire, timer output labeled as P105.TD_ CASCADE_ START.DN becomes **True,** and P105_CONVEYOR_ CMD will turn ON starting P105 conveyor in this simple material handling conveyor system application.

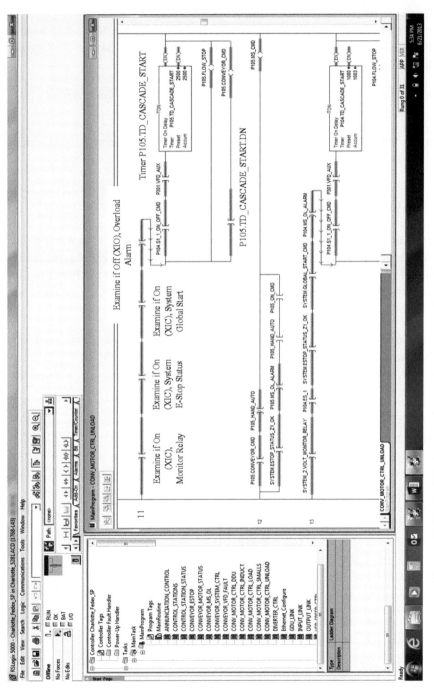

Figure 10.10: PLC Relay Ladder Logic Programming Example—Cascade start timer.

Example 10.2

Consider the logic associated with Timer P105.TD in Rung # 11 of the PLC relay ladder logic program shown in Figure 10.10 and answer the following questions:

a)   Which "register" of field on Rung #11 represents the *time elapsed* while the timer is timing?

b)   What is the status of the timer? How many seconds remain on the clock before the timer times out?

c)   What is the function of Rung #11 branch consisting of the "P105.TD_ CASCADE_START.DN" bit and the "P105 Conveyor _CMD" coil?

(a) Answer: As we examine the right side of Timer P105.TD in Rung # 11, we see the register or field labeled "Preset 2500." This register holds the timer preset time of 2500 milliseconds. The register directly below the Preset register is labeled "Accum 2500." This field can be interpreted as: Accumulated or elapsed time = 2500 milliseconds, or 2.5 seconds.

(b) Answer: As observed in part (a), the accumulated or elapsed time = 2500 milliseconds. Since the accumulated or elapsed time is equal to the Preset time, the timer has timed out, and "0" milliseconds remain.

(c) Answer: As we examine the right side of Timer P105.TD in Rung # 11, and observe the branch with two elements, the "P105.TD_CASCADE_ START. DN" bit and the "P105 Conveyor _CMD" coil, we note the fact that P105 Conveyor is commanded to turn ON when the P105.TD_CASCADE_ START. DN contact or bit is closed, or True. In other words, after the accumulated time reaches the preset time of 2500 milliseconds, the "**Done**" coil labeled as— **(DN)**—, turns ON or becomes **True**. This results in the closure (or conversion to **True** state) of the timer **Done** bit: "--| |--" 105.TD_ CASCADE_ START. DN, and with the three XIC and one XIO conditions satisfied (**True**), the P105 Conveyor _CMD" coil is energized to turn on P105 Conveyor.

Self-assessment Problems and Questions—Chapter 10

1.   Answer the following questions pertaining to the branch circuit shown in the schematic diagram below:
      (a) What is the maximum voltage the power distribution system for this branch circuit rated for?

(b) How many wires and phases is the power distribution system for this motor branch circuit rated for?

(c) What would be the proper rating for the branch circuit disconnect switch?

(d) What should the solid-state overload device be set for at commissioning of this branch circuit?

2. Determine the sizes/specifications of the following components in the branch circuit shown below using the Bus®Table and information included in the questions:

   (a) Conductor size.

   (b) Conduit size.

   (c) Overload setting based on the 115% NEC® stipulation.

   (d) Disconnect switch size, safety and fusible.

3. Consider the wiring diagram for the 75-hp motor shown in Figure 10.6 and answer the following questions based on the control logic explained in this chapter:

   a) What would be the likely outcome if the START switch is depressed while the motor is operating?

480 V Bus, 3 φ, 4W, 600 A, 60 Hz,

b) What would be the outcome if START and STOP switches are depressed simultaneously?

c) What would be the likely outcome if the main disconnect switch for the MCC is opened?

d) Can the motor be stopped if the motor starter latching contact "welds" shut due to overheating?

4. Consider the logic associated with Timer P105.TD in Rung # 11 of the PLC relay ladder logic program shown in Figure 10.10 and answer the following questions:

a) What would be status of the timer **Done** bit: "--| |--" 105.TD_ CAS-CADE_ START. DN, if the XIC System E-Stop bit turns "**False**" 1 second after Timer P105.TD in Rung # 11 is turned on?

b) Would the "P105 Conveyor _CMD" bit—commanding P105 Conveyor to turn ON be **True** if and when the system E-Stop switch is engaged/pressed?

Chapter 11

Electrical Power Rate Schedules And Electrical Energy Cost Savings Opportunities

Introduction

The purpose of this brief chapter is to introduce the reader to three pronged approach associated with comprehensive electricity cost reduction. This chapter allows readers an opportunity to enhance their knowledge about various electrical power rates or contracts. Better awareness of electrical energy cost structure allows a utilities engineer, plant engineer, or a facilities manager an opportunity to make calculated and informed decisions associated with electrical energy cost minimization. We will also explore some electrical cost reduction ideas. We will explore the concepts of EPC and ESCOs as a means for pursuing electrical energy intensity improvement projects with little or no capital investment.

THREE PRONGED APPROACH TO ELECTRICAL ENERGY COST REDUCTION*

Most energy programs—adopted for residential, commercial, or industrial settings—focus, mainly, on *energy conservation*. However, in order to develop a more comprehensive and complete energy program, one should view an electrical energy intensity improvement program as a three pronged approach. Those three prongs or pillars are as follows:

I. **Energy Conservation:** This facet consists of the application of high efficiency electrical equipment, i.e. premium efficiency electric motors, energy efficient lighting systems, etc.

II. **The Supply Side:** Exploration, evaluation and application of local electrical power generating systems that are cost competitive with the grid.

---

*Finance and Accounting for Energy Engineers by S. Bobby Rauf

III. **Energy Cost Rate:** This approach focuses, primarily, on exploration and financial evaluation of electrical power schedules and contracts that offer the lowest overall cost rates for commercial and industrial applications.

ELECTRIC UTILITY RATE SCHEDULES[*]

Measurement, verification and computation of utility bills are key elements in understanding, planning for and managing the consumption of various utilities. Conservation of utilities is an essential and strategic component of overall utility cost reduction effort. Its primary impact is through enhancement of utility productivity; or, in the strict energy realm, through maximization of energy productivity. The other vital determinants of the overall cost of utilities are contract and rate structure. In a given billing month, once the utility usage has been recorded, it must be processed using the rate schedule or structure the contract calls for.

The thrust of our discussion, in this chapter, will entail introduction of various rate schedules and the impact of major rate schedules on the composition of the electric bill in the industrial and residential realm. The residential rate structures and computation methods are relatively straightforward. Consequently, our discussion on residential electric bills will be brief. Some of the contract categories and rate schedules are as follows:

- Industrial
- Commercial
- Residential
- Municipal or Co-ops
- OPT, or Time of the Day
- HP, or Hourly Pricing
- Interruptible

Most utility companies, around the United States, offer rate schedules similar to the ones stated above. However, in many instances the names of the rate schedules may vary from one utility to another. Despite the apparent semantic differences among rate schedules offered by various utilities, the cost components, cost tiers and rate application mechanisms are similar. The essence and format of electrical rate schedules are being introduced in this text using Duke Energy® Company as an example.

---

[*]*Finance and Accounting for Energy Engineers* by S. Bobby Rauf

Duke® is one the largest electric power holding company in the United States, supplying and delivering energy to approximately 7.2 million U.S. customers. Duke® has approximately 57,700 megawatts of electric generating capacity in the Carolinas, the Midwest and Florida. Duke® is headquartered in Charlotte. Duke's® service area consists of approximately 104,000 square miles in the Southeast and Midwest.

Details on Duke® Energy electric power rates are available at the following Duke Energy® Website:

http://www.duke-energy.com/rates/north-carolina.asp

Some of the rates schedules offered by Duke® are listed in Table 11.1 below, with the more common ones highlighted.

Three of the most common schedules are explained in more detail below through excerpts from the Duke Energy® website. These schedules are also more pertinent to the topics introduced in this text. The reader is advised to regard the following excerpted rate schedule information as *reference information* intended to *introduce* the reader to typical electrical billing terminology and general format of the electrical power bills in the residential and industrial sectors. The rates quantified in the discussion that follows represent a "snapshot" at the time this text was authored, and only the current prevailing rates should be used in the practice of engineering, energy or facilities management.

Schedule RS (NC)

The RS stands for Residential Service. This specifics of this schedule, as published by Duke®, apply to the State of North Carolina, only.

This schedule and associated rates are available only to residential customers in residences, condominiums, mobile homes, or individually-metered apartments which provide independent and permanent facilities complete for living, sleeping, eating, cooking, and sanitation.

Type, or Specification, of (RS) Service

Under this schedule, Duke Energy®, agrees to furnish 60 Hertz service through one meter, at one delivery point, at one of the following approximate voltages, where available:

— Single-phase, 120/240 volts
— Or, 3-phase, 208Y/120 volts
— Or other available voltages at the Company's (Duke's®) option.

Table 11.1: Some Rate schedules offered by Duke® Energy

Rate Code	Description
RS	**Residential Service**
RE	Residential Service, Electric Water Heating and Space Conditioning
ES	Residential Service, Energy Star
RT	**Residential Service, Time-of-Use**
WC	Residential Service, Water Heating, Controlled/Submetered
SGS	Small General Service
OPT-G	**Optional Power Service, Time of Use General Service**
BC	General Service, Building Construction Service
LGS	**Large General Service**
FL	General Service, Floodlighting Service
OL	General Service, Outdoor Lighting Service
GL	Government Lighting Service
PL	General Service, Street and Public Lighting Service
NL	Nonstandard Lighting Service
TS	General Service, Traffic Signal Service
I	**Industrial Service**
OPT-E	Optional Power Service, time of Use, Energy-Only (Pilot)
OPT-H	**Optional Power Service, Time of Use, General Service, High Load Factor**
OPT-I	**Optional Power Service, Time of Use, Industrial Services**
HP	**Hourly Pricing for Incremental Load**
PG	Parallel Generation
REPS	Renewable Energy Portfolio Standard Rider
PM	Power Manager Load Control Service Rider
NM	Net Metering Rider (NM)
SCG	Small Customer Generator Rider (SCG)
IS	**Interruptible Power Service Rider**
SG	**Standby Generator Control Rider**
EC	Economic Development Rider
ER	Economic Redevelopment Rider
PSC	Power Share Call Option
PP-N	Purchased Power Non-Hydroelectric
PP-H	Purchased Power Hydroelectric

— Motors in excess of 2 H. P., frequently started, or arranged for automatic control, must be of a type to take the minimum starting current and must be equipped with controlling devices approved by the Company (Duke®).

Where three-phase and single-phase service is supplied through the same meter, it will be billed in accordance with the rates listed below. Where three-phase service is supplied through a separate meter, it will be billed on the applicable **General Service schedule**:

I.   Basic Facilities Charge per month: **$ 9.90**

II. Energy Charges

- For the billing months of July—October, for all kWh used per month: **9.2896¢** per kWh*

- For the billing months of November—June, for all kWh used per month: **9.2896¢** per kWh*.

Note that the two rates stated above may not always be the same.

*For complete details and exemptions, the reader is encouraged to visit Duke Energy® rate schedule website, referenced above.

Schedule OPT-I (NC)

This rate schedule represents an ***optional power industrial service*** that is based on time of use. Therefore, it is often referred to as **Time of Use Industrial Service.**

This rate schedule, offered by Duke Energy®, as specified in this text, is available in North Carolina Only. Furthermore, this rate schedule is available only to establishments classified as "Manufacturing Industries" by the Standard Industrial Classification Manual published by the United States Government, and where more than 50% of the electric energy consumption of such establishment is used for its ***manufacturing processes***.

The specifications of this rate schedule or service, as stipulated by Duke Energy®, are as follows:

- The Company (Duke Energy®) will furnish 60 Hertz service through one meter, at one delivery point, at one of the following approximate voltages, where available:

— Single-phase, 120/240 volts, 120/208 volts, 240/480 volts or other available single-phase voltages at the company's option; or

— 3-phase, 208Y/120 volts, 460Y/265 volts, 480Y/277 volts; or

— 3-phase, 3-wire, 240, 460, 480, 575, or 2300 volts; or

— 3-phase, 4160Y/2400, 12470Y/7200, or 24940Y/14400 volts

• Motors of less than 5 H.P. may be single-phase. All motors of more than 5 H.P. must be equipped with starting compensators. The Company (Duke Energy®) reserves the right, when in its opinion the installation would not be detrimental to the service of the Company, to permit other types of motors.

RATES UNDER OPT—I SCHEDULE

As evident below, due to the fact that this rate schedule is a function of time and seasons, rate structure under the OPT-I schedule is multi-dimensional and more complex. The rate (or charge) components of an OPT-I based electric power bill, at the time this text was authored, were as shown in Table 11.2.

Table 11.2: Time of Use, OPT-I Rate Schedule offered by Duke® Energy

RATE: I.	Basic Facilities Charge per month	$39.79
II. Demand Charge	**Summer Months**	**Winter Months**
A. On-Peak Demand Charge	**June 1 – September 30**	**October 1 – May 31**
For the first 2000 kW of Billing Demand per month, per kW	**$14.0767**	**$8.2981**
For the next 3000 kW of Billing Demand per month, per kW	**$12.8972**	**$7.1075**
For all over 5000 kW of Billing Demand per month, per kW	**$11.7067**	**$5.9064**
B. Economy Demand Charge	**$1.1448**	**$1.1448**

III. Energy Charges	All Months
A. All On-Peak Energy Per Month, Per kWh	**5.7847¢**
B. All Off-Peak Energy, Per Month, Per kWh	**3.4734¢**

On-Peak and Off-Peak Hours, for Schedule OPT-I, are classified as follows:

On-Peak Hours:
Summer Months
June 1-September 30, Monday-Friday
On-Peak Period Hours
1:00 p.m. -9:00 p.m.

Winter Months
October 1-May 31, Monday-Friday
On-Peak Period Hours
6:00 a.m.-1:00 p.m.

Off-Peak Hours:
- All weekday hours not included under On-Peak hours, and all Saturday and Sunday hours.
- All hours for the following holidays:
  — New Year's Day, Memorial Day, Good Friday, Independence Day, Labor Day, Thanksgiving Day, Day after Thanksgiving, and Christmas Day.

Determination of Billing Demand

A. The **On-Peak Billing Demand** each month shall be the largest of the following:
1. The maximum integrated thirty-minute demand during the applicable summer or winter on-peak period during the month for which the bill is rendered.
2. Fifty percent (50%) of the Contract Demand (or 50% of the On-Peak Contract Demand if such is specified in the contract)
3. 15 kilowatts (kW)

B. **Economy Demand:** To determine the Economy Demand, the larger of
1. The maximum integrated thirty-minute demand during the month for which the bill is rendered; or
2. 50% of the Contract Demand shall be compared to the On-Peak Billing Demand as determined in A. above.

If the demand determined by the larger of B.1 and B.2 above exceeds the On-Peak Billing Demand, the difference shall be the Economy Demand.

Power Factor Correction

When the average monthly power factor of the Customer's power requirements is *less than 85 percent*, the Company may correct the integrated demand in kilowatts for that month by multiplying by *85 percent* and dividing by the average power factor in percent for that month.

HOURLY PRICING OPTION/SCHEDULE

If your business is expecting incremental loads and has the ability to manage consumption patterns on a day ahead, hourly basis, Duke Energy® can offer electric energy cost savings through rate schedule HP, also referred to as *Hourly Pricing*. This opportunity is available to customers who have the ability to be responsive to fluctuations in hourly prices (use more energy when prices are low, use less energy when prices are high).

The HP schedule is available to non-residential establishments with a minimum contract demand of 1000 kW who qualify for service under the Duke Energy® rate schedules LGS, I, OPT-G, OPT-H, OPT-I, or PG, at the Company's (Duke Energy®) option on a voluntary basis. The maximum number of customers on the system to be served under this schedule is one hundred fifty (150). Service under this Schedule is available to contracting customer in a single enterprise, located entirely on a single, contiguous, premises.

AREAS OF OPPORTUNITY FOR ELECTRICAL ENERGY COST SAVINGS

There are several areas of potential cost savings available to electrical power consumers, beyond the ones discussed in this text, so far. The reader is encouraged to see organized and comprehensive approach to energy cost reduction, supported by illustrative case studies, in *Finance and Accounting for Energy Engineers, By S. Bobby Rauf*. Some areas of opportunity, related specifically to electrical rate schedules, are listed below:

Schedule or Rider SG, On-Site Generation

This program is designed to offer large industrial or commercial consumers an opportunity take advantage of on-site emergency power generation assets they already own. Most emergency power generators have to be tested, under load, periodically, for predictive maintenance, preventive maintenance and personnel training purposes. Why not get compensated or credited for this routine, but necessary, exercise by the power company? That is where Schedule or Rider SG offers the larger consumers an opportunity to do so.

Therefore, if a large consumer owns emergency standby generators and can make their capacity available for use by Duke Energy® during times of system emergencies, monthly credits are offered through rate rider SG.

Schedule or Rider IS, Interruptible Power Service

The interruptible power service schedule, IS, is available for nonresidential customers receiving concurrent service from the Company (Duke Energy®) on Schedules LGS, I, HP, OPT-G, OPT-H or OPT-I, served under continually effectively agreements for this Rider made prior February 26, 2009. Under this schedule, the customer agrees, at the Company's (Duke Energy's®) request, to reduce and maintain his load at a level specified in the individual contract. The Company's request to interrupt service may be initiated at any time the Company has capacity problems.

Contracts for interruptible power service are accepted by Duke Energy on the basis of successive contracts, and each contract must specify an interruptible, integrated demand of not more than 50,000 KW to be subject to these provisions. The Company can limit the acceptance of contracts to a total of 1,100,000 KW of Interruptible Contract Demand on all non-residential schedules on the total system.

Duke Energy® reserves the right to test the provisions of this Rider once per year if there has not been an occasion during the previous 12 months when the Company requested an interruption. Duke Energy® gives advance notice of any test to customers served under this Rider. In return for participation in this schedule, Duke issues credits to the participating customers based on a specific formula that can be reviewed at Duke Energy® Website.

Tips on Utility Rate Schedules and Contracts

Its, ultimately, the consumer's responsibility to maintain vigilance and awareness of opportunities and incentives available through the power company that serves the region. The following tips can prove to be useful to the consumer for minimizing the *cost* of electrical energy:

1. Electric Utility Company is not obligated to notify facilities about availability of more favorable rates or "schedules." Some utility companies, for example, Duke Energy®, assign Account Managers to certain segments of their market. In order to maximize the utility of the consumer's relationship with the assigned Account Manager, the consumer's representative, namely the energy or utilities engineer, must note the following:
   — Utility company Account Managers will, at times, advise their larger accounts of favorable rate schedules as the customers' demand and usage changes. However, it is good practice to explore suitable contract alternatives, when some of the following changes are experienced in load characteristics and profile:

— Addition or removal of loads that constitute a substantial percentage of facility's overall load.
— Addition of highly reactive loads, i.e. large motors, transformers. This might impact the facility's overall power factor and energy consumption.

2. Look for newfound **flexibility in load schedules**. For example "Off-Peak" unloading of rail cars, in industries where mass transport of raw materials is required.

3. Look for addition of **onsite generation**; Standby or Cogeneration assets.

4. Be vigilant of any change in facility's operation constraints. For instance:
    — Could the facility tolerate power interruption, with some advance notice?
    — Could facility participate in, on-line, diesel generator testing.

5. **Renewable Energy Portfolio Standard (REPS) Adjustment**:
    The monthly bill shall include a REPS Adjustment based upon the revenue classification:

    Commercial/Governmental Classification—**$3.22/month**

    Industrial/Public Authority Classification—**$32.20/month**

    Upon written request, only one REPS Adjustment shall apply to each premise serving the same customer for all accounts of the same revenue classification. If a customer has accounts which serve in an auxiliary role to a main account on the same premise, no REPS charge should apply to the auxiliary accounts regardless of their revenue classification (see Annual Billing Adjustments Rider BA).

6. **Transformation Discounts:** When Customer owns the step-down transformation and all other facilities beyond the transformation which Company (Duke Energy) would normally own, except Company's metering equipment, the charge per kW of on-peak billing demand will be reduced as follows:

    Transmission Service: $0.75/kW

    Distribution Service: $0.5/kW

For customer to qualify for the Transmission Service Transformation Discount, customer must own the step-down transformation and all other facilities beyond the transformation which Company (Duke Energy®) would normally own, except Company's metering equipment, necessary to take service at the voltage of the 69 kV, 115 kV, or 230 kV transmission line from which Customer received service.

ENERGY PERFORMANCE CONTRACTING AND ESCO OPPORTUNITIES*

EPC, energy performance contracting, is a service vehicle for provision of energy conservation or energy productivity enhancement services. In most cases, EPC type contracts and projects involve turnkey service. Turnkey service is defined as a comprehensive service provided by a vendor or contractor that begins with definition and scope of an energy project and ends with project start up, commissioning and subsequent verification of energy savings. Projects covered by EPC range from simple energy conservation efforts such as replacement of inefficient lighting systems to more complex projects that address the supply side of the energy equation through renewable energy systems.

While EPC and ESCO alternatives are often adopted for substantial energy conservation or renewable energy projects, initiatives that leverage lower cost power contracts can be included in the overall energy project portfolio. EPC projects and contracts often include guarantees that the savings produced by a project will be sufficient to finance the full cost of the project.

The guarantee to fund the project through the savings generated by the project is what distinguishes **EPC** projects from **non-EPC**, or owner funded energy projects. Even though it appears counter intuitive, often EPC projects are not limited to energy conservation or energy capacity enhancement projects, instead, the breadth of their scope includes water conservation, sustainable materials and operations.

ESCO stands for Energy Service Company. One way to understand the relationship and distinction between EPC and ESCOs is to think of EPC, Energy Performance Contracting, as a *process* and the ESCOs as *entities that implement the EPC process*. ESCOs can provide a full range of services required to complete an energy project. Such services, often, include the following or a combination, thereof:

- Energy audit
- Design engineering

---

*Finance and Accounting for Energy Engineers by S. Bobby Rauf

- Construction management and supervision
- Facilitation or provision of project financing
- Start-up and commissioning
- Operations and maintenance
- Monitoring and verification of energy savings

Historically, the inception of EPC could be traced as far back as the early eighties. In the pre-1985 era, ESCOs were established, as a part of the DSM, Demand Side Management, efforts to provide personnel and equipment resources to the utilities as they strived to meet the energy conservation mandates imposed by federal and state governments.

From the mid 1980s through 2003, ESCOs, and EPC industry as a whole, have seen substantial ebb and flow in growth, acceptability and revenue. Over this period, some of the ESCOs have transformed and some have grown either organically or accretively, through consolidation. This evolution within the EPC domain was influenced, favorably, by the state and federal governments. The successes of the EPC industry, in the 1994-2002 period could, to a certain extent, be attributed, to studies by LBNL, Lawrence Berkley National Laboratory, and NAESCO, National Association of Energy Service Companies, that highlighted the EPC successes and encouraged the state and federal governments to promote EPC. Another important event that could be credited for EPC growth and successes in the 1994-2002 period was the formulation and implementation of the IPMVP, or International Performance Measurement and Verification Protocol. The IPMVP provided standard methods for documenting project savings and provide commercial lenders the confidence to finance EPC projects.

As plausible, the EPC industry was impacted unfavorably by the ENRON debacle, as ENRON was a significant player in the ESCO market. The ENRON collapse coupled with the uncertainty about the deregulation of the electric utility industry impeded the growth of EPC in the 2002 to 2004 period.

The 2004-2006 period showed 20% growth in the EPC industry with comparable projected growth trend. The 20% growth, and subsequent upward trend can be attributed to volatility in the energy market and the increasing energy prices. Other contributors to the heightened interest in the EPC are state and federal mandates, inadequate capital and maintenance budgets for federal and state facilities. Growing awareness of the greenhouse gas emissions and realization of the fact that large scale, sustained, remediation is needed in this regard has made EPC more attractive at local, state and federal levels, and in the private sector.

While ESCOs, in response to customer requests, are constantly adding new measures and services to their project portfolios, electrical and non-electrical, they are not to be construed as stewards for technological research, development and marketing in the energy domain. ESCOs and their clients tend to be fairly conservative and risk averse in selection of technologies for projects. Due to the fact that the cost of most ESCO projects are paid from energy savings, often secured with financial guarantees, ESCOs and their clients tend to lean in favor of proven technologies.

Energy performance contracting involves distinct skills and areas of expertise in the following two areas:

— Energy procurement
— Commercial law

Both of these disciplines involve risk management, risk allocation, isolating benefits, and option analysis. The following list could serve as a checklist for energy managers, utilities engineers and facilities managers as they consider EPC and ESCOs in formulation of the energy program strategy:

✓ EPC, or Energy Performance Contracting is one way to finance and implement energy conservation projects.

✓ Remuneration sought by the EPCs is typically included in the overall cost of the project.

✓ Initial investment, maintenance cost, energy cost (over the life of the project) monitoring and training cost, are—and should be—included in the overall cost of the project.

✓ The energy and cost savings produced by the project need to be sufficient to cover all project costs over the term of the contract.

✓ EPC project contracts, typically, span a period of 10 years or more.

Some of the benefits associated with the EPC and ESCO approach for implementation Energy Programs and Strategies are as follows:

☐ EPCs save company capital for projects that lack financial justification.

☐   EPCs fund energy conservation projects from savings generated by the project.

☐   EPCs reduce repair and maintenance costs caused by inadequate, aging, or obsolete equipment, electrical or non-electrical.

☐   EPCs, and energy conservation projects in general, provide secondary benefits i.e. increased employee productivity, safe and more comfortable working environment. For example, energy efficient lighting projects/programs, computer/PLC based automated HVAC systems and computer/PLC based EMS, Energy Management Systems.

☐   Improve the environment and conserve energy resources.

ELECTRICAL ENERGY RELATED MEASURES TYPICALLY INCLUDED IN ESCO OR EPC ENDEAVORS

Additional details on some of the mainstream technical measures, typically included in ESCO or EPC driven projects, are recounted for reference below:

•   Lighting: Replacement of inefficient lighting systems with energy efficient lamps, energy efficient ballasts and optimally designed light fixtures. Examples of such measures include:
    —   Replacement of mercury lamps with higher efficacy sodium vapor lamps
    —   Replacement of incandescent and, in some cases, florescent lamps with high efficacy LED, Light Emitting Diode, lamps. An LED is constructed similar to a regular diode with the exception of the fact that when an LED is forward biased, holes and the electrons combine at the p-n junction releasing light energy, or photons. Symbol for a light emitting diode is shown below:

LED

    —   Replacement of older florescent lamps and fixtures with high efficiency T-8 or T-5 florescent light systems.
    —   Replacement of lighting systems, that are inadequately designed by

today's standards, with light systems that are designed with emphasis on important factors such as lighting efficacy (lumens/watt), CU, coefficient of utilization (Lumens Reaching the Work Plane/Total Lumens Generated). See Chapter 12.

• HVAC, heating, air conditioning and ventilation: Optimization and improvement of chilled water systems, replacement of lower efficiency and high maintenance HVAC Systems with HVAC systems that carry high Energy Star Rating, utilize high efficiency chillers, use green technologies, i.e. geothermal, solar, thermal storage, etc. Convert manual HVAC control systems to BMCS, Building Management Control Systems, or Direct Digital Control Systems*.

• Control systems: Control systems incorporating effectively designed and optimally applied control architecture in energy usage and energy generating systems. These control systems employ cutting edge—yet proven—sensors, transducers and other control devices for field application. Furthermore, these control systems are driven by CPUs, Central Processing Units, or computers and PLCs, Programmable Logic Controllers that offer the latest improvements in hardware, firmware, application software and HMI, Human Machine Interface, options.

• Building Envelope Improvements: Measures in this category include infrastructure improvements related to the building envelope or exterior.

• Cogeneration and CHP: Cogeneration and CHP, Combined Heat and Power, measures address the supply side of the energy equation, through production of electricity while catering to other needs such as steam required for production processes. Measures in this category can include the following:
  — Topping Cycle Cogeneration System: In topping cycle cogenerating systems, electrical power is generated at the top, or beginning, of the cogeneration cycle.
  — Bottoming Cycle Cogeneration System: In bottoming cycle cogenerating systems, electrical power is generated at the bottom, or tail end, of the cogeneration cycle.
  — Combined Cycle Cogeneration System: In the combined cycle co-

---

Thermodynamics Made Simple for Energy Engineers, by S. Bobby Rauf generating systems, electrical power is generated at the top and bottom segment of the cogeneration cycle.

Since the combined cycle systems employ both the topping cycle feature as well as the bottoming cycle feature, they offer higher efficiency in the production of electricity.

•   Demand Response Measures: Demand response measures are projects or actions undertaken to avert the need for electrical power generating capacity expansion. Demand response measures are also referred to as DSM, or Demand Side Management, measures. DSM is an important tool to help balance supply and demand in electricity markets, to reduce price volatility, to increase system reliability and security. This enables the utility industry to rationalize investment in electricity supply infrastructure and to reduce greenhouse gas emissions. Examples of these measures include the following:

   —   Energy efficiency enhancement technologies, management practices or other strategies in residential, commercial, institutional or governmental arena that reduce electricity consumption.

   —   Demand response or load management technologies, management practices or other strategies in residential, commercial, industrial, institutional and governmental arena that shift electric load from periods of peak demand to periods of off-peak demand, including pump storage technologies.

   —   Industrial by-product technologies consisting of the use of a by-product from an industrial process, including the reuse of energy from exhaust gases, steam or other manufacturing by-products that are used in the direct production of electricity.

Figure 11.1 is Quantity (Q)—Price (P) graph. This graph shows the effect of *demand response* on *demand elasticity*. The inelastic demand in the electrical power market place is represented by curve **D1**. The high price **P1** associated with the *inelastic* demand **D1** is extrapolated off the point of intersection of the supply curve **S** and the demand curve **D1**. When demand response measures are introduced, demand becomes *elastic*. The elastic demand is represented by curve **D2**. The point of intersection of elastic demand curve **D2** and the supply curve **S** precipitates in a substantially **reduced market price P2**.

It is estimated that a 5% lowering of the demand would result in a 50% reduction in price during peak hours, as demonstrated in the California Electricity Crisis of 2000-2001*.

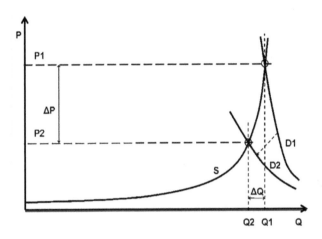

Figure 11.1. Demand elasticity and the effect of demand response.

Other studies, such as the two studies sponsored by Carnegie Mellon in 2006†, examined the impact of demand response measures on demand elasticity and price.

The price reduction can be explained by the fact that operators generally plan to use the least expensive, or lowest marginal cost, generating capacity first, and use additional capacity from more expensive plants as demand increases.

Renewable Energy Measures: Renewable energy is, mostly, derived from natural resources such as sunlight, wind, rain, tides, and geothermal heat; these are energy sources, or forms of energy, that are renewable and replenished naturally. Of course, the energy harnessed from these renewable natural sources is often in form of **electrical energy**.

---

*The Power to Choose—Enhancing Demand Response in Liberalised Electricity Markets Findings of IEA Demand Response Project, Presentation 2003.

†CEIC-07-01 "Demand Response and Electricity Market Efficiency," Kathleen Spees and Lester Lave. CEIC-07-02 "Impacts of Responsive Load in PJM: Load Shifting and Real Time Pricing" Kathleen Spees and Lester Lave.

Some of the more proven renewable energy technologies are as follows:

- **Hydroelectric Power**

- **Biomass Energy**
  — Biomass Heat
  — Biomass Power

- **Solar Energy**
  — Solar Heat Energy
  — Solar Photovoltaic Electrical Energy

- **Wind Energy**

- **Geothermal Energy**
  — Geothermal Heat
  — Geothermal Power

- **Ocean or Tidal Energy**

Published data show that approximately 21% of the total US electrical energy is derived from renewable sources. This 21% of the overall electricity generated in the US can be segmented as shown in Figure 11.3. This segmentation shows that:

  — Almost 80 GW, Gigawatts, of renewable electrical energy is derived from *conventional hydroelectric dams*.

  — Almost 40 GW, Gigawatts, of renewable electrical energy is derived from *wind turbines*.

  — *Solar* comes in 3rd, at almost 10 GW

  — The rest of renewable electrical energy is derived from *biomass and geothermal sources.*

Water and Sewer: Consumption of water and discharge into sewage systems, at face value, may not appear to have a direct relationship with the electrical energy. However, as one examines the process behind the reclamation and

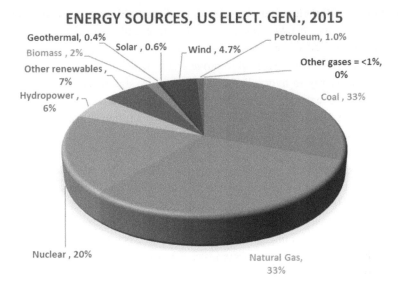

Figure 11.2: Sources of U.S. Electricity Generation. Based on 2015 data published by EIA, U.S. Energy Information Administration

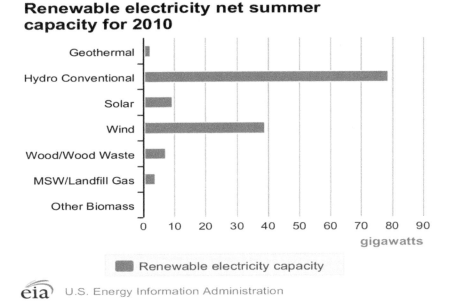

Figure 11.3: Renewable electricity generation breakdown—U.S.

distribution of potable water, the vital and indispensable role that energy plays becomes palpable. It takes energy, and therefore dollars, to drive the pumps to collect untreated water. The filtration and sanitization phases of water purification require hydraulic head to move water, which amounts to conversion of electrical energy to hydraulic head. In addition, significant and continuous amount of electrical energy is needed to store and pump water to consumers at a standard head. Hence, it's obvious that if the consumption of water is reduced, energy is conserved and the demand for electrical energy abates.

Sustainable Materials and Associated Operations: Mining, transportation, refinement, treatment, sizing, packaging, marketing, sale, distribution and application of all types of materials—pure metals, alloys, non-metallic substances, polymers and myriad synthetic substances—require energy in all phases of production. Even recycling of materials, as "green" as it is, requires energy. As we assume this perspective, it becomes easy to see how conservation in the use of materials can have an impact on energy demand.

Self-assessment Problems and Questions—Chapter 11

1.   HP, or hourly pricing, program is a standard feature in all OPT, or Time of Use, schedules.
      A. True
      B. False

2.   The energy charge rate structure with Electrical OPT schedule is:
      A. Flat, year round
      B. Tiered
      C. Exponential
      D. Is a function of time and season.

3.   A large industrial electricity consumer set the peak demand in July's billing month at 40 megawatt. The demand rate structure is same as that included in Duke Energy® OPT-I, Time of Use, rate schedule, as shown in Table 11.2. Determine the demand cost for the month.
      A. $367,000
      B. $505,000

    C. $407,000

    D. $476,579

4.   Calculate the ***energy charge*** for the month of July considered in Problem 3 if all of the energy is consumed during On-Peak hours, for 10 hours per day. Assume that there are 30 days in the billing month and that the load factor is 1, or 100%.

    A. $902,000

    B. $416,808

    C. $2,064,187

    D. None of the above

5.   Which of the following statements describes the role of EPC and ESCO most accurately?

    A. The terms EPC and ESCO are synonymous

    B. EPC is method for implementing energy projects and ESCOs are entities that offer this alternative.

    C. EPC is required by Department of Energy, ESCOs are not.

    D. None of the above

Chapter 12

Illumination and Lighting System Design

Introduction

Electricity is the lifeline of most non-solar, artificial, light produced and used by mankind. Substantial portion of electricity consumed in the world can be attributed to the generation of light. Most electrical energy conservation programs include lighting improvement as a cornerstone of the overall, comprehensive energy conservation effort. This is, especially, true when a program is evaluated with consideration to the ROI, return on investment, payback period and NPV, Net Present Value Analysis*. Many large energy productivity improvement project portfolios depend on short payback, high ROI and high NPV lighting improvement measures to "float" the more capital intensive measures like energy efficient chillers and energy efficient air compressors. Therefore, understanding of illumination and efficacies of various lighting systems is an important and integral part of electrical engineering.

In this chapter, we will examine fundamental concepts of illumination. We will gain familiarization with illumination and lighting principles, and we will learn basic considerations pertaining to lighting system design.

In the interest of facilitating continuous learning and updating of reader's knowledge on the subject of lighting, it should be noted that IESNA, Illumination Engineering Society of North America, provides a vital leadership role in the illumination arena. IESNA is a non-profit organization which develops and publishes lighting standards. IESNA also collaborates with ICI, International Commission on Illumination to promote uniformity of illumination standards worldwide. The reader is encouraged to stay abreast of IESNA publications to stay current on the evolving illumination standards.

---

*Finance and accounting for energy engineers, by S. Bobby Rauf.

LIGHTING TERMS, CONCEPTS AND STANDARDS

Solid angle: A solid angle is an angle represented by a given area on the surface of a sphere. It can also be defined as a two-dimensional angle subtended, at a point, by an object, in three dimensional space. Solid angle is denoted by symbol ω. The units for solid angle are steradians, **sr.**

Luminous flux: Luminous flux is a measure of the amount of light that passes through a certain area of cross-section ΔA. The symbol used to denote luminous flux is Φ. The units for luminous flux are **lumens**. One lumen is defined as the amount of luminous flux produced by a light source when it emits **one candela** of luminous intensity over a solid angle of **one steradian**.

A lumen is an **SI** unit. As apparent from Table 12.1, luminous flux can also be measured in watts per steradian, or **w/sr**; which implies that luminous intensity can be quantified in watts.

The formula for Luminous Flux Φ is:

$$\Phi = 4\pi . I \qquad \qquad \textbf{Eq. 12.1}$$

Where,

I = Luminous intensity from a uniformly radiating point source
Φ_t = Total luminous flux originating from the source
4π = Solid angle in steradians

Luminous intensity: Luminous intensity is a measure of the source intensity as seen by the eye. The symbol used to denote luminous flux is I. If light is perceived as emanating out, three dimensionally, from a point source, luminous intensity could be conceived as flow of the light radiation on per unit solid angle basis.

The formula for luminous intensity is as follows:

$$I = \frac{d\Phi}{d\omega}$$

$$I = \frac{\Phi}{\omega} = \frac{\Phi}{\dfrac{A}{r^2}} = \frac{\Phi_t r^2}{4\pi r^2} = \frac{\Phi_t}{4\pi} \qquad \qquad \textbf{Eq. 12.2}$$

Where,

I = Luminous intensity from a uniformly radiating point source
Φ = Luminous flux, in lumens
Φ_t = Total luminous flux originating from the source

$$\omega \quad = \quad \text{Solid angle in steradians (sr)} = \frac{A}{r^2} \qquad \textbf{Eq. 12.3}$$

A = Area illuminated, in ft^2
r = Distance of the work plane from the light source, in ft

Luminous intensity is measured in **candelas**. One candela can also be defined, mathematically, as follows:

$$1 \text{ cd} \quad \frac{1 \text{ lm}}{\text{sr}}$$

Illuminance: Illuminance defines the *quantity* of light that falls on a given *surface area*. The area can be real area —pertaining to a real surface —or it can be a virtual area in space. Illuminance is denoted by **E**. The formula for computing illuminance is as follows:

$$E = \frac{\Phi_t}{A} = \frac{\Phi_t}{4\pi r^2} \qquad \textbf{Eq. 12.4}$$

Where,

Φ_t = Total luminous flux originating from an omnidirectional light source
A = Spherical area illuminated by the omnidirectional light source
r = Distance the light source and the illuminated spherical area
E = Illuminance

The units for illuminance, in the SI realm are "**lux**." The units for illuminance, in the US realm are **ft-c**, or foot candles.

One lux can be defined as:

$$1 \text{ lux} = \frac{\text{Number of lumens}}{\text{m}^2} \quad \text{or} \quad \frac{\text{Lm}}{\text{m}^2}$$

One ft-c can be defined as:

$$1 \text{ fl-c} = \frac{\text{Number of lumens}}{\text{ft}^2} \quad \text{or} \quad \frac{\text{lm}}{\text{ft}^2}$$

Illuminance is also referred to as ***illumination*, *brightness*** or, ***irradiance*,** and as such, it could be quantified in terms of **Watts/m²**.

While the more common illumination entities are introduced and discussed in this chapter, Table 12.1 shows symbols, units and conversion factors for many common and uncommon illumination entities.

Example 12.1

A two (2) lumen omnidirectional light source is located 5 meters from a 1-meter-diameter target as shown in the diagram below. Determine the following illumination parameters for this scenario:

a)　　Luminous intensity of the source.
b)　　Luminous flux emitted by the source and received by the target
c)　　Illuminance at the given distance of 5 meters.

Table 12.1: Symbols, units and conversion factors in illumination

Symbols, Units and Conversion Factors for Common Illumination Entities				
Entity/Quantity	Symbol	Metric/SI Units	Alternative Units	Explanation
Illuminance	**E**	lux	lm/m^2	lm (lumens)
Irradiance	**E**	W/m^2	-	W (Watts)
Radiometric Illuminance	$\mathbf{E_e}$	W/cm^2	-	W (Watts)
Photometric Illuminance	$\mathbf{E_v}$	Phot	lm/cm^2	Phot = Photons
Luminous Intensity	**I**	cd (candela)	Lumens/sr	sr (steradians)
Luminous Efficacy	**K**	lm/watt	-	Lumens/watt
Luminance	**L**	cd/m^2	-	Candela per unit area
Radiance	**L**	$W/sr.m^2$	-	watts/steradians/meter²
Luminance Existence	**M**	lm/m^2	-	Lumens/sq. meter
Luminance Flux Density	**M**		lm/ft^2	Lumens/sq. foot
Luminous Flux	**φ**	cd (candela)	Watt/sr	watts/steradians
Radiant Intensity	**I**	cd (candela)	Watt/sr	watts/steradians
Transmittance	**T**	-	-	ϕ_T/ϕ_i
Radiant Energy Density	**ω**	J/m^3		joules/cubic meters
Luminous Efficacy	**V**	-	-	V = Joules/mass

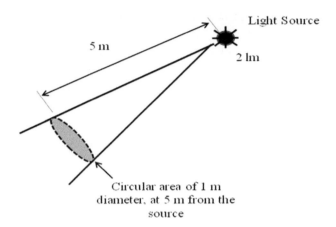

Light Source

5 m

2 lm

Circular area of 1 m
diameter, at 5 m from the
source

Solution

a)   Luminous intensity of the source.

The formula for luminous intensity is as follows:

$$I = \frac{\Phi}{\omega}$$

And, ω = solid angle, in steridians = $\dfrac{A}{r^2}$

Where,

    A = Area illuminated, in m²
         = π.(radius of illuminated circle)²
         = π. (1/2 m)² = **0.785 m²**
    r = Distance of the work plane from the light source = **5 m**
    I = Luminous intensity from a uniformly radiating point source
    Φ = Luminous flux at the source, in lumens

Therefore,

$$\omega = \frac{A}{r^2} = \frac{0.785 \text{m}^2}{5 \text{m}^2} = 0.0314 \text{ sr}$$

And,

$$I = \frac{\Phi}{\omega} = \frac{2}{0.0314} = 6369 \text{ lm/sr}$$

b)   Luminous flux emitted by the source and received by the target

The formula for Luminous Flux Φ is:

$$\Phi = 4.\pi.I$$

Where,
   I = Luminous intensity as calculated above = 63.69 lm/sr
   4π = Solid angle in steradians

   $\Phi = 4.\pi \text{ sr.}(I) = 4.\pi \text{ sr.}(63.69 \text{ lm/sr}) = 800 \text{ lm}$

c)   Illuminance at the given distance of 5 meters.

$$E = \frac{\Phi_t}{A} = \frac{\Phi_t}{4\pi r^2} = \frac{2}{4.(3.14).(5m)^2} = 0.0064 \text{ lm}$$

Where,
   Φ_t = Total luminous flux originating from an omnidirectional light source
   A = Spherical area illuminated by the omnidirectional light source
   r = Distance between the light source and the illuminated spherical area

LUMINOUS EFFICACY OR LUMINOUS EFFICIENCY

Luminous efficacy is also referred to as luminous efficiency or lamp efficacy. Luminous efficacy is defined as the ratio of luminous flux of a light source
to the amount of power required by the light source. Since the units for luminous flux are lumens and the electrical power is measured in watts, the units for
luminous efficacy are lm/W. Luminous efficacy or efficiency is denoted by the
symbols "η" or "K," and is defined, mathematically, as:

$$\eta = K = \frac{\Phi}{P}$$

Where,
   Φ = Flux, in lumens
   P = Electrical power demanded by the light source or lamp

Luminous efficacy is used extensively to compare energy productivity*
of various light sources. Luminous efficacies of some common light sources or
lamps are listed in Table 12.2.

Table 12.2: Efficacy values for common light sources

Source	Power	Efficacy
	Watts	lm/W
High Pressure Sodium Vapor Lamp	400/1000	125/130
Low Pressure Sodium Vapor Lamp	180	180.00
Mercury Fluorescent Lamp	400	58.00
Mercury free Fluorescent Lamp	40	80.00
Mercury Vapor Lamp	100/1000	36/65
Metal Halide	400/1000	85/100
Tungsten Lamp (Incandescent)	25/100/1000	10/16/22

The significance of luminous efficacy and its role in the design of lighting
systems is illustrated through Example 12.2

Example 12.2:
A 40 watt fluorescent lamp, with a rated luminous efficiency factor (**K**) of
80 lm/W, is mounted to the ceiling of an office as shown in the diagram below.
The purpose of this light fixture is to provide adequate lighting for work being
performed on the desk. Assuming that the interior design of the office allows
the lamp to illuminate hemispherically, determine the illumination level on the
desk in the office.

15 ft
(4.57m)

2.33 ft
(0.71m)

*Finance and Accounting for Energy Engineers, by S. Bobby Rauf

Solution

Illumination level and illuminance, **E**, are synonymous. As stated earlier in this chapter, Illumination level can be defined, mathematically, as follows:

$$E = \frac{\Phi}{A} \qquad\qquad \textbf{Eq. 12.4}$$

Where,

Φ = Luminous flux

A = Spherical area illuminated, in ft² or m²

In this case,

Φ = Luminous flux = K (lm/W).(Power Rating of the Lamp in Watts)

= 80 (lm/W).(40 Watts)

= 3200 lm

Since the area over which the light is radiated is "hemispherical' and not spherical:

$$A = \frac{1}{2} .(\text{Area of a Sphere}) = \frac{1}{2} .(4\pi r^2) = 2\pi r^2$$

$$E = \frac{\Phi}{2\pi r^2}$$

Where,

r = Radius of the hemisphere

= Height of the lamp from the top of the desk = 15 ft or 4.57 m

Therefore, illumination level in foot-candles (US Unit System) would be:

$$E_{ft-c} = \frac{\Phi_{Lumens}}{2\pi r^2} = \frac{\Phi_{Lumens}}{2\pi(15\,ft)^2} = \frac{3200\,lm}{2\pi(15\,ft)^2}$$

$$= 2.264\ \text{lm/ft}^2$$

$$= 2.264\ \text{ft-c}$$

And, the illumination level in lux (SI Unit System) would be:

$$E_{Lux} = \frac{\Phi_{Lumens}}{2\pi r^2} = \frac{\Phi_{Lumens}}{2\pi(4.572m)^2} = \frac{3200\,lm}{2\pi(4.572m)^2}$$

$$= 24.36\ \text{lm/m}^2$$

$$= 24.36\ \text{lx}$$

Example 12.3:

A special application lamp is rated 500 W. Using a photometer, the lamp's light output is measured and recorded to be 8000 lumens, at the lamp. Determine the lamp's luminous efficacy, K.

Solution

Given:

P = 500 W

Φ = 8000 lumens

$$\eta = \text{Lumanious Efficacy} = K = \frac{\Phi}{P} = \frac{8000 \text{ lm}}{500 \text{ W}}$$

$$= 16 \ \frac{\text{lm}}{\text{W}}$$

IESNA RECOMMENDED ILLUMINATION LEVELS AND LIGHTING EFFICACY COMPARISON

IESNA, Illumination Engineering Society of North America has made numerous significant contributions as a non-profit lighting standards, guidelines and recommendations development body. Some of those contributions are as follows:

1) Recommended illuminance levels
2) Lighting design
3) Efficacy comparison between various lamps, or light sources

Recommended Illuminance or Illumination Level

Comprehensive illuminance guidelines for common indoor venues and specific tasks are shown in Table 12.3. These indoor venues, tasks and illuminance ranges are classified into categories A, B, C, D, E, F, G, H, and I. The range of illuminance levels, allocated to each category is stated in lux as well as fc (foot-candles). The illuminance levels range from low, moderate to high.

An illuminance level is selected for a specific activity or task based on an aggregate weighting factor derived from a combination of age of occupants, specific activity, pace or frequency, venue, level of importance, and background or surface reflectance. The weighting factors for these attributes are listed in Tables 12.4 and 12.5.

Table 12.3: IESNA illumination guidelines

IESNA Illuminance Categories and Values for Generic Indoor Activities			
Activity	**Category**	**lux**	**fc**
Public spaces with dark surroundings	A	20 - 30 - 50	2 - 3 - 5
Simple orientation for short tasks/visits	B	50 - 75 - 100	5 - 7.5 - 10
Working spaces with occasional visual tasks	C	100 - 150 - 200	10 - 15 - 20
Visual tasks involving high contrast or large size	D	200 - 300 - 500	20 - 30 - 50
Visual tasks involving medium contrast or small size	E	500 - 750 - 1000	50 - 75 - 100
Visual tasks involving low contrast or very small size	F	1000 - 1500 - 2000	100 - 150 - 200
Visual tasks involving low contrast or very small size, over a prolonged period	G	2000 - 3000 - 5000	200 - 300 - 500
Performance of very prolonged exacting visual tasks	H	5000 - 7500 - 10000	500 - 750 - 1000
Performance of very special visual tasks of extremely low contrast	I	10000 - 15000 - 20000	1000 - 1500 - 2000

Table 12.4 lists the weighting factors for categories A through C. If the nature of the task or activity falls within categories A through C, age of the subjects (or occupants) is used in combination with reflectance level (%) in the room to determine the aggregate weighting factor. If the aggregate weighting factor is +2, High illuminance number is selected. If the aggregate weighting factor is -2, Low illuminance number is selected. For aggregate scores of -1, 0 and +1, middle illuminance value is selected. This method for determination of recommended illuminance level based on age and reflectance factors is illustrated through example 12.4.

Example 12.4:

The lobby of a government building is visited by individuals with a mean age of 59. The reflectance in the lobby is, approximately, 75%. These individuals sign a log and collect supplies once each day. Determine illuminance level the lobby's lighting system should be designed for.

Table 12.4: IESNA illumination level weighting guidelines, categories A through C.

For Categories A through C			
Room & Occupant's		**Weighting Factor**	
Characteristics	**-1**	**0**	**1**
Age	Under 40	40 - 55	Over 55
Average Room Surface Reflectances	>70%	30% - 70%	<30%

Solution

Based on the problem statement, this activity would be classified under **Category C** on Table 12.3, i.e. "working spaces with occasional visual tasks."

• From Table 12.4, the weighting factor associated with age 56 and over is +1.

• The weighting factor for reflectance of 70% or greater is -1.

• Therefore, the total weighting factor for this scenario would be = +1 + (-1) = 0.

According to Table 12.1 and the rules stated earlier, the aggregate weighting factor of "0" would point to the use of middle lux or lumen value under **Category-C.** The illuminance range in this category is 100 – 150 – 200 lux or 10 – 15 – 20 fc.

Therefore, the recommended illuminance would be **150 lux or 15 fc.**

If the nature of a task or activity falls in categories D through I, Table 12.5 is used to assess the weighting factor. When a situation is classified in categories D through I, the age of the subjects (or occupants) is used —in combination with speed of activity, desired accuracy of the task, and reflectance level (%) —to determine the aggregate weighting factor. If the aggregate weighting factor is +2 or +3, high illuminance number is selected from Table 12.3. If the aggregate weighting factor is -2 or -3, Low illuminance number is selected. For aggregate scores of -1, 0 and +1, middle illuminance value is selected. This method for determination of recommended illuminance level based on age, nature of activity and reflectance factors is illustrated through example 12.5.

Table 12.5: IESNA illumination level weighting guidelines, categories D through I.

For Categories D through I			
Room & Occupant's		**Weighting Factor**	
Characteristics	**-1**	**0**	**+1**
Age	Under 40	40-55	Over 55
Imp. of Speed & Accuracy	Not Imp.	Imp.	Critical
Reflectance of Task Background	>70%	30 - 70%	<30%

Example 12.5:

An upscale custom watch-maker needs to assess the minimum illumination level in the final assembly room of its facility. The mean age of the craftsmen in this room is 56. The craftsmen are expected to work an average of 10 hours per day. The accuracy of the task is critical. The walls and the ceilings offer a measured reflectance 40%. Calculate the required illumination level.

Solution

Upon examination of Table 12.3, the most suitable classification of this task would be **Category-G,** which entails "**Visual tasks involving low contrast or very small size, over a prolonged period.**"

As explained earlier, weighting factors for categories D through I are assessed in accordance with Table 12.5.

*   From Table 12.5, the weighting factor associated with **age 56** and over is **+1**.

*   Since **accuracy** is **critical** in the watch assembly task, the weighting factor for this task attribute would be **+1**, as well.

*   The **reflectance** is given as **40%**, therefore, the weighting factor contribution in this category, in accordance with Table 12.5, would be **0**.

*   The total weighting factor for this scenario is = +1 +1 + (0) = **+2**.

According to the rules stated earlier, and the illuminance ranges stated in Table 12.3, the aggregate weighting factor of "+2" would point to the use of **High** lux or lumen value under category **G**. The illumination range for Category **G** is 2000 – 3000 – 5000 lux, or 200 – 300 – 500 fc. The highest illumination level in Category G is 5000 lux or 500 fc. Therefore, the recommended illuminance level in this watch assembly room is **5000 lux or 500 fc.**

When the nature of activity being conducted in a specific illuminated space is well defined but other attributes such as age, speed of activity, size of work elements, contrast and reflectance levels are not well defined and can safely be assumed to be mid-range, the process of determining the illumination level can be simplified through the use of Table 12.6, below. This table is a simplified derivative of Table 12.3. Table 12.6 is derived from Table 12.3 by assuming that all weighting factor defining attributes are mid-range. This assumption yields total weighting factor for each category to be zero, which in turn results in selection of mid-range illumination level for each category.

Alternative, Rae Method

An alternative to the IESNA method is the Rae Method, named after M. S. Rae, who first introduced this approach in 1988. The Rae method results in target illumination levels comparable to the ones derived through the IESNA method.

Table 12.6: Illumination level guidelines for **important** tasks performed by **40- to 55-year-old** individuals with **30-70% task background reflectance.**

IESNA Illuminance Categories and Values for Generic Indoor Activities			
Activity	Category	lux	fc
Public spaces with dark surroundings	A	30	3
Simple orientation for short tasks/visits	B	75	7.5
Working spaces with occasional visual tasks	C	150	15
Visual tasks involving high contrast or large size	D	300	30
Visual tasks involving medium contrast or small size	E	750	75
Visual tasks involving low contrast or very small size	F	1500	150
Visual tasks involving low contrast or very small size, over a prolonged period	G	3000	300
Performance of very prolonged exacting visual tasks	H	7500	750
Performance of very special visual tasks of extremely low contrast	I	15000	1500

LAMBERT'S LAW

Lamberts law, also called the *cosine law*, establishes the relationship between the illuminance on a surface, or target, the illuminance at the source, and the angle "θ" the inclined source radiation (light) portends with respect to the target surface's normal vector. Where, normal vector would be defined as a unit vector that is perpendicular to the target surface.

Figure 12.1: Illustration of Lambert's Law

Figure 12.1 illustrates a scenario where light travelling from a source, at a luminous level of E_1, falls on a target area that lies at an angle θ with respect to the inclined path of the light. In such a case, Lambert's law could be interpreted to state that the illuminance on any surface varies as the cosine of the angle of incidence, θ. Therefore, Lambert's Law, can be stated, mathematically, as:

$$E_2 = E_1 \cos\theta \qquad\qquad \textbf{Eq. 12.5}$$

Where,

E_1 = Illuminance at the source or at a specified point along the path of the light.

E_2 = Illuminance, at a point on a surface, whose normal vector portends an angle θ with respect to the inclined path of the light from the light source.

Notice that according to Lambert's Law, and **Eq. 12.5**, if the light source is directly above, perpendicular, or orthogonal to the target surface, θ, would be "0," and

$$E_2 = E_1 \cos\theta = E_1 \cos(0) = E_1$$

Or,

$$E_2 = E_1$$

ILLUMINANCE INVERSE SQUARE LAW

Within the realm of physics, inverse square law defines the relationship of extending force or radiating energy to the distance. In illumination domain, inverse square law stipulates the relationship between illuminance, illuminance intensity, and the distance. Illuminance on a surface varies in accordance with the **illuminance inverse square law.** In other words, according the inverse square law, Illuminance is inversely proportional to the distance between the source and the surface. This law can be stated, mathematically, as follows:

$$\text{Iluminance, in lx} = E = \frac{I}{d^2} \qquad\qquad \textbf{Eq. 12.6}$$

Where,

E = Illuminance expressed in lx, or lux
I = Illuminance intensity expressed in cd, candela
d = Distance between the luminaire and the surface being illuminated, measured in "m" or meters.

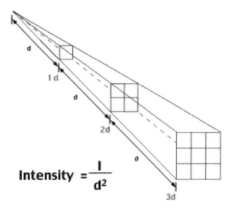

$$\text{Intensity} = \frac{I}{d^2}$$

Figure 12.2: Illustration of the Inverse Square Law

The inverse square law applies only to situations where the distance **d** is at least five times the max dimension of the light source. The inverse square law can be illustrated graphically as shown in Figure 12.2. Eq. 12.7 is an extension of Eq. 12.6 and it can be used to assess the illumination contribution, by the *same* source at multiple points along the work surface.

$$E_1 r_1^2 = E_2 r_2^2 \qquad\qquad \textbf{Eq. 12.7}$$

Where,

$E_1 =$ Illuminance, in lx or cd, due to the source at a distance r_1.

$E_2 =$ Illuminance, in lx or cd, due to the source at a distance r_2.

Example 12.6:

Consider the scenario depicted in Example 12.2 and assume that the manufacturer's specifications show tested illuminance of the lamp, at 1.0 m (3.28 ft), to be 900 lx. Determine the amount of illuminance, E, at the top of the desk surface.

Solution

Apply the single source special case interpretation of the inverse square law in form of Eq. 12.7:

$$E_1 r_1^2 = E_2 r_2^2$$

In this case,

$E_1 = 900$ lx

$r_1 = 1.0$ m

$r_2 = 4.57$ m

$E_2 = ?$

Then, by rearranging Eq. 12.7:

$$E_2 = \frac{E_1 r_1^2}{r_2^2} = \frac{(900 \text{ lx}).(1\text{m})^2}{(4.57)^2} = 43.1 \text{ lx}$$

ILLUMINANCE COSINE-CUBED LAW

The illuminance cosine-cubed law can be viewed as a combination of the Lambert's law and the inverse square law. As described earlier, Lamberts law, also referred to as the *cosine law*, establishes the relationship between the illuminance on a surface, or target, the illuminance at the source, and the angle "θ" the inclined source radiation (light) portends with respect to the target surface's normal vector. Equation 12.5 is the mathematical representation of this law. On the other hand, the inverse square stipulates the relationship between illuminance and the target distance. Mathematical representation of this inverse square law is embodied in the form of Eq. 12.6.

Combination of these laws, and their respective equations, yields **Eq. 12.8**, which is a mathematical representation of the cosine-cubed law. In order to facilitate a comparison of these laws, their respective equations are restated below.

$$E_2 = E_1 Cos\theta \qquad\qquad \text{Eq. 12.5}$$

$$\text{Illuminance, in lx} = E = \frac{I}{d^2} \qquad\qquad \text{Eq. 12.6}$$

Combination of Eqs. 12.5 and 12.6 yields

$$\text{Illuminance} = E = \frac{I}{d^2} Cos\theta \qquad\qquad \text{Eq. 12.8}$$

When light is radiating toward a target at an angle of inclination θ, the geometry of its path can be portrayed as shown in Figure 12.3.

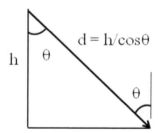

Figure 12.3: Geometry of the inclined light path

Substitution of **d = h/cosθ**, from Figure 12.3 into Eq. 12.8, yields Eq. 12.9, or the **cosine-cubed law:**

$$\text{Illuminance} = E = \frac{I}{h^2} Cos^3\theta \qquad\qquad \text{Eq. 12.9}$$

Example 12.7:

Proposed layout of the lighting system for a work space is depicted in the figure below. The luminous intensity, I, for lamps X, Y and Z is 450 cd. Determine the following:

a)   Illuminance $E_{Z\text{-}C}$, at point C, due to light source Z.
b)   Illuminance $E_{Z\text{-}B}$, at point B, due to light source Z.

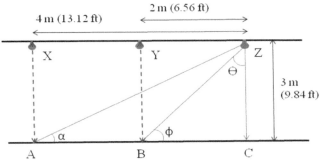

Geometry of the inclined light path

Solution

a)   There are multiple alternative methods for solving this problem. One approach is to simply apply the cosine-cubed law. Since point C lies directly below source Z, angle $\theta = 0$. Apply Eq. 12.8:

$$\text{Illuminance} = E = \frac{I}{h^2}\ Cos^3q \qquad\qquad \textbf{Eq. 12.9}$$

$$\text{Illuminance} = E = \frac{I}{h^2}Cos^3\theta$$

$$E_{Z\text{-}C} = \frac{I}{h^2}Cos^3\theta = \frac{450\ cd}{3^2}Cos^3(0)$$

$$= \frac{450\ cd}{(3m)^2}(1) = 50\ cd/m^2\ \text{or}\ 50\ lx$$

b)   In this case, the angle of the angle of the light path under consideration is not zero. Therefore, before the law of cubed cosine can be applied, the angle θ must be determined.

Applying Pythagorean theorem to the triangle **ZBC**:

$$\theta = \text{Tan}^{-1}\left(\frac{\text{BC}}{\text{ZC}}\right) = \text{Tan}^{-1}\left(\frac{2}{3}\right) = 33.7°$$

Then, by applying the cosine-cubed law, or Eq. 12.9:

$$E_{Z-B} = \frac{I}{h^2}\text{Cos}^3\theta = \frac{450 \text{ cd}}{3^2}\text{Cos}^3(33.7°)$$

$$= \frac{450 \text{ cd}}{(3m)^2}(0.576) = 28.8 \text{ cd/m}^2 \text{ or } 29 \text{ lx}$$

LIGHTING DESIGN CONSIDERATIONS

Lighting system design is driven primarily by the illuminance requirement at the point of interest. Factors that must be considered in lighting design are as follows:

- Type of light source
- Nature of task, or tasks, being performed
- Illuminance contributed by direct light
- Illuminance contributed by reflected light
- Illuminance contribution by delighting measures, when lumen method is used
- Interior only light when zonal cavity method is used

DAYLIGHT BASED DESIGN CONSIDERATIONS

As implied in the list of factors above, lighting systems design varies depending on the lighting approach employed. When daylight is included among the light sources in a lighting system, the following points must be noted:

- Daylight consists of unique spectra and light-level distributions. This distinguishes the characteristics and performance of daylight from those exhibited by artificial light sources.

- Daylight varies seasonally

- Standard power distributions of daylight, allow illuminance from the sun to be determined by factoring in the following factors:
  — Site location
  — Time of desired illumination
  — Solar position
  — Skylight

— Ground light contributions
— Human response factors

Daylight Factor Method

The daylight factor method is used with known illuminance distributions. These distributions include the standard three daylight illuminations. The data for calculating the illuminance is given in a tabular format. The daylight factor, DF, is a ratio of the illuminance at a point to the illuminance from the unobstructed sky. This definition of daylight factor, DF, can be stated mathematically as follows:

$$DF = \frac{E_x}{E_s} \qquad \textbf{Eq. 12.10}$$

Where,

E_x = Illuminance at a point x
E_s = Illuminance from the unobstructed sky or another light source

Daylight factor can also be defined as the sum of sky component, externally reflected illuminance, and internally reflected illuminance. This definition of daylight factor can be stated in form of an Eq. as follows:

$$\textbf{DF = SC + ERC + IRC} \qquad \textbf{Eq. 12.11}$$

Where,

SC = Sky Component
ERC = Externally reflected component, in lumens, emitted by each light source
IRC = Internally reflected illuminance component.

Eq. 12.11 can be used to compute the DF value for a daylight system. The DF value thus defined can be substituted in Eq. 12.10, in conjunction with known source illuminance value E_s, to determine the estimated illuminance, E_x, at specific point x.

The Lumen Method

The lumen method accounts for total lumens projected by a luminaire, derated by the coefficient of utilization factor, CU, for a specific lighting scenario. The CU factor accounts for the following lighting system parameters:

- Efficiency of the luminaire
- Distribution of light or shape of the light distribution area
- Reflective properties of the space

$$E_i = \frac{(L_{Total}).(CU)}{A_W}$$ **Eq. 12.12**

Where,

E_i = Initial, unmitigated, illuminance in the work plane.

L_{Total} = Total illuminance emitted by the light source; based on manufacturer's specs.

A_w = Area in the work plane to be lit

CU = Coefficient of utilization; either given or computed using the zonal method described below.

As a light source ages, it experiences a loss in its luminance due to various factors, including dirt accumulation on the lamp, lens, reflector, etc. Such degradation in the illuminance is accounted through computation of maintained illuminance, E_m as shown in Eq. 12.13.

$$\text{Maintained Illuminance} = E_M = \frac{(L_T).(CU).(LLF)}{A_W}$$ **Eq. 12.13**

Where,

L_T = Lumens (initial) emitted by the light source

CU = Coefficient of utilization

LLF = Light loss factor used to derate the initial illuminance

A_W = Area, in the work plane, to be illuminated

When a lighting system consists of multiple lamps or light sources, the maintained illuminance can be computed through Eq. 12.14, below.

$$\text{Maintained Illuminance} = E_M = \frac{(N_T).(L_{E/L}).(CU).(LLF)}{A_W}$$ **Eq. 12.14**

Where,

N_L = The number of lights or lamps

$L_{E/L}$ = Lumens emitted by each light source or lamp

CU = Coefficient of utilization for the lighting system
LLF = Light loss factor
A_W = Area, in the work plane, to be illuminated

There are various methods for determining, accessing or computing the CU values for specific lighting system scenarios. A thorough discussion of CU computation methods is beyond the scope of this text. Use of tables to identify the CU value for particular situations is illustrated through Example 12.8. This example utilizes Table 12.7.

Table 12.7: Coefficient of utilization

Coefficient of Utilization (%)				
Wall Reflectance (%)	50	0	50	0
Base Reflectance (%)	90	90	80	80
Cavity Ratio				
0.2	86	82	77	72
0.6	**80**	73	71	63
1.0	75	62	67	55

When using Table 12.7 for determination of CU, certain parameters are expected to be known or available. These parameters include (1) reflectance of the ceiling or floor, commonly referred to as the base reflectances, (2) the wall reflectance, and (3) cavity ratio.

Example 12.8:
Determine the CU, coefficient of utilization, for a lighting scenario where the wall reflectance is known to be 50%, the base (ceiling or floor) reflectance is estimated to be 90%, and the cavity ratio is 0.6.

Solution
Identify the column in Table 12.7 that represents the given wall reflectance of 50% and base reflectance of 90%. Then, identify the row in the table that represents the given cavity ratio of 0.6. The CU value for this lighting scenario would be represented by the table entry that lies at the point of intersection of the identified column and row as shown below. This entry is 80%. Therefore, the CU value for this scenario would be **82%**.

Coefficient of Utilization (%)				
Wall Reflectance (%)	50	0	50	0
Base Reflectance (%)	90	90	80	80
Cavity Ratio				
0.2	86	82	77	72
0.6	80	73	71	63
1.0	75	62	67	55

Chapter 12—Self-assessment Problems and Questions

1. Consider the lighting scenario described in Example 12.1. Determine
    the impact on the luminous flux at the 1 m circular target if the distance
    increased to 10 m. See the diagram below.

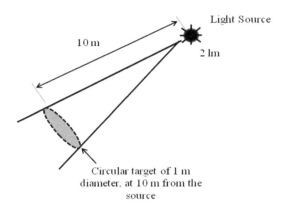

Circular target of 1 m
diameter, at 10 m from the
source

2. A 180W low pressure sodium vapor lamp is being used to illuminate an
    exterior space. Determine the luminous flux that would be emitted by this
    lamp.

3. Consider the scenario depicted in Example 12.2 and assume that the
    manufacturer's specifications show tested illuminance of the lamp, at 3.0
    ft, to be 1000 lx. Determine the amount of illuminance, E, at the floor
    elevation, directly below the lamp.

4.   Consider the situation stated in Example 12.7. Proposed layout of the lighting system for the work space is shown below. The luminous intensity, I, for lamp **Y** is 700 cd. The luminous intensity for lamps **X** and **Z** is 600 cd. Determine the following:

   a)   Illuminance E_{Y-B}, at point B, due to light source Y.
   b)   Total Illuminance, E_B, at point B, due to light sources X, Y and Z.

Chapter 13

Batteries

Introduction

Invention of battery is credited to Alessandro Volta, dating back to 1800. Volta constructed the first battery from stack of copper and zinc plates, separated by brine-soaked paper disks, which could produce a steady current for a considerable length of time. See Figure 13.1. Volta named his invention "Voltaic Pile." At the time, however, Volta did not fully recognize the fact that the voltage was due to chemical reactions, with a finite life. He thought that his cells were an inexhaustible source of energy, and that the corrosion effects he noted at the electrodes were a mere nuisance. It was Michael Faraday who, later in 1834, proved that charge and energy held by the battery was indeed limited and exhaustible.

By: Luigi Chiaesa, September 28, 2018

Alessandro Volta demonstrating his pile to French emperor Napoleon Bonaparte, Circa 1801

Figure 13.1. "Voltaic Pile," the first battery, by Alessandro Volta. Alessandro pictured demonstrating his invention to Napoleon Bonaparte.

Ever since the invention of the first battery, scientists and engineers have relied on its unique and valuable, portability and energy storage, characteristics. Over the last two centuries, since the invention of the first battery, scientists and engineers have focused on the size, portability, mass, capacity, safe charging and discharging characteristics of batteries. Today, in the 21st century, we find batteries to be vital, integral and indispensable part of our daily lives. Batteries play a crucial role in our phones, computers, watches, light sources, kitchen appurtenances, cordless tools, health care equipment, our automobiles, aircrafts and space crafts. Batteries are a vital link in making renewable energy sources more practical, viable and versatile. As the 21st century all-electric automobiles—also known as EV's, or Electric Vehicles—have already proven, battery powered electric motors cannot just replace the environmentally unfriendly, depletable fuel dependent, internal combustion engine, they can, in many ways, *outperform* it. The key component—and sometimes a significant constraint—in electric vehicles, and many other types of portable or transportable electric equipment, is the battery.

Therefore, to complete the introduction of non-electrical engineers to electrical engineering, in this chapter we will explore basic, yet important, facets of battery, and its sustainable and safe operation. We will introduce the reader to the different types of commercially available batteries. We will examine the construction of common batteries, the charging and discharging characteristics of secondary batteries, influence of temperature on the performance of batteries, and more. While a significant portion of this chapter is devoted to fundamentals of battery technology, in this chapter, the reader will get an opportunity to learn about the not so obvious charging and discharging performance characteristics of batteries that can be determining factors in their selection, application and optimal performance. In this text, in depth discussion of the chemistry, charging and discharging characteristics of batteries will focus, mainly, on the common lead-acid (SLI) battery.

As is the case with all of the electrical engineering topics introduced in the earlier chapters of this text, battery technology related engineering and physics concepts, principles and quantitative analytical methods will be explained and illustrated with the help of pictures, diagrams, and sample analytical problems. As before, the reader will have the opportunity to reinforce their learning by attempting the self-assessment problems and questions at the end of this chapter. The solutions to the self-assessment problems are made available in Appendix A.

BATTERY CLASSIFICATIONS

Overall, contemporary batteries can be classified into two major categories:

- Primary Batteries
- Secondary Batteries

Within these two major classifications listed above, batteries can be specified or classified into subcategories. The subcategories differ based on the following attributes:

- Construction, chemistry or chemical composition, maintenance free or non-maintenance free characteristics, etc.
- Physical size, weight, volume and derived ratios, i.e. Specific Energy, Specific Power, and Energy Density
- Voltage, Amp-hour (energy) capacity, C-rating
- Discharging characteristics
- Charging characteristics.

A few commercially available primary and secondary batteries are depicted in Figure 13.2. Their voltages, energy (Amp-hour) ratings at specific discharge current levels are listed in latter sections of this chapter.

PRIMARY BATTERIES

Most primary batteries are of dry cell construction. With the exception of some especially designed rechargeable alkaline batteries, most primary batteries are not rechargeable. Most primary batteries are typically discarded when discharged. Therefore, primary batteries are sometimes referred to as "One-cycle" batteries. Primary batteries can be subdivided into two major categories:

- Zinc-Carbon, or Zinc-manganese dry cell batteries
- Alkaline batteries, or alkaline-manganese dry cell batteries

Cross-section of a typical Zinc-Carbon, or Zinc-manganese dry cell primary battery is shown in Figure 13.3.

Alkaline batteries are referred to as "alkaline" batteries because of the fact that the chemically active substance—the electrolyte—in these batteries is an alkaline, or a base, as compared to the acidic electrolyte used in zinc-carbon, or zinc-manganese primary dry cell batteries. This chemically active substance,

Duracell ® Primary Batteries

Lead Acid, SLI, Secondary, Battery

Nokia® Lithium Ion Secondary
Battery, By Kristoferb.

Figure 13.2. Example of common primary batteries

or electrolyte, in alkaline batteries is typically potassium hydroxide instead of
the acidic ammonium chloride, or zinc chloride electrolyte, used in zinc-carbon
primary batteries. Some alkaline primary batteries *are* rechargeable.

Secondary Batteries

Secondary batteries are batteries that generally can be recharged. Albeit,
the number of discharge-recharge cycles—before the battery is retired—are finite
and do vary based on battery types and depth of discharge (DOD). The concept of
DOD is explained later in this chapter.

Some of the more commonly used, commercially available, secondary
batteries are as follows:

1. Brass cap or the (+) terminal, connected to the carbon rod
2. Plastic seal
3. Expansion space
4. Porous cardboard
5. Zinc can, or the (-) electrode, with high concentration of electrons
6. Inert center carbon rod and black carbon paste constitute the (+) electrode system
7. Manganese dioxide
8. Chemical mixture, zinc-chloride or ammonium chloride

Figure 13.3. Typical dry cell primary battery construction

- Li-ion, or Lithium Ion
- Ni-Cad, or Nickle Cadmium
- Lead – Acid

Li-Ion Battery

The lithium-ion battery is sometimes abbreviated as LIB. Li-ion is a general term for several different lithium-battery chemistries that are used for different applications. i.e.:

(LCO, LMO, NMC, LFP, LTO, LNA). In lithium-ion batteries, lithium ions (Li^+) move through the electrolyte, from the negative electrode to the positive electrode while the battery is being discharged, or in use. However, when a Li-Ion battery is being charged, the Li^+ ions move from the positive electrode to the negative electrode. This migration of Li^+ ions to the negative electrode, during the charging phase, eventually results in saturation of the negative electrode with positive charge and positive potential; which, when connected to load, can drive holes (electrons in the opposite direction) through the load, thus comprising the flow of electric current.

Lithium-ion batteries are commonly used as rechargeable batteries in consumer electronics and other common electric applications. Some of the strengths of Li-Ion batteries are as follows:

- They offer high energy density (Wh/L, or Watt-hours per Liter of volume). The concept of energy density is explained later in this chapter. For now, note that higher energy intensities are more desirable, in that, high energy density batteries deliver more energy while occupying less space. Note that a battery that contains more energy can perform more work before it needs recharging. In other words, a high energy density will operate those consumer electronics longer between charges while occupying less space—thus allowing those consumer electronics to be more compact in size.

- The memory effect is insignificant. In other words, the recharging performance of Li-Ion battery is relatively independent of the magnitude and profile of discharging.

- They exhibit low self-discharge. Self-discharge is the non-work producing energy discharge that occurs while the battery is not in use.

Because of the strengths cited above—in addition to their popularity among consumer electronics—Li-Ion batteries show a strong demand in the following applications:

- Military sector
- All-electric and hybrid automobiles
- Aerospace applications
- Energy storage, for base load needs, in renewable energy systems, i.e. wind turbines, PV farms, etc.
- Cordless tools
- Consumer electronics, i.e. cell phones, cameras, lap-top PC's,
- As replacements for the lead acid batteries in general applications that have, traditionally, been dominated by lead-acid batteries, i.e. golf carts, scooters, ATV's, mass transit equipment, like busses, etc.

Lithium-ion batteries can be dangerous and can pose a safety hazard under certain circumstances due to the fact that they contain flammable electrolyte, under pressure. Therefore, the testing standards for Li-Ion batteries are more stringent than those for acid-electrolyte, or Lead Acid batteries; tests that span over a broad range of operating conditions, specific to the type and construction of batteries. This is in response to reported accidents and failures which have resulted in battery-related recalls by some battery manufacturers.

Major Li- Ion batteries, in use today, are listed below:

- **Li-FePO$_4$**: Also known as Lithium Phosphate Battery, or LFP battery, with "LFP" standing for "lithium ferro-phosphate. LiFePO$_4$ batteries have a nearly constant discharge voltage. In other words, the voltage stays close to 3.2V during discharge until the cell is exhausted. This permits the LFP cell to deliver virtually full power until it is discharged. This attribute greatly simplifies or even eliminates the need for voltage regulation circuitry. The nominal 3.2V output of an LFP cell, allows four cells to be placed in series for a nominal voltage of 12.8 V; which is close to the nominal voltage of a six-cell lead-acid battery. This advantage, coupled with good safety characteristics of LFP batteries, makes them a good potential replacement for lead-acid batteries. On the other hand, LFP batteries have a somewhat lower energy density than the more common $LiCoO_2$ design found in consumer electronics, but the LFP does offer longer lifetimes, better power density, expressed in W/L, or Watts per liter of overall battery volume. $LiFePO_4$ is a natural mineral of the olivine family (triphylite) and is used as a cathode material. Because of LFP batteries' low cost, non-toxicity, the natural abundance of iron, its excellent thermal stability, its safety characteristics, electrochemical performance, and specific energy capacity (170 $mA \cdot h/g$), it has a respectable market share and acceptance. LFP's key barrier to initial commercialization was its intrinsically low electrical conductivity. This problem has, to a great extent, been overcome by reducing the particle size and the coating of the $LiFePO_4$ particles with conductive materials such as carbon. LFP batteries are now being used in industrial products by major corporations including Black and Decker's DeWalt brand, the Fisker Karma, Daimler, Cessna and BAE Systems.

- **Li-Po:** Also known as Lithium Polymer Battery. The difference between Li-Po batteries and other Li-Ion batteries is in the type of electrolyte used. In lithium ion polymer batteries, a solid polymer electrolyte is used. This polymer may be a solid or a semi-solid gel. While other Li-Ion batteries consist of liquid electrolytes.

- **Li-Mn$_2$O$_4$:** Also known as Lithium Manganese Oxide Battery. In Li-Mn2O4 batteries, manganese oxide as cathode material. The $Li-Mn_2O_4$ architecture forms a three-dimensional "spinel" structure that improves ion flow on the electrode. This results in lower internal resistance and improved current flow. Another advantage of spinel is high thermal stability and enhanced safety. However, the cycle and operating life of Li-Mn2O4 batteries are relatively limited. The low internal cell resistance of Lithium Manganese Oxide battery enables fast charging and discharging. It is also possible to apply one-second load pulses of up to 50A with Li-Mn$_2$O$_4$ battery. Continuous high load or current in Manganese Oxide battery can cause heat buildup. An operating—discharging or charging—temperature of 80°C (176°F) or higher can result in catastrophic failure of this battery. Li-manganese is used in power tools, medical instruments, as well as hybrid and electric vehicles.

- **Li-CoO$_2$:** Also known as Lithium Cobalt Battery. Li-CoO$_2$ battery's high specific energy makes it the popular choice for mobile phones, laptops and digital cameras because of its high specific energy (Wh/kg). The Li-CoO$_2$ battery consists of a cobalt oxide cathode and a graphite carbon anode. The cathode has a layered structure and during discharge, lithium ions move from the anode to the cathode. The flow of lithium ions reverses on charge. The drawback of Li-cobalt is a relatively short life span, low thermal stability and limited load capabilities. Nominal cell voltage of Lithium Cobalt Battery if 3.7 volt.

Lead-Acid or SLI Battery

Lead-Acid battery is also known as the "SLI" battery. The name SLI stands for Starter, Light and Ignition. The Lead-Acid or SLI battery is designed to deliver an instantaneous burst of energy to loads such as the motor starter of an automobile. Once the internal combustion engine of the automobile starts, the alternator—or the onboard DC generator of the automobile—assumes the task of recharging the lead-acid battery and also provides DC power and energy to all onboard electrical and electronic components.

The SLI battery is designed with a high density of thin electrode plates resulting in a large electrode surface area for fast electrochemical reaction to provide the needed instantaneous burst of energy. This distinctive design of the SLI battery is also conducive to faster charging characteristics. However, this distinctive property of the SLI battery does carry with it the vulnerability of this design to deep and permanent discharging.

Lead—Acid batteries can be divide into two major categories based on their construction and application:

- Deep Cycle Battery
- Regular SLI Battery

Deep Cycle Lead-Acid Battery vs. Regular SLI Battery

The deep cycle battery is designed for endurance, longer and deeper discharge cycles without substantial depreciation of life. Since deep cycle batteries are designed to work reliably even when regular charging is not viable, they offer a better charge or energy storage alternative for renewable energy storage applications, as compared to the regular SLI batteries. From construction perspective, the key distinction between the deep cycle battery and the SLI battery is that the former consists of thicker cell or electrode plates. The thicker battery plates resist corrosion during extended charge and discharge cycles.

Deep-cycle lead-acid batteries generally fall into two distinct categories:

- Valve-regulated lead-acid (VRLA),
- Flooded (FLA)

VRLA Battery: The acronym VRLA stands for Valve Regulated Lead-Acid battery The VRLA battery is also known as a Sealed Lead Acid (SLA), battery or the Maintenance Free Battery. Even though the VRLA batteries do not require constant maintenance, their characterization as "Maintenance Free" is somewhat of a misnomer. The VRLA batteries still require cleaning and regular functional testing. The VRLA batteries are widely used in large portable electrical devices, and off-grid power systems; where large amounts of storage are needed. The VRLA batteries, in such applications, in some cases, offer a lower cost alternative as compared to the Li-Ion batteries. As implicit from its classification in the secondary battery category, the VRLA or SLA battery is rechargeable. The VRLA or SLA batteries can be sub-classified further as:

- Sealed VR Wet Cell Batteries
- AGM Batteries
- Gel Batteries

AGM (Absorbed Glass Mat) batteries feature fiberglass mesh between the battery plates which serves to contain the electrolyte. Gel cells add silica dust to the electrolyte, forming a thick putty type substance with the consistency of a gel. The Gel cells or Gel Batteries are sometimes referred to as "silicone batteries." Both designs offer advantages and disadvantages compared to conventional batteries, the sealed VR wet cells, as well as each other.

VRLA Gel and AGM batteries offer several advantages compared with VRLA flooded wet cell lead acid and conventional standard lead-acid batteries. Because the Gel and AGM batteries don't contain low viscosity liquids, they can be mounted in any position, since the valves only operate in response to over-pressure conditions. Gel and AGM battery chemical system is designed to

be recombinant with little or no emission of gases in case of overcharge. Therefore, when AGM and Gel batteries are applied, room ventilation requirements
are reduced as no acid fume is emitted during normal operation. In reality, even
with flooded cell applications, gas emissions are of little consequence in all but
the smallest confined areas, and these wet cell batteries pose little threat to residential users, who can benefit from the durability and the lower cost, in dollars
per kWh. Other advantages of gel cells or gel batteries are as follows:

•   In case of damage, with AGM's or Gel Cells, the volume of free electro-
        lyte that could be released would be significantly less than the potential
        release from a damaged flooded (wet) lead-acid battery.

•   The AGM's or Gel Cell batteries require little or no maintenance. While,
        with flooded lead-acid batteries the level of electrolyte needs to checked
        and replenished with distilled water, periodically. However, note that the
        wet cell batteries can be maintained by automatic or semi-automatic wa-
        tering systems. The need for watering is dictated often by overcharging.

Construction, Composition and Operation of a
Common Lead Acid Battery

General construction of a common Lead-Acid battery is depicted in
Figure 13.4. The term "flooded" is used in the Flooded Lead Acid batteries
because in this type of battery the electrode plates are completely submerged
in the electrolyte. The electrolyte level must be maintained above the tops of
plates. The minimum electrolyte level is marked on the battery case. As shown
by the charging phase reaction in Eq. 13.6, some of the electrolyte water is
decomposed and depleted during the charging cycles. Therefore, regular maintenance of flooded batteries requires inspection of electrolyte level and addition
of water. Major modes of failure of deep-cycle batteries are loss of the active
material due to the decomposition and shedding of the plates, and the corrosion
of internal grid that supports the active material. The limiting factor with capacity of a deep cycle battery is the decline in the concentration of the electrolyte.
Because the deep cycle battery electrode plate mass is substantial, electrode
plate attrition is seldom the cause of its failure.

As shown in Figure 13.4, the components actively participating in the
electrochemical reaction of a lead acid battery are:

•   Cathode constructed from solid lead peroxide, or, PbO_2.
•   Anode constructed from relatively porous or "spongy" lead, or Pb.

- Aqueous solution of sulfuric acid, H_2SO_4.

The nominal voltage in an individual Lead-Acid, battery cell is 2.1 volts.

Figure 13.4. Lead-Acid battery and the flow of current during battery discharging phase.

A charged battery represents a state of chemical "imbalance." In the charged state, the cathode consists of lead peroxide, PbO_2, and the anode consists of "spongy" lead, Pb. These two electrodes—cathode and anode—are immersed in an aqueous solution of sulfuric acid or H_2SO_4. The aqueous solution of sulfuric acid disperses into positively charged H^+ "cations" and negatively charged HSO_4^- anions.

During the discharging mode of a lead acid battery, the electro-chemical reaction that occurs at the lead (negative) terminal or anode, is as follows:

$$HSO_4^- + Pb \rightarrow PbSO_4 + H^+ + e^- \qquad \text{Eq. 13.1}$$

Furthermore, during the discharge mode, the electrons released through the electrochemical reaction at the anode travel through the closed electrical circuit, as shown in Figure 13.4; thus establishing the flow of current through the load. These electrons arrive at the cathode and engage in the electrochemical reaction stated in Eq. 13.3. The flow of current through the load delivers energy to the load, which can converted to work or simply heat, in accordance with Eq. 13.2.

$$\text{Energy} = \text{Work} = W = I^2 \times R \times t \qquad\qquad \text{Eq. 13.2}$$

Where,

   W = Work or Energy can be measured in N-m, or Joules, and can be convert-
            ed into ft-lbf, or BTU's.
   R = Resistance offered by the load
   t = Time, measured in seconds; representing the time span over which the
            battery is used, or duration of discharge phase of the battery.

The electro-chemical reaction that occurs during the discharging mode of a lead acid battery, at the lead (negative) terminal or anode, is shown in Eq. 13.3, below.

$$HSO_4^- + PbO_2 + 3H^+ + e^- \rightarrow PbSO_4 + 2H_2O \qquad\qquad \text{Eq. 13.3}$$

Overall, combined, reaction at the two electrodes:

$$2HSO_4^- + Pb + PbO_2 + 2H^+ \rightarrow 2PbSO_4 + 2H_2O + \text{Energy} \qquad \text{Eq. 13.4}$$

During the discharging mode of the battery, and as a result of the electrochemical reactions shown in Eq. 13.1 and Eq. 13.3, the amount of H_2SO_4 is gradually depleted and replaced by H_2O; the lead peroxide, PbO_2, cathode and the lead, Pb, anode transform progressively into lead sulfate, $PbSO_4$. In other words, the chemical differential between the anode and the cathode diminishes or neutralizes. This reduction of chemical differential is characteristic of the discharging of the battery and loss of energy.

Because the concentration of H_2SO_4 is depleted and replaced by H_2O during the discharging process of a battery, one method for assessing the level of charge or energy left in the battery consists of testing the concentration and specific gravity of the electrolyte in the battery fluid by using a hydrometer. In other words, the state of charge, or SOC, of a lead-acid battery can be measured by testing the specific gravity of the battery fluid. The specific gravity of a fully charged battery—with high concentration of H_2SO_4 in the electrolyte—is substantially higher than 1.0. The specific gravity of the electrolyte would be expressed in form of Eq. 13.5.

$$\text{Specific Gravity of the Electrolyte} = \frac{\text{Density of the Electrolyte}}{\text{Density of Water}} \qquad \text{Eq. 13.5}$$

The density of electrolyte with high concentration of H_2SO4 is substantially higher than the density of pure H_2O, or approximately, 1.3. As substantial concentration of H_2SO_4 is replaced by H_2O during the discharging phase, the specific gravity of the electrolyte approaches 1.1.

It is important to note that since such hydrometer based testing for the level of charge left in the battery requires opening of each cell, and the hazard of exposure to a strong acid, the voltage measurement method—especially under load—is preferred.

Since a Lead-Acid battery is a secondary battery and it, of course, can be recharged. Recharging of a lead acid battery, essentially, involves a reaction that is the reverse of the discharging reaction; the charging electrochemical reaction is shown in Eq. 13.6. The charging phase is also illustrated in Figure 13.5.

$$2PbSO_4 + 2H_2O + Energy \rightarrow 2HSO_4^- + Pb + PbO_2 + 2H^+ \qquad \text{Eq. 13.6}$$

Figure 13.5. Lead-Acid battery and the flow of current during battery charging phase.

During the charging phase, electrons are driven by the battery charger into to the terminal labelled "Anode." In order to avoid confusion, it is best to ignore the polarity of the battery terminals at this point in time, and to focus attention on the dynamics of the charge flow, the current flow and the electro-chemical reaction shown in Eq. 13.6. As apparent from the chemical reaction equation Eq.13.6, during the charging phase, the battery charger pumps elec-trons—and, therefore, energy—into the battery via the "anode." This injection of electrons at the "anode," the positive holes at the "cathode," and energy from the battery charging system, results in the restoration of the anode to lead (Pb) and the cathode to lead per oxide (PbO_2). So, in essence, the charging process reconstitutes the electrochemical differential between the cathode and the an-ode as the battery is fully charged.

From battery sustainability point of view, it is important to note that when lead acid batteries are fully discharged and kept in the discharged state for prolonged periods of time, the lead sulfate ($PbSO_4$) compound formed on the electrodes (anode and cathode) crystalizes and hardens. An attempt to recharge a battery in such a state is often unsuccessful.

Distinctive discharging and charging characteristics of batteries lead to two major sub-categories within the lead acid secondary class of batteries, namely, the SLI Battery and the Deep Cycle Battery.

Having introduced some of the major families of batteries, and their gen-eral characteristics, let's examine some of the key parameters that are used to compare the performance of various batteries, quantitatively.

Specific Energy

Specific Energy is energy expressed on per unit mass basis. Therefore, the units for assessment of specific energy of batteries would be a conventional unit of energy per unit mass. Common unit for quantification of electrical energy is watt-hours, or Wh, often expanded to kilowatt-hours, or kWh, Megawatt-hours, or MWh, etc. Common unit for specification of battery specific energy is Wh/kg. Specific energy is also referred to as the "gravemetric energy density," not to be confused with "energy density"—which is measured in Wh/L, as dis-cussed later.

Examples Specific Energy:

- Li-Po, or Lithium polymer cell battery's specific energy is approximately 180 Wh/kg. This specification can, conversely, be interpreted as 5.6 kg mass of battery per 1kWh of energy capacity of the battery.

- $LiFePO_4$, or Lithium phosphate cell battery's specific energy is approxi-mately 100 Wh/kg. Conversely, it could be said that a Lithium Phosphate cell battery with an energy capacity of 1 kWh would have a mass of 10 kg.

*   Pb-H_2SO_4, or lead-acid secondary battery's specific energy is approximately 40 Wh/kg. Conversely, it could be stated that a Lead-Acid battery with an energy capacity of 1 kWh would have a mass of 25 kg.

Because energy delivered by a battery, in Wh (or kWh, MWh, etc.)—as explained later in the discussion of the "C" ratings of batteries—can vary based on the rate of discharge, the specific energy of a battery is not a constant or permanent characteristic.

Energy Density

Energy density is energy expressed on per unit volume basis. Therefore, the units for assessment of energy density of batteries would be a conventional unit of electrical energy per unit volume. Therefore, the units for energy density would be Watt-hours per Liter, or Wh/L, or kWh/L, etc.

Examples of Energy Density

*   **Li-Po**, or Lithium polymer cell battery's energy density is, approximately, 440 Wh/L. Conversely, a Li-Po would occupy 2.3 L of volumetric space for each kWh of energy capacity.

*   **LiFePO$_4$**, or Lithium phosphate cell battery's energy density is, approximately, 220 Wh/L. Conversely, a Lithium polymer cell battery capable of holding and delivering 1 kWh of electrical energy would occupy a volume of approximately 4.6 Liters.

*   **Pb-H_2SO_4**, or lead-acid secondary battery's energy density is, approximately, 85 Wh/L Conversely, a Lithium polymer cell battery capable of holding and delivering 1 kWh of electrical energy would occupy a volume of approximately 11.8 Liters.

It is important to note that just as specific energy of a battery can vary with non-nominal, or abnormal demand, so can energy density, due to that fact that energy delivered by a battery, in Watt-hours, can vary based on the rate of discharge. Therefore, the energy density of a battery is not a constant or permanent fixed characteristic. In addition to specific energy and energy density, another important performance parameter of batteries is specific power.

Specific Power

Specific Power is power (in Watts) supplied by the battery, expressed on per unit mass (kg) basis. Therefore, the units for specific power would be Watts per kilogram, or W/kg.

Examples Specific Power:

- **Li-Po**, or Lithium polymer cell battery's specific power is, approximately, 300 W/kg. Conversely, a Lithium Polymer battery capable of delivering 1 kW of power would have a mass of 3.33 kg.

- **LiFePO$_4$**, or Lithium Iron Phosphate cell battery's specific power is, approximately, 240 W/kg. Conversely, a Lithium Iron Phosphate battery capable of delivering 1 kW of power would have a mass of 4.17 kg.

- **Pb-H$_2$SO$_4$**, or lead-acid secondary battery's specific power is, approximately, 180 W/kg. Conversely, a lead-acid battery capable of delivering 1 kW of power would have a mass of 5.55 kg.

Comparison of Electrical Energy and Charge Storage Systems

Electrical charge or energy storage devices are not limited to batteries. Other practical and comparable electrical charge and energy storage devices include fuel cells and capacitors. Comprehensive discussion on the fuel cells is not included in this text. However, to facilitate a brief comparison between various, mainstream, energy storage systems, fuels cells can be defined as systems that, similar to batteries, convert energy from chemical form to electrical form. The chemical substance being hydrocarbon fuel, such as methane, or CH_4 and oxygen or air. In a methane (or alkane) based fuel cell, the fuel is "reacted" with oxygen in the air, and steam to produce ions that move between two electrodes, the anode and the cathode. As we learnt earlier in this text, movement of charge or charged particles constitutes current. The current in the case of fuel cells is DC, or direct current. The DC current is then converted to AC current through an inverter. Of course, unlike the electrolyte in batteries, in fuel cells the key source of energy is the hydrocarbon fuel. In addition, unlike the electrolyte in batteries, the hydrocarbon fuel in the fuel cells is consumed and must be replenished. Like the recharging phenomenon in secondary batteries, the replenishment of hydrocarbon fuel serves a supply of energy, storage of energy and subsequent conversion and availability of energy in electrical form.

Among capacitors, as energy storage systems, there are Double-Layered Capacitors, Ultra-Capacitors, and Aluminum-Electrolytic Capacitors. The electrical energy storage in a capacitor, as introduced earlier in this text, can be quantified as $\frac{1}{2}CV^2$. Where, "C" is the capacitance of the capacitor, measured in Farads, and "V' is the voltage of the capacitor, in volts. Having introduced the mainstream contemporary batteries, with prior introduction to capacitors, and the brief introduction to fuel cells above, let's examine Figure 13.6 to delineate the differences, strengths and weaknesses between these various charge storage devices. The chart depicted in Figure 13.6 contrasts the charge or energy storage devices discussed in this chapter on the basis of their specific energy (in Wh/kg) and specific power (in W/kg).

Energy or Charge Storage Device Comparison—
Specific Energy Perspective

From *specific energy* (kWh/kg) perspective, the charge storage devices compared in Figure 13.6 can be ranked in descending specific energy capacity as follows:

1. Fuel Cell
2. Batteries: Lead-Acid, Ni-Cad and Li-Io
3. Capacitors: Double-Layer Capacitors, Ultra-Capacitors, and Aluminum-Electrolytic Capacitors

Energy or Charge Storage Device Comparison—
Specific Power Perspective

From specific power (W/kg) perspective, the charge storage devices compared in Figure 13.6 can be ranked in descending specific energy capacity as follows:

1. Capacitors
    a. Aluminum-Electrolytic Capacitor
    b. Ultra-Capacitors, and
    c. Double-Layer Capacitors

2. Batteries:
    a. Li-Ion
    b. Ni-Cad and
    c. Lead-Acid

3. Fuel Cells

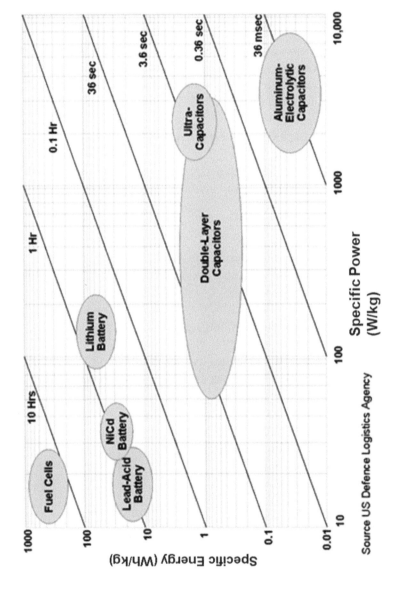

Figure 13.6. Specific power and specific energy comparison of common electrical energy storage devices.

Energy or Charge Storage Device Comparison—
Duration of Energy Supply and Storage Perspective

From ***Duration*** of Energy Supply (Hours, Minutes, Seconds, or Milliseconds) perspective, the charge storage devices compared in Figure 13.6 can be ranked in descending energy supply duration as follows:

1. Fuel Cells
2. Batteries: Lead-Acid, Ni-Cad and Li-Ion
3. Capacitors:
   a. Double-Layer Capacitors
   b. Ultra-Capacitors and
   c. Aluminum-Electrolytic Capacitors

Specific Energy and Energy Density Comparison between
Lead-Acid and other Contemporary Batteries

Having compared the specific energy and specific power characteristics of major energy and charge storage devices, let's narrow or focus to the specific energy and energy density characteristics of just batteries. In order to compare the specific energy and energy density characteristics of batteries, let's examine Figure 13.7. The chart depicted in Figure 13.7 compares: Lead-Acid, Ni-CAD, or Nickle Cadmium, NMH, or Nickle Metal Hydride, and Li-Ion batteries. The specifications, such as "cylindrical," "can," or "prismatic" relate to the physical design and construction of the various batteries. As ostensible from Figure 13.7, Li-Ion, Li-Po, and Li-Phosphate batteries out-perform all other batteries insofar as the specific energy and energy density attributes are concerned; of course, at a price. In other words, if cost is not a determining factor, and the objective is to compact the maximum amount of energy in the least amount of volume, while adding the least amount of mass or weight, then the best alternative would be one of the Lithium secondary batteries. Note that, because of slight cost advantage in some cases, in some applications, and in some markets—despite the obvious disadvantages associated with low specific energy and energy density—the Lead-Acid batteries may be more suitable, particularly, in applications that require large amounts of energy storage. Examples of such applications include: some UPS (Uninterruptible Power Supply Systems) systems and some renewable energy systems like photovoltaic farms, wind turbines, etc. However, even in these large battery applications, as shown in the Cost Comparison sections below, Li-Ion batteries are gaining ground over their Lead-acid counterparts.

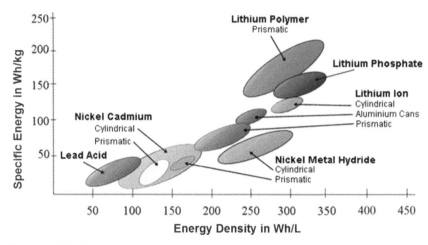

Figure 13.7. Specific energy and energy density of mainstream secondary batteries.

Example 13.1

A battery must be selected to operate on-board electronics on a NASA rover for an exploratory mission to Mars. If the energy required is 1000 Wh (1kWh) and the weight of the battery must not exceed 6 kg, based on specific energy considerations, which of the following battery systems would be suitable?

a.   Li-Po, or Lithium Polymer
b.   Lead-acid
c.   LiPO4

Solution/Answer: As stated in the specific energy discussion:

"Li-Po, or Lithium polymer cell battery's specific energy is approximately 180 Wh/kg. This specification, as shown below, can conversely be interpreted as 5.6 kg mass of battery per 1kWh of energy capacity of the battery."

$$\frac{1}{180 \text{ Wh/kg}} \Rightarrow 0.00556 \text{ kg/Wh}; \therefore 1000 \text{ Wh of energy would entail:}$$

Mass = (0.00556 kg/Wh) . (1,000 Wh) ≈ 5.6 kg

Therefore, a Li-Po battery, that offers a specific energy of 180 WH per kg of mass would be the best choice because a Li-Po battery capable of supplying 1,000 Wh, or 1 kWh, of energy and would at a mass less than 6 kg.

Note: Based on charts in Figures 13.6, Figure 13.7, and other information provided in this chapter, a Lead-Acid battery system would have a mass of, approximately, 25 kg and a $LiPO_4$ battery would have a mass of, approximately, 10 kg. In addition, more important from the physical application point of view, because this problem entails battery application in a spacecraft—which is designed to undergo three dimensional motion—the Lead-Acid battery would not offer a viable alternative.

COST COMPARISON OF MAJOR SECONDARY BATTERIES

Cost of major secondary, rechargeable, batteries has continued to trend down over the last two decades. Table 13.1 compares estimated cost of mainstream rechargeable batteries commonly used in consumer electronics, tools and electric automobiles—also referred to as EV's, or Electric Vehicles. Car maker GM® revealed at their 2015 Annual Global Business Conference that they expected the price of EV (Li-ion cell) batteries to drop to $145 per-kilowatt-hour in 2016—substantially lower than other analyst's earlier cost estimates. GM also expects a cost of US $100 per kWh by the end of 2021. However, reader is advised to bear in mind that cost of batteries—or energy storage devices, in general, especially, those that have not reached technological maturity—is volatile, dynamic and in state of flux. Therefore, the only most current and verifiable costs should be used in formal work. The cost data provide herein is for relative comparison purposes, only.

Battery Type	Year	Cost ($/kWh)
Li-Ion	2016	145
Li-Ion	2014	200-300
Li-Ion	2012	500-600
Nickel Metal Hydride	2004	750
Nickel Metal Hydride	2013	500-550
Lead Acid	2013	256.68

Table 13.1: Mainstream Battery Cost Estimate in $/kWh

Universal Symbol for Battery:

Most batteries consist of a stack of cells connected in series. Therefore, as shown in Figure 13.8, the universal battery symbol depicts a stack of cells in a battery. From another perspective, the individual cells within the overall battery are the "building blocks" of the battery. The more the number of cells in a battery, the more the overall voltage of the battery, and generally, the more the energy storage capacity. As introduced in earlier chapters of this text, and later in this chapter, the symbol for battery depicted in Figure 13.8 is used to represent a battery in electrical circuits and circuit analysis.

Figure 13.8. Standard symbol for a multi-cell battery

As evident, application of batteries beyond the ordinary applications such as toys, remote controls watches, cell phones and other myriad portable electronic devices, is limitless and transformative, especially with the focus on renewable energy technologies and the need to cater to base load.

Discharging Characteristics of Batteries
An *ideal* battery would always deliver the same amount of energy and amp-hours regardless of the rate of discharge or the number of amps, at a constant voltage, as depicted in Figure 13.9.

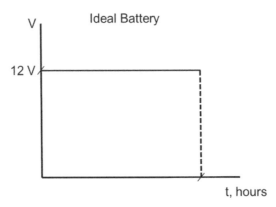

Figure 13.9. Discharge profile of an *ideal* battery

The performance of *real* batteries, however, could be explained more accurately through the voltage versus time graph shown in Figure 13.10. As shown in Figure 13.10, when the rate of discharge, or amperage (current), is higher, the operating voltage of the battery drops in conjunction with the decrease in the amount of energy the battery is capable of delivering before cutting off.

Figure 13.10. Discharge profile of real battery

Depth of Discharge

Depth of discharge, or DOD, is an indication of the amount of energy or charge depleted. The SOC, or State of Charge, on the other hand is the amount of energy or charge left in the battery. In percent, the relationship between DOD and SOC can be expressed as:

DOD (%) = 100% – SOC (%)

Or, SOC (%) = 100% – DOD (%)

Advantages and Disadvantages, Lithium Ion vs. Lead Acid

Charging
- A lithium ion could be charged in around one hour versus, approximately, eight hours for a lead acid battery. However, charging of a lithium ion battery would be difficult at sub-zero temperatures; its temperature would need to be raised through the use of an external heater before it can be charged.

- Lithium polymer cells, with a nominal voltage of 3.7 volts, can experience catastrophic failure if charged to 4.7 volts or greater. Therefore, robust protection design is required to avoid unsafe charging of lithium ion batteries. Lead acid batteries, in contrast, are more tolerant of overcharging. While sudden overcharging of lead acid batteries can also be damaging, it is less likely to result in a catastrophe.

Self-discharge Rates

*   A Lithium ion battery can lose approximately 10% of its charge over a month. While, a lead acid battery could lose as little as 5%.

Specific Power, Specific Energy and Power Density

*   A lithium ion battery carries *higher specific energy* (Wh/kg) than a lead acid battery; a ratio of greater than 4:1, in favor of lithium ion battery.

*   A lithium ion battery offers *higher energy density* (Wh/L) than a lead acid battery; a ratio of greater than 5:1, in favor of lithium ion battery.

*   A lithium ion battery offers *higher specific power* (W/kg) than a lead acid battery.

Life Span of Batteries

Life span of a battery depends on the following factors:

*   **Temperature**: Batteries operated within temperature ranges specified by the manufacturer tend to outlast those that are operated at extreme temperatures, outside the specified range.

*   **Storage**: Batteries stored within temperature ranges specified by the manufacturer tend to outlast those that are operated at extreme temperatures.

*   **Treatment**: Frequent and high depths of discharge (DOD) tend to shorten the lifespan of batteries.

C—Rating of Batteries

C-Rating of a battery relates the capacity of the battery to the rate of discharge. C-Rating is also used to correlate the battery capacity to the charging rate. For example, C_{20} rating of a battery would be the energy capacity of the battery if it were to deliver a constant current for a period of 20 hours. The C-Rating of a battery also marks the duration *after* which the battery operation must be ceased. In other words, if a C_{20} battery is not cut-off after continuous operation of 20 hours, it can be damaged permanently.

Additionally, a C_{20} battery capacity rating of 7Ah would imply that this battery would deliver 350mA of current over a period of 20 hours. In other words:

Current Delivered by a 7 Ah C_{20} Battery = 7 Ah/20 h = 0.350 A or 350 mA.

So, a C_{20} battery rated at for nominal 7 Ah would *not* be capable of delivering 7 Amps for a period of 20 hours. Instead, as shown in the Table 13.2, a nominal 7 Ah C_{20} battery would deliver 7 Amps for a period of only 1 hour. Therefore, a more complete and explicit rating of a battery should include specification of the discharge rate of the battery in addition to the "C" rating. So, as indicated in Table 13.2, a more complete specification of a 7 Ah, C_{20}, battery, being applied in a 1 A discharge mode would be: Nominal 7 Ah, 1 C_{20}. For addition clarification, let's examine the 7 Ah C_{20} battery in a 0.14 A discharge mode. A complete and proper C-rating specification of the battery, in this case, would be **7Ah, 0.02 C_{20}**. Note that the coefficient of the C-rating represents the "Discharge Rate of the Battery," as stated in the second column of Table 13.2. From the second column in Table 13.2, we also see that when a nominal 7 Ah, C_{20} battery is discharged at only 0.14 Amps, it can sustain the demanded current and wattage for up to 50 hours—a period that is 50 times longer than the "20—hour" safe discharge specification of the battery. So, in this 0.14 A discharge case, the battery discharge rate—or the coefficient of C_{20}—would be 1/50 or 0.02. As apparent from further examination of Table 13.2, batteries, almost always, will last longer than the "C" rating when discharged at a rate lower than the specified "Discharge Rate" of the battery.

Nominal Amp Rating, in Ah	Discharge Rate of a C_{20} Battery	C-Rating of the Battery in hours	Current Draw in A	Time to Complete Discharge or Cut-off, in hours
7	0.02	20	0.14	50
7	0.05	20	0.35	20
7	0.1	20	0.7	10
7	0.2	20	1.4	5
7	0.5	20	3.5	2
7	1	20	7	1
7	2	20	14	0.5

Table 13.2: Voltage, and energy capacity of common primary batteries

The C_{20} rating is more common among smaller batteries. The larger batteries, commonly used for "off-the-grid" applications, are rated C_{100}, or have C-rating of 100, and are rated based on 100 hour discharge.

Battery Discharge Rate and Nominal Capacity

The performance of a battery, in terms of its nominal capacity, is a function of the discharge rate. Relationship of battery's nominal capacity and the rate at which it is discharged is illustrated in Figure 13.11. The graph in Figure 13.11 represents a 12 V, 7 Ah, C_{20}, battery. As apparent from the graph, the performance of the battery, or the nominal capacity of the battery, would be 100% of the rated value if it is discharged at 0.05 C_{20} rate. As the discharge rate is increased to 0.5 C_{20} and beyond, the nominal capacity and performance of the battery declines; delivering only 34.4% of rated capacity at 2 C_{20}.

Figure 13.11: Battery discharge rate and its nominal capacity

Temperature and Battery Discharge Characteristics

The performance of a battery, when in use or being discharged, is influenced by temperature as shown in Figure 13.12. Figure 13.12 depicts the performance of a 12 V, 7 Ah, 0.5 C_{20} battery relative to temperature. A battery, typically, exhibits optimal 100% performance at standard temperature of approximately 20°C (68°F). The graph also shows that at a sub-zero temperature of –0°C (23°F), the capacity of the battery declines, substantially, delivering only 80% of the rated energy.

Figure 13.12. Discharge performance of a 12 Volt, 0.5 C_{20} 7Ah battery relative to temperature

Power and Energy Performance Assessment of Battery

The power and energy assessment of battery will be illustrated through Example 13.2, which is based on the battery performance versus temperature information stated in the previous section.

Example 13.2:

The data pertaining to a 7Ah 0.5 C_{20} battery is as follows:

Rated Battery Charge: 7 Ah (Amp-hours)
Rated Discharge Period: 20 hours (C_{20})
Discharge Current, based on the 0.5 coefficient of the 0.5 C_{20}: 3.5 A.
Battery Voltage: 12 V

Determine the following:

(a) Power and energy drawn or demanded from the battery at 20°C (68°F)
(b) Power and energy drawn or demanded from the battery at -5°C (23°F)

Solution:

Based on the specifications of the 12 V, 7Ah, 0.5 C_{20} battery stated above, and the graph depicted in Figure 13.12:

(a)   Power, at 20°C (68°F):
          In W: 12 V x 3.5 A = 42 W

        Energy, at 20°C (68°F):
          In Ah: 3.5 A x 2 h = 7 Ah
          In Wh: 12 V x 3.5 A x 2 h = 84 Wh

(b)   Power, at -5°C (68°F):
          In W: 0.8 x 12 V x 3.5 A = 33.6 W

        Energy, at -5°C (68°F):
          In Ah: 0.8 x 3.5 A x 2 h = 5.6 Ah
          In Wh: 0.8 x 12 V x 3.5 A x 2 h = 67.2 Wh

Safe Discharge Ratings of Batteries

Safe discharge ratings of batteries are somewhat analogous to the concept of "Service Factor" discussed earlier in this text, in that, if a battery is rated eight times the nominal "Ah" rate, it has been tested, under extreme circumstances, to discharge at eight times the nominal current, safely. So, in the example of nominal 7 Ah, C_{20} battery stated above, a safe discharge rating of "8" imply that the battery is capable of safely discharging at the rate of:

$$8 \text{ x } C_{20}\text{-Rating, or, } 8 \text{ x } 7 \text{ Ah nominal, or } 56 \text{ A}$$

Note that as a battery discharges, its voltage decreases slightly.

Power and Energy (Ah) Formulas and Numerical Examples:

Example 13.3:

How much does the energy capacity of a D-cell drop when the drain current is increased from 10 mA to 100 mA?

Solution:

From the Table 13.3, the capacity at 10mA:

C = (10x10-3 A) . (500 h) = 5000x10-3 A-h

The capacity at 100 mA is:

C = (100x10-3 A) . (45 h) = 4500x10-3 A-h

The decrease in capacity is:

Battery Type	Voltage Rating in Volts	Drain Current mA	Service Capacity in hours	Energy Capacity in mAh
AAA	1.5	2	290	580
	1.5	10	45	450
	1.5	20	17	340
AA	1.5	3	350	1050
	1.5	15	40	600
	1.5	30	15	450
C	1.5	5	430	2150
	1.5	25	100	2500
	1.5	50	40	2000
D	1.5	10	500	5000
	1.5	50	105	5250
	1.5	100	45	4500

Table 13.3: Voltage, and energy capacity of common primary batteries

$$1 - \frac{4500 \times 10^{-3} \text{ A-h}}{5000 \times 10^{-3} \text{ A-h}} = 0.1$$

Battery Equivalent Circuit

Batteries possess and exhibit internal resistance which results in lower output voltage as current increases. The internal resistance of a battery is labeled as R_{int} in the battery equivalent circuit shown on Figure 13.13 below. The internal resistance of a battery increases with the age of the battery and the frequency and magnitudes of the DOD, Depth of Discharge. This is the reason why the terminal voltage of a battery declines with the age of the battery.

Figure 13.13. Battery equivalent circuit

Example 13.4

A 1.5 volt battery with an internal

resistance of 0.3 Ω is connected to a load of 1 Ω.

(a)   What is the load current?
(b)   What is the power dissipated in the battery resistance?
(c)   What is the power delivered to the load?

Solution:
(a)   Load current:
$$V = IR$$

$$I = \frac{V}{R} = \frac{V}{R_{int} + R_{load}}$$

$$= \frac{1.5V}{0.3\ \Omega + 1\ \Omega} = 1.15\ A$$

(b)   Power dissipated in the internal resistance of the battery:

The power is determined by

$$P = I^2 R_{int}$$
$$= (1.15\ A)^2(0.3\ \Omega) = 0.397\ W$$

(c)   Power delivered to the load:

$$P = I^2 \cdot R_{load} = (1.15\ A)^2 \cdot (1\ \Omega) = 1.32\ W$$

Battery Voltage Regulation
Voltage regulation of a battery is an indication of how much voltage drop is encountered in a battery application, or in a battery circuit, when a battery is supporting a load.

Voltage Regulation of a Battery: VReg

$$\%\ VReg = \frac{V_{NL} - V_{FL}}{V_{FL}} \cdot 100\% \qquad \text{Eq. 13.7}$$

In Figure 13.13, V_{NL}, the no-load voltage, would be the voltage (V_{term}) measured at the terminals before the load R_{load} is connected, with no current flowing in the circuit. The no-load voltage V_{NL} will also be the rated voltage specified on the battery, V_{bat}, i.e. 1.5 V for AAA, AA, C, and D cell primary batteries, as indicted in Table 13.3. Once the load, R_{load}, is connected into the circuit, the current "I" begins to flow and a voltage drop, $V_{R\text{-}int}$, develops across the internal resistor, R_{int}, of the battery. This results in a finite reduction in the terminal voltage. This lower terminal voltage with the current flowing in the circuit is V_{FL}.

Application of KVL in the battery circuit, underload, yields:

$$+V_{term} - V_{Bat} + V_{R\text{-}int} = 0$$
$$\text{Or, } V_{term} = V_{Bat} - V_{R\text{-}int} = V_{FL}$$
$$\text{Since } V_{Bat} = V_{NL}, \; V_{term} = V_{VL} = V_{NL} - V_{R\text{-}int}$$

Therefore, the voltage regulation equation can be expanded as follows:

$$\% \, VReg = \frac{V_{NL} - V_{FL}}{V_{FL}} \cdot 100\% = \frac{V_{NL} - (V_{NL} - V_{R\text{-}int})}{V_{Bat} - V_{R\text{-}int}} \cdot 100\%$$

$$\text{Or, } \% VReg = \frac{I.R_{int}}{V_{Bat} - I.R_{int}} \cdot 100\% \qquad \qquad \text{Eq. 13.8}$$

Example 13.5

What is the full load voltage across the battery terminals in Example 13.4? What is the voltage regulation of the battery with the full load of 1 Ω in the circuit?

Strategy: Apply KVL in the circuit and then use Eq. 13.8 to assess the % Voltage Regulation.

Application of KVL in the battery circuit, underload, yields:

$$+V_{term} - V_{Bat} + V_{R\text{-}int} = 0$$
$$\text{Or, } V_{term} = V_{Bat} - V_{R\text{-}int} = 1.5 - (I).(R_{int}) = 1.5 - (1.15A).(0.3)$$
$$\therefore V_{term} = 1.155 \, V$$

$$\% \, VReg = \frac{I.R_{int}}{V_{Bat} - I.R_{int}} \cdot 100\% = \frac{(1.15 \, A).(0.3 \, \Omega)}{1.5 \, V - (1.15 \, A).(0.3 \, \Omega)} \cdot 100\% = 29.87\%$$

Note: The regulation in this example is relatively high due to the high current demand from the 1.5 V battery.

Peukert's Law

Although we have explored Lead-Acid battery discharging characteristics in foregoing sections of this chapter, we would be remiss if didn't discuss Peukert's Law contributions toward understanding of Lead-Acid battery operation. Peukert's law describes the relationship between the discharge rate and the available capacity, SOC, of a Lead-Acid battery. According to Peukert's law, as the rate of discharge increases, the battery's available capacity decreases. Note: ***Peukert's law holds true only if a battery is discharged a constant discharge current and constant temperature.*** If the exponent constant "**k**" was equal to "1," the delivered capacity would be independent of the current. For a lead–acid battery the value of **k** is typically between **1.1** and **1.3**. The Peukert constant varies according to the age of the battery, generally ***increasing*** with age.

The approximate change in capacity of rechargeable lead-acid batteries for different **k** values, and at different rates of discharge, can be assessedusing Eq. 13.9 and Eq. 13.10.

$$C_p = I^k t, \qquad\qquad\qquad\qquad\text{Eq. 13.9}$$

$$t = H \left(\frac{C}{IH} \right)^k \qquad\qquad\qquad\qquad\text{Eq. 13.10}$$

Where,

  H = Rated discharge time (in hours),
  C_p = is the capacity at a one-ampere discharge rate, expressed in Ah
     (Ampere-hours)
  C = Rated capacity at that discharge rate (in ampere-hours),
  I = Actual discharge current (in amperes),
  k = The Peukert constant (dimensionless),
  t = Actual time to discharge the battery (in hours).

Example 13.6

A battery is rated 100 A-h, when discharged over 20 hours. The battery is connected to a circuit that draws 10 amps. If the Peukart's exponent of the battery is approximately 1.4, the amount of time the battery can sustain the current in the circuit is most nearly:

  (A)   10hour
  (B)   20 hours

(C)   7.5 hours
(D)   25.5 hours

Solution:
Apply the following Peukart's Law equation, Eq. 13.10:

$$t = H \left(\frac{C}{IH} \right)^k$$

Where,

H = Rated discharge time (in hours) = 20 hours
C = Rated capacity at that discharge rate (in ampere-hours) = 100 A-h
I = Actual discharge current (in amperes) = 10 A
k = The Peukert constant (dimensionless) = 1.4
t = Actual time to discharge the battery (in hours) = ?

$$t = H \left(\frac{C}{IH} \right)^{(k)} = 20 \left(\frac{100}{(10) \cdot (20)} \right)^{(1.4)} = 7.58 \text{ hours}$$

Answer: **(C)**

Peukert's exponent can be determined algebraically. Peukert's constants for some of the more common batteries are listed in Table 13.4.

$$\frac{Q}{Q_o} = \left(\frac{T}{T_o} \right)^{\left(\frac{k-1}{k} \right)}$$

Eq. 13.11

Where, Q and Q_o represent charging or discharging in times T and T_o, respectively.

Another commonly used form of the Peukert's law is:

$$\frac{Q}{Q_o} = \left(\frac{I}{I_o} \right)^a$$

Eq. 13.12

Where, exponent or constant "a" can be defined mathematically as:

$$a = \left(\frac{k-1}{2-k} \right)$$

Eq. 13.13

Table 13.4: Voltage, and energy capacity of common primary batteries

α	k	Description/Comments
0	1	Full charge regardless of the applied current, i.e. an ideal battery
0.1	1.090909	VRSLAB AGM batteries
0.2	1.166667	VRSLAB AGM batteries; available with a different α
0.25	1.200000	Gelled
0.3	1.230769	Gelled; available with a different α
0.33	1.249998	Flooded Lead-acid battery
0.8	1.444444	Flooded Lead-acid battery; available with a different α
0.9	1.473684	Additional Example of α and k for a Flooded Lead-acid Battery
0.4	1.285714	Regular Lithium Ion battery
0.5	1.333333	Diffusion Control, Cottrell-Warburg
0.75	1.428571	Additional Example of α and k for Diffusion Control, Cottrell-Warburg Battery

Charging of Secondary Batteries

Charging of secondary batteries is a function of voltage, current, time, chemistry and the design of the battery. In the ideal world, one would wish to charge a battery to its full capacity, in the least amount of time, safely, without negatively impacting the life of the battery. In the real world, however, one has to choose between capacity of recharge, the speed of recharge, physical constraints—i.e. size, weight, energy density, specific power, specific energy, power density, etc.—and economics.

Gassing Voltage and Maintenance of Lead Acid Secondary Batteries

Gassing voltage is the voltage at which electrolysis begins to electrically break down water, H_2O, into hydrogen, H_2, and Oxygen, O_2. The gassing voltage for a lead acid battery is, approximately, 2.4 volts per cell. A typical 12 V lead acid SLI battery consists of six (6) cells, therefore, its gassing voltage is approximately 14.4 volts. Because hydrogen and oxygen are highly flammable, they must be vented safely during the charging process. A certain amount of gassing is expected and it serves to circulate the electrolyte. However, excess gassing depletes the water and increases the concentration of electrolyte, H_2SO_4. Therefore, as a regular maintenance requirement, distilled water must be added to the battery, periodically, to sustain proper electrolyte concentration. This periodic maintenance requirement can be circumvented through "maintenance free" battery design, wherein, by design, the vented hydrogen and oxygen gases are recombined to form water, thus maintaining the electrolyte concentration.

Charge rate of a battery is a function of the voltage applied and its capacity, or state of charge, SOC. Note that, with constant internal resistance of the battery, charging voltage determines the charging current. In other words, higher voltage would result in higher charging currents and accelerated rate of electrolysis, or gassing, at or greater than gassing voltage. Safe and optimal charging of a battery requires that the battery voltage not exceed gassing voltage for prolonged periods of time. If the battery capacity is substantially depleted, or its SOC is low, for instance 10%, it should be charged at lower voltage and low charging current to avoid prolonged gassing. If a battery is already at a high SOC, say 90%, aggressive charging would likely drive the voltage up to or beyond the 14.4 V electrolysis voltage level, resulting in higher current and subsequent gassing. It is for this reason that most battery charging systems are designed with voltage and current control systems.

A battery charging system can be designed to meet one of the following functional criteria, or a combination of both:

Single Step Constant Current Charging

In a single step *constant current* charging process, voltage is ramped up gradually as the battery is charged, or the state of charge increments from low SOC to high SOC. However, the rate of increase of the voltage, as depicted by the slope of the voltage line, in Figure 13.14, is controlled in such a fashion that the charging current remains constant at a low

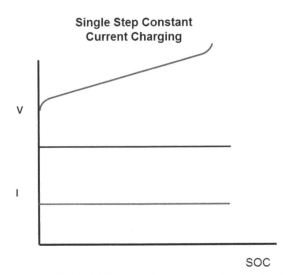

Single Step Constant Current Charging

Figure 13.14. Voltage and current profile in single step constant current charging process magnitude. While this type of charging approach is safe, it can prolong the charging or recharging time.

Multi-step Constant Current Charging

The slow charging rate shortcoming of the single step constant current charging approach can be overcome with the addition of more controls, such that a voltage is ramped up, initially, at a high rate. But, as the voltage approaches the gassing level, it is reduced or stepped down by a finite amount to lower the current level as shown in Figure 13.15. As shown in Figure 13.15, this closed looped control approach is repeated until the battery has reached 100% SOC.

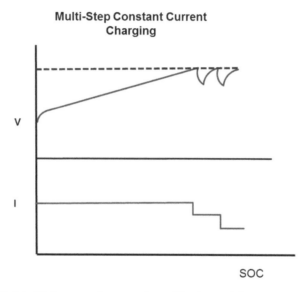

Figure 13.15. Voltage and current profile in multi-step constant current charging process

Constant Voltage Charging

A constant voltage charging approach, as shown in Figure 13.16, is relatively simple and can be implemented at low cost through the incorporation of a voltage regulator. With the application of a constant voltage, at a magnitude just below the gassing voltage, the current would decline as the battery gets charged, or as the SOC increases. However, this simplicity and cost advantage is coupled with the risk of battery drawing a large, unsafe, amount of current when it is deeply discharged, or at a very low SOC level.

Constant Voltage Charging

V

I

SOC

Figure 13.16. Voltage and current profile in multi-step constant current charging process

Modified Constant Voltage Charging

Single Step Constant Current Charging, Multi-Step Constant Current Charging and Constant Voltage Charging approaches, in isolation, present disadvantages. However, a hybrid approach that leverages the desirable characteristics of all three the charging approaches discussed above, is the approach employed by most contemporary secondary battery chargers. This approach is referred to as the "Modified Constant Voltage Charging," or, "Fast Charging." The principle behind the modified constant voltage charging method is that it combines the benefits of constant voltage and constant current approaches to yield versatile and robust battery charging systems. See Figure 13.17.

As depicted in Figure 13.17, the modified constant charging process can be categorized into three stages:

- Stage I—Constant Current Stage
- Stage II—Constant Voltage Stage
- Stage III—Constant voltage stage

Stage I—Constant Current Stage: The *constant current* stage pertains to the charging phase when the battery at low SOC. The modified constant voltage charging phase begins with a gradual increase in voltage, while keeping he current constant and below the gassing voltage level and at or below the maximum current allowable by the battery manufacturer. This, controlled, low level of

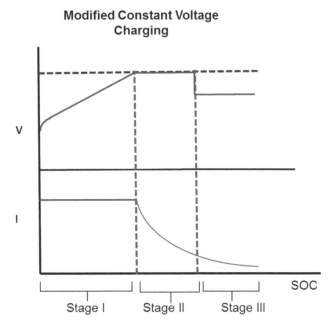

Figure 13.17. Modified Constant Voltage Charging Process

current ensures that the battery does not overheat. This stage, typically, takes up almost *half* of the required battery charge time. This stage is also referred to the "bulk stage" because the battery receives bulk of its charge in this stage.

Stage II—Constant Voltage Stage: In this stage, the battery charger switches to *constant voltage* mode and allows the current to ramp down as the battery absorbs more and more charge. That is the reason why this stage is sometimes referred to as the "Absorption Stage," or the "Topping Charge Stage." Since the battery has already achieved a substantial magnitude in voltage when it reaches this stage, the voltage differential between the battery and the charger is much smaller and the current driven through the battery continues to decline. The battery voltage at this stage is usually just below the gassing voltage, or approximately, 14.4 V for a typical lead acid battery.

Stage III—Float Voltage Stage: The battery enters the float voltage stage when it is almost saturated with charge or energy, and it only needs a small amount of "trickle charge" to keep the battery in fully charged (100% SOC) state. In typical lead acid batteries, at this stage, the charges switches to a constant voltage of, approximately, 13.8 V.

Batteries—*The Future*

Recent advances in nano-technology holds a promising future for lithium-ion. Egg-line nanoparticles of sulfur present an option; improving energy transfer and hugely increasing capacity. Use of silicon nanoparticles in the Li-Ion battery anode is also expected to boost their performance.

Research into the overcoming current challenges with lithium-ion tech—such as reducing the tendency for lithium to gather around the battery electrodes—is underway.

Work is underway on Lithium-air technology. This is an offshoot of lithium-ion batteries, and one that could significantly increase energy density. In essence, at this vantage point in time, it appears that Li-ion technology will be around as viable and effective electro-chemical energy storage technology for years to come.

Chapter 13—Self-assessment Problems and Questions

1. What type of battery would be most suitable for the following applications:
   a. UPS, Uninterruptable Power Supply System
   b. Starting an automobile
   c. Electric camp light
   d. Remote Wind turbine
   e. Remote solar PV farm

2. In Example 13.1, if the volume of space reserved for the battery is 3 liters or less, based on energy density considerations, which of the given battery designs would be suitable?
   a. Li-Po, or Lithium Polymer
   b. Lead-acid
   c. LiPO4

3. It has been disclosed that the battery being selected for the mission to Mars, as outlined in problem 2, must be capable of serving as an emergency source of power for 1 kW of ignition power without overheating. However, the mass of the battery must not exceed 3.5 kg. Based on the specific power data provided in this chapter, which of the three batteries would meet the specifications?

4.   A LiFePO4 cell battery is discharged at three times the nominal discharge rate. Will the energy density of the battery:
      (a) Remain 220 Wh/L?
      (b) Be greater than 220 Wh/L?
      (c) Be less than 220 Wh/L?
      (d) None of the above?

5.   A Li-Po, or Lithium polymer cell battery, is discharged at half the nominal discharge rate. Will the specific energy of the battery be:
      (a) 180 Wh/kg?
      (b) Greater than 180 Wh/kg?
      (c) Less than 180 Wh/kg?
      (d) None of the above?

6.   A double-layered capacitor is being considered as an energy storage device to in an electrical system. The amount of initial energy needed by the system is 10 Wh. Based on the specific energy and specific power chart in Figure 13.6, the approximate weight of the double-layered capacitor would be:
      (a) 1kg
      (b) 10 kg
      (c) 100 kg
      (d) 1000 kg

7.   If, in problem 8, the maximum weight allowed for the charge storage system is 1kg, based on the specific energy and specific power chart in Figure 13.6, which charge storage system would be the most suitable?
      (a) Lead acid battery
      (b) Double-layered capacitor
      (c) Ultra-capacitor
      (d) None of the above

Appendices

Appendix A

This appendix includes the solutions and answers to end of chapter self-assessment problems and questions.

CHAPTER 1—SOLUTIONS

1.   In an AC system, a voltage source $V(t) = 120Sin(377t + 0°)$ volts, rms, sets up a current of $I(t) = 5Sin(377t + 45°)$ amps, rms. Calculate the maximum values of voltage and current in this case.

Solution
According the Eq. 1.2:

$$V_p = V_m = \sqrt{2}V_{RMS} = \sqrt{2}V_{EFF}$$

$$\therefore$$

$$V_p = \sqrt{2}V_{RMS} = \sqrt{2}(115) = 163 \text{ V}$$

According the Eq. 1.5:

$$I_p = I_m = \sqrt{2}I_{RMS} = \sqrt{2}I_{EFF}$$
$$\therefore \quad I_p = \sqrt{2}I_{RMS} = \sqrt{2}(5) = 7.1 \text{ A}$$

2.   A phase conductor of a transmission line is one mile long and has a diameter of 1.5 inch. The conductor is composed of aluminum. Calculate the electrical resistance of this conductor.

Solution
        Solution Strategy: Since the resistivity value of aluminum, as stated in Chapter 1, is in metric or SI unit system, the length and diameter specifications stated in this problem must be streamlined in metric units before application of Eq. 1.9 for determination of resistance in Ohms (Ωs).

L = 1 mile = 1,609 m
Diameter = 1.5 inch = 0.0381m; ∴ **R** = Radius = D/2 = 0.019 m
A = Area of cross-section = $\pi.\text{R}^2$ = (3.14).(0.019)^2 = 0.00113m^2
$\rho_{aluminum}$ = 28.2 n Ωm = 28.2x10^{-9} Ωm; given in Chapter 1

$$R = r.\frac{L}{A}$$ **Eq. 1.9**

$$\therefore \quad R = \rho.\frac{L}{A} = 28.2 \times 10^{-9}\left(\frac{1609}{0.00113}\right) = 0.039\Omega$$

3.   What is the resistance of the following circuit as seen from the battery?

Solution
      No current will flow through the two 4 Ω resistors, the two 3 Ω resistors, or the 7 Ω resistor. The current finds the path of least resistance through the highlighted short circuit segment of the circuit. Therefore, the circuit reduces to one 6 Ω in series with two 12 Ω in parallel.

$$R = 6\Omega + \frac{(12\Omega).(12\Omega)}{(12\Omega + 12\Omega)}$$
$$R = 6\Omega + 6\Omega = 12\Omega$$

4.   Consider the RC circuit shown in the diagram below. The source voltage is 12V. The capacitor voltage is 2V before the switch is closed. The switch is closed at t = 0. What would the capacitor voltage be at t = 5 sec?

Solution

This particular case represents a capacitor charging scenario. Given the value of **R, C,** $v_c(0)$ and the source voltage V, Equation 1.18 allows us to calculate the voltage after and elapsed time "**t,**" during the capacitor charging phase.

$$v_c(t) = v_c(0)\, e^{-\frac{t}{RC}} + V(1 - e^{-\frac{t}{RC}})$$ **Eq. 1.18**

Given:

R = 1 MΩ = 1,000,000
C = 5µF = 5x10⁻⁶F
$v_c(0)$ = 2 V = Voltage across the capacitor at t = 0
$v_c(t)$ = Voltage across the capacitor, at a given time t = ?
V = Voltage of the power source = 12V
t = 5 sec
τ = RC circuit time constant = **R.C**
       = (1,000,000)(5x10⁻⁶⁾ = 5 sec

Substitute the given and derived values in Eq. 1.18:

$$v_c(t) = v_c(0) \cdot e^{-\frac{t}{RC}} + 12.(1 - e^{-\frac{t}{RC}})$$

$$= (2V).(e^{-\frac{5}{5}}) + (12).(1 - e^{-\frac{5}{5}})$$

$$= (2V).(0.368) + (12).(0.632)$$

$$= 0.736V + 7.585V$$

$$= 8.32V$$

5.   Determine the equivalent capacitance for the DC circuit shown in the circuit diagram below if C_1 = 5µF and C_2 = 10µF.

Solution

Application of Eq. 1.21 to the two capacitor series circuit shown in the given circuit diagram yields:

$$C_{EQ} = \frac{C_1 C_2}{C_1 + C_2}$$

$$C_{EQ} = \frac{(5x10^{-6})(10x10^{-6})}{(5x10^{-6}) + (10x10^{-6})}$$

$$= 3.33x10^{-6} = 3.33\,\mu F$$

6.   Determine the equivalent capacitance for the DC circuit shown below if this circuit consists of twenty 100µF capacitors in series.

Solution

Apply Eq. 1.22 to "n" series capacitor circuit shown in diagram:

$$C_{EQ} = C_{EQ\text{-}n} = \frac{C}{n}$$

Where, **n** = 20 and **C** = 100µF

$$C_{EQ} = \frac{100\,\mu F}{20}$$

$$= 5\,\mu F$$

7. Determine the equivalent capacitance in series and parallel combination circuit shown below. The capacitance values are: C_1 = 10μF, C_2 = 10μF, C_3 = 20μF, C_4 = 20μF.

Solution

The capacitors in this circuit that lend themselves to linear combination are C_3 and C_4. Therefore, the combined capacitance, C_{34}, would be:

$$C_{34} = C_3 + C_4 = 20\mu F + 20\mu F = 40\mu F$$

Then, by applying Eq. 1.24 to this special hybrid capacitor combination case:

$$C_{EQ} = \frac{C_1 C_2 C_{34}}{C_1 C_2 + C_2 C_{34} + C_1 C_{34}}$$

$$C_{EQ} = \frac{(10x10^{-6})(10x10^{-6})(40x10^{-6})}{(10x10^{-6})(10x10^{-6}) + (10x10^{-6})(40x10^{-6}) + (10x10^{-6})(40x10^{-6})}$$

$$C_{EQ} = 4.44\mu F$$

8. Assume that the circuit in problem 6 is powered by a 60 Hz AC source instead of the DC source. Determine the total capacitive reactance, X_c, seen by the AC source.

Solution

If the DC source is replaced by an AC source, the circuit would appear as follows:

As computed in problem 6, the combined or net capacitance contributed to the circuit by the parallel and series network of capacitors is $C_{EQ} = 4.44\mu F$. Then, by applying Eq. 1.26:

$$X_c = \frac{1}{\omega C} = \frac{1}{2\pi f C}$$

Where,

      f = frequency = 60 Hz

      $C_{EQ} = 4.44\mu F$

$$X_c = \frac{1}{\omega C} = \frac{1}{2\pi f C} = \frac{1}{2(3.14)(60)(4.44 x 10^{-6})} = 597\Omega$$

9.   Consider the series RL circuit shown in the diagram below. The source voltage is 12V, R = 10Ω and L = 10mH. The switch is closed at t = 0. What would be magnitude of current flowing through this circuit at t = 2ms?

Solution

      In most series RL cases, the current value at a certain time "t" can be predicted through **Eq. 1.31**.

$$i_L(t) = i_R(t) = i(0)e^{-\frac{R}{L}t} + \frac{V}{R}(1 - e^{-\frac{R}{L}t})$$

Given:

      t = 2×10^{-3}s

      L = 10×10^{-3}H

      R = 10Ω

      V = 12 V

      i(0) = 0

$$i_L(t) = (0)e^{-\frac{10}{0.01}(0.002)} + \frac{12}{10}(1 - e^{-\frac{10}{0.01}(0.002)})$$

$$i_L(t) = \frac{12}{10}(1 - e^{-\frac{10}{0.01}(0.002)})$$

$$i_L(t) = (1.2)(1 - 0.135) = 1.04 \text{ A}$$

10. Consider the series RL circuit given in problem 9, in discharge mode, with voltage source removed. Parameters such as R = 10Ω and L = 10mH, are the same. The switch has been closed for long period of time, such that the current has developed to the maximum or steady state level 1.04 A. How much time would need to elapse for the current to drop to 0.5 A after the switch is opened.

$$R = 10\Omega \qquad L = 10\text{mH}$$

Solution

Apply series RL current equation, **Eq. 1.31**.

$$i_L(t) = i_R(t) = i(0)e^{-\frac{R}{L}t} + \frac{V}{R}(1 - e^{-\frac{R}{L}t})$$

Given:

t = ?
L = 10 x 10⁻³H
R = 10Ω
V = 0
i(0) = 1.04A
$i_L(t)$ = 0.5A

$$i_L(t) = (0.5) = (1.04)e^{-\frac{10}{0.01}t} + (0)(1 - e^{-\frac{10}{0.01}t})$$

$$0.5 = (1.04)e^{-\frac{10}{0.01}t}$$

$$0.481 = e^{-\frac{10}{0.01}t}$$

$$\ln(0.481) = \ln(e^{-\frac{10}{0.01}t})$$

$$-.7324 = -1000t$$

$$t = 0.00073s \text{ or } 0.73 \text{ ms}$$

11.   Determine the equivalent inductance L_{EQ} for three parallel inductor DC circuit shown in the diagram below if $L_1 = 2mH$, and $L_2 = 5mH$ and $L_3 = 20mH$.

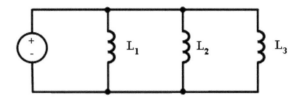

Solution

Apply Eq. 1.36 to compute L_{EQ} for the three parallel inductor circuit.

$$L_{EQ} = \frac{L_1 L_2 L_3}{L_1 L_2 + L_2 L_3 + L_1 L_3}$$

$$= \frac{(2mH)(5mH)(20mH)}{(2mH)(5mH) + (5mH)(20mH) + (2mH)(20mH)}$$

$$= 1.33mH$$

12.   Calculate the net or total inductance as seen from the 24V source vantage point in the circuit shown below.

$L_1 = 10mH$

Solution

Focus on the parallel combination of L_2, L_3, and L_4, first. Apply Eq. 1.36 to calculate the equivalent inductance L_{234} for the three parallel inductors:

$$L_{234} = \frac{L_2 L_3 L_4}{L_2 L_3 + L_3 L_4 + L_1 L_4}$$

$$= \frac{(20\text{mH})(30\text{mH})(40\text{mH})}{(20\text{mH})(30\text{mH}) + (30\text{mH})(40\text{mH}) + (20\text{mH})(40\text{mH})}$$

$$= 9.23\text{mH}$$

This reduces the circuit as shown below:

Inductors L_1 and L_{234}, in this reduced circuit, lend themselves to a linear combination. Therefore, the equivalent inductance L_{EQ} for the entire parallel and series inductor hybrid circuit would be:

$$L_{EQ} = L_1 + L_{234} = 10\text{H} + 9.23\text{H} = 19.23 \text{ H}$$

13. Assume that the circuit in Problem 12 is powered by a 60 Hz AC source. Calculate the inductive reactance, X_L, as seen by the AC voltage source.

Solution

If the DC source is replaced by an AC source, the circuit would appear as follows:

L_{EQ}, as seen by the AC voltage source, is shown in the simplified equivalent circuit below:

$$L_{EQ} = 19.23\text{mH}$$

As computed in Problem 12, the combined or net inductance contributed to the circuit by the parallel and series network of inductors is $L_{EQ} = 19.23$ mH. Then, by applying Eq. 1.37, the inductive reactance, $X_{L\text{-}EQ}$ as seen by the AC voltage source V_{AC}, would be:

$$X_{L\text{-}EQ} = \omega.L = (2\pi f).L_{EQ}$$
$$= 2(3.14)(60\text{Hz})(19.23\text{mH})$$
$$= 7.25 \ \Omega$$

CHAPTER 2—SOLUTIONS

1.   Determine the following for the DC circuit shown below if $R_1 = 5\Omega$, $R_2 = R_3 = 10\Omega$, and $R_4 = R_5 = 20$:
   a) Current flowing through resistor R_1
   b) Voltage across resistor R_5

Solution
   a) R_{eq} was derived in Example 1.2 as follows:
Combination of R_2 and $R_4 = R_{2,4} = R_2 + R_4 = 10\Omega + 20\Omega = 30\Omega$
Combination of R_3 and $R_5 = R_{3,5} = R_3 + R_5 = 10\Omega + 20\Omega = 30\Omega$

Combination of $R_{2,4}$ and $R_{3,5}$ =

$$R_{2-5}= \frac{(30\Omega).(30\Omega)}{(30\Omega+30\Omega)} = \frac{900}{60} = 15\Omega$$

$$R_{eq} = R_1 + R_{2-5} = 5\Omega + 15\Omega = 20\Omega$$

$$R_{eq} = R_1 + R_{2-5} = 5\Omega + 15\Omega = 20\Omega$$

Current through R_1 would be the same as the current through the 12V supply:

$$I = \frac{V}{R_{eq}} = \frac{12V}{20\Omega} = 0.6A$$

b) One method for determining V_{R5}, voltage across R_5, is to first calculate V_{R2-5}, the voltage across the combined resistance of resistances R_2, R_3, R_4, and R_5. Then, by applying voltage division, calculate V_{R5}:

According to Ohm's law:
$$V_{R2-5} = I.(R_{2-5}) = (0.6A).(15\Omega) = 9V$$

Then, by applying the voltage division rule:

$$V_{R5} = (9V).(\frac{R_5}{R_5 + R_3})$$

$$= (9V).(\frac{20\Omega}{20\Omega+10\Omega})$$

$$= (9V).(0.67) = 6V$$

2. What is the current through the 6 Ω resistor?

Solution

Simplify the circuit.

3 Ω in parallel with 6 Ω = 2 Ω

2 Ω in series with 4 Ω = 6 Ω

$$i = \frac{6\text{ V}}{6\text{ }\Omega} = 1\text{ A}$$

$$R_{parallel} = 3\text{ }\Omega$$

$$R_{total} = 3\text{ }\Omega + 6\text{ }\Omega = 9\text{ }\Omega$$

$$i = (1\text{ A})\left(\frac{3\text{ }\Omega}{3\text{ }\Omega + 6\text{ }\Omega}\right) = 1/3\text{ A}$$

3. Find the current through the 0.5 Ω resistor.

Solution

The voltage sources around the left loop are equal to the voltage drops across the resistances.

$$20\text{ V} - 19\text{ V} = 0.25\text{ }\Omega\ i_1 + 0.4\text{ }\Omega\ (i_1 - i_2)$$

The same is true for the right loop.

$$19\text{ V} = 0.4\text{ }\Omega\ (i_2 - i_1) + 0.5\text{ }\Omega\ i_2$$

Solve two equations and for two unknowns, using the simultaneous equation method:

$$0.65\text{ }\Omega\ i_1 - 0.4\text{ }\Omega\ i_2 = 1\text{ V}$$

$$-0.4\text{ }\Omega\ i_1 + 0.9\text{ }\Omega\ i_2 = 19\text{ V}$$

$$i_1 = 20\text{ A}$$

$$i_2 = 30\text{ A}$$

The current through the 0.5Ω resistor is 30A.

4.   Determine the value of currents I_1, I_2 and I_3 in the circuit shown below if the voltage source V_3 fails in short circuit mode. The specifications of all components are listed in the table below:

R_1	10Ω
R_2	2Ω
R_3	3Ω
R_4	4Ω
R_5	7Ω
R_6	3Ω
R_7	5Ω
V_1	20V
V_2	5V
V_3	12V

Solution

    Two noteworthy observations are in order before formulation of the three equations Necessary for the derivation of the three unknown currents:

(1)   Even though the stated value of voltage for source V3 is 12V, the value used in the formulation of second and third loop equations would be 0V because, in this scenario, voltage source V3 is assumed to have failed in short circuit mode.

(2)   As a matter of simplification, series resistors R2 and R7 are combined into one resistor R2-7

$$R_{2\text{-}7} = R_2 + R_7 = 2\Omega + 5\Omega = 7\Omega$$

The revised, simplified, would then be:

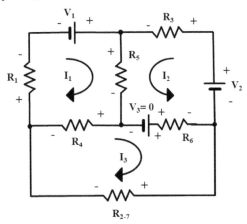

The three simultaneous equations derived by applying KVL to loops 1, 2 and 3, as described in Example 2.5, are:

$$21I_1 + 7I_2 - 4I_3 = 20$$
$$7I_1 + 13I_2 + 3I_3 = 5$$
$$-4I_1 + 3I_2 + 14I_3 = 0$$

As in Example 2.5, apply the Cramer's rule to solve for the three unknown currents I_1, I_2 and I_3. The augmented matrix thus developed would be:

$$\begin{vmatrix} 21 & 7 & -4 & \vdots & 20 \\ 7 & 13 & 3 & \vdots & 5 \\ -4 & 3 & 14 & \vdots & 0 \end{vmatrix}$$

The coefficient matrix, denoted as **A**, would be:

$$\begin{vmatrix} 21 & 7 & -4 \\ 7 & 13 & 3 \\ -4 & 3 & 14 \end{vmatrix}$$

The determinant of the coefficient matrix, denoted as $|\mathbf{A}|$, would be:

$|\mathbf{A}| = 21\{(13 \times 14) - (3 \times 3)\} - 7\{(7 \times 14) - (-4 \times 3)\} - 4\{(7 \times 3) - (4 \times 13)\} = \mathbf{2571}$

The determinant of the substitutional matrix, $\mathbf{A_1}$, for determining the value of I_1, is denoted as $|\mathbf{A_1}|$, and

$$\mathbf{A_1} = \begin{vmatrix} 20 & 7 & -4 \\ 5 & 13 & 3 \\ 0 & 3 & 14 \end{vmatrix}$$

$|\mathbf{A_1}| = 20\{(13 \times 14) - (3 \times 3)\} - 7\{(5 \times 14) - (0 \times 3)\} - 4\{(5 \times 3) - (0 \times 13)\} = \mathbf{2910}$

The determinant of the substitutional matrix, $\mathbf{A_2}$, for determining the value of I_2, is denoted as $|\mathbf{A_2}|$, and

$$\mathbf{A_2} = \begin{vmatrix} 21 & 20 & -4 \\ 7 & 5 & 3 \\ -4 & 0 & 14 \end{vmatrix}$$

$|\mathbf{A_2}| = 21\{(5 \times 14) - (0 \times 3)\} - 20\{(7 \times 14) - (-4 \times 3)\} - 4\{(7 \times 0) - (-4 \times 5)\} = \mathbf{-810}$

The determinant of the substitutional matrix, $\mathbf{A_3}$, for determining the value of I_3, is denoted as $|\mathbf{A_3}|$, and

$$\mathbf{A_3} = \begin{vmatrix} 21 & 7 & 20 \\ 7 & 13 & 5 \\ -4 & 3 & 0 \end{vmatrix}$$

$|\mathbf{A_3}| = 21\{(13 \times 0) - (3 \times 5)\} - 7\{(7 \times 0) - (5 \times -4)\} + 20\{(7 \times 3) - (-4 \times 13)\} = \mathbf{1005}$

Applying the Cramer's rule, the unknown variables, currents I_1, I_2 and I_3, can be calculated by dividing the determinants of substitutional matrices $\mathbf{A_1}$, $\mathbf{A_2}$ and $\mathbf{A_3}$, respectively, by the determinant of the coefficient matrix $\mathbf{A}$.

Therefore,

$$I_1 = \frac{|A_1|}{|A|} = 1.132 \text{ A}$$

$$I_2 = \frac{|A_2|}{|A|} = -0.315 \text{ A}$$

$$I_3 = \frac{|A_3|}{|A|} = 0.391 \text{ A}$$

Note: The negative sign for I_2 indicates that the counterclockwise direction assumed for this current is incorrect and that the correct direction of the flow of current in loop 2 is clockwise.

5. Use current division to determine the value of current I_1 in the circuit below:

Solution

We must to determine the value of source current **I**, first. In order to determine the value of current **I** flowing through the source and the two 5Ω resistors, we must consolidate all resistors into an equivalent resistance R_{EQ} and then apply the Ohm's law.

$$R_{EQ} = 5\Omega + \left(\frac{(10\Omega).(10\Omega)}{(10\Omega)+(10\Omega)} \right) + 5\Omega = 5\Omega + 5\Omega + 5\Omega = 15\Omega$$

$$I = \frac{V}{R_{EQ}} = \frac{12V}{15\Omega} = 0.8 \text{ A}$$

Apply current division formula in form of **Eq. 2.7**

$$I_{10\Omega} = I_1 = \left(\frac{R_{parallel}}{R_{total}} \right).I = \left(\frac{10\Omega}{10\Omega + 10\Omega} \right).(0.8A) = 0.4A$$

6.   Using Kirchhoff's Voltage Law, calculate the current circulating in the series resistor network below:

Solution

This problem is similar to the Example 2.2, with following exceptions:

1)   The two voltage sources are driving the current the same direction, i.e., clockwise.
2)   There are four resistors in series instead of three.

Using the strategy described in Example 2.2 and preparing the circuit for KVL application, the circuit would appear as follows:

Apply the Ohm's law to define the voltages, or voltage drops, across the four resistors. **Note:** since all four of the resistors are in series, we ***could*** combine them into a single R_{EQ} before applying KVL. However, in this case we will keep resistors separate just to maintain consistency with the approach adopted in Example 2.2.

$$\therefore \ V_{R1} = IR_1 = 10(I), \ \ V_{R2} = IR_2 = 20(I),$$
$$V_{R3} = IR_3 = 10(I), \ \text{and} \ V_{R4} = IR_4 = 30(I)$$

With all voltages—voltage source, voltage load and voltage drops across the resistors—identified and their polarities noted, apply KVL by "walking" the annotated circuit beginning at the cathode or negative electrode of the voltage source, V_{1s}. Add all voltages, with respective polarities, as you make a complete circle around the circuit, in the clockwise direction.

$$\Sigma V = 0$$
$$- 18V + 10(I) + -6V + 20(I) + 10(I) + 30(I) = 0$$
$$70(I) = 24V$$
$$\text{Or,} \ I = \frac{24}{70} = 0.343 \ A$$

Ancillary exercise: Verify the derived value of current through the alternative, R_{EQ} and the Ohm's law method, as illustrated in Approach 1 of the solution for Example 2.2.

7.   Determine the value of voltage source **current** in the parallel circuit below using KCL, Kirchhoff's Current Law.
Ancillary question: If one of the 5Ω resistors is removed (or replaced with an open circuit) and the other one is replaced with a short circuit, what would be the source current?

Solution
    KCL is applied to the given circuit after the node has been identified, circuit has been annotated with voltage designation, voltage polarity, branch currents and current directions. See circuit diagram below:

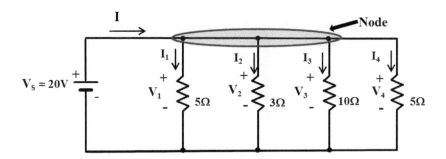

Subscribing to the definition of a node as a point where three or more conductors merge, the shaded segment in the diagram above is designated as the node for this circuit. Then, before applying KCL to determine the source current—using the Ohm's law—define the individual currents through each of the resistors, in terms of the specific resistance values and the voltages around them:

$$I_1 = \frac{V_1}{R_1} \quad I_2 = \frac{V_2}{R_2} \quad I_3 = \frac{V_3}{R_3} \quad I_4 = \frac{V_4}{R_4}$$

Since all of the resistors are in parallel with the voltage source,

$$V_1 = V_2 = V_3 = V_4 = V_s = 20V$$

Therefore,

$$I_1 = \frac{20}{5} = 4A, \quad I_2 = \frac{20}{3} = 6.67A, \quad I_3 = \frac{20}{10} = 2A$$

$$I_4 = \frac{20}{5} = 4A$$

Then, application of KCL at the designated node yields the following equation:

$$I = I_1 + I_2 + I_3 + I_4$$

Or,

$$I = 4A + 6.67A + 2A + 4A = 16.67A$$

Ancillary Question: If one of the 5Ω resistors is open-circuited and the other one short-circuited, the parallel resistor network would appear as follows:

The highlighted segment in the circuit above represents the short circuit that replaces R_1. Since R_1 is replaced by a short circuit—regardless of the disposition of other circuit elements—it becomes a ***path of least resistance*** for the entire circuit. In other words, the voltage source is short-circuited. Interpreted in terms of the Ohm's law, this would mean:

$$I = \frac{V_s}{0} = \infty$$

Since infinite current is not practical, this means that a very large amount of current would pass through the shunt or short circuit resulting in a catastrophic failure (burning or melting) of the short circuiting conductor, the interconnecting wires or a fault in the voltage source.

CHAPTER 3—SOLUTIONS

1.   A plating tank with an effective resistance of 100 Ohm is connected to the output of a full-wave rectifier. The AC supply voltage is $340V_{peak}$. Determine the amount of time, in hours, it would take to perform 0.075 faradays worth of electroplating?

Solution

Therefore, the amount of charge transfer, in Coulombs, in this electro deposition case, would be:

$$q = \left(\frac{96{,}487 \text{ Coulombs}}{1 \text{ Faraday}}\right) \cdot (0.075 \text{ Faraday}) = 7237 \text{ Coulombs}$$

As explained in Example 3.1, we are interested in the amount of time it takes to transfer a known amount of charge, so, rearrangement of Eq. 3.1 results in:

$$t = \frac{q}{I}$$

The next step entails determination of the DC voltage and current produced by the full wave rectification of 340 V_{peak} AC; which is the same as 340V_{max}.

$$V_{max} = \sqrt{2}.\, V_{rms} = 340 \text{ V}$$

$$V_{DC} = 2.\left(\frac{V_{max}}{\pi}\right) = 2.\left(\frac{340}{3.14}\right) = 216 \text{ V} \qquad \text{Eq. 3.2}$$

And,

$$I_{DC} = \frac{V_{DC}}{R} = \frac{V_{Ave}}{R} = \frac{216}{100} = 2.16\text{A} \qquad \text{Eq. 3.4}$$

Then, application of Eq. 3.1 yields:

$$t = \frac{7237 \text{ Coulombs}}{2.16 \text{ Coulombs/sec}} = 3350 \text{ seconds}$$

Or,

$$t = \frac{3350 \text{ seconds}}{3600 \text{ second/hr}} = 0.93 \text{ hours}$$

2. Determine the source current I_{rms} in the AC circuit below.

Solution

Solution strategy: Convert the given inductance value of L = 4 mH into its equivalent inductive reactance X_L. Convert the given capacitance value C = 800µF into its equivalent capacitive reactance X_C. Convert the given "black box" impedance of $Z=10\angle45°$ into the rectangular form. Combine R, X_L, X_C and Z, linearly, to determine Z_{EQ}. Then apply Ohm's Law to calculate the source current

I. Note that the given polar version of the "black box" impedance **Z=10∠45°Ω** can be converted into the equivalent rectangular form through Pythagorean Theorem, as explained in this chapter, or through a scientific calculator to:

$$Z = 10\angle 45° = 7.07 + j7.07\Omega$$

$$\therefore Z_{Eq} = R + Z_C + Z_L + Z$$

$$= R - jX_C + jX_C + Z$$

$$= R - j\frac{1}{\omega C} + j\omega.L + Z$$

$$= R - j\frac{1}{(2).(\pi).(f).(C)} + j(2).(\pi).(f).(L) + Z$$

$$= 5-j\frac{1}{(2).(3.14).(60).(800\times10^{-6})} + j.(2).(3.14).(60).(4\times10^{-3}) + 7.07 + j7.07$$

$$= 5 - j\frac{1}{0.301} + j1.51 + 7.07 + j7.07$$

$$= 5 - j3.32 + j1.51 + 7.07 + j7.07$$

$$= 12.07 + j5.26 = 13.2\angle 23.55°\Omega$$

Current "**I**" calculation:

$V(t)$, in rms form $= 110 \angle 30°V_{rms}$ as determined in the chapter examples.

Then, according to Ohm's Law, $I = \dfrac{V}{Z_{Eq}} = \dfrac{110 \angle 30°}{13.17\angle 23.55}$

$\therefore$ The source rms current, $I = 8.35 \angle 6.5°A$

3.   Calculate the impedance **Z_EQ** as seen by the AC voltage source in the circuit below:

$R_1 = 10\Omega$

$C_1 = 10\ \mu F$

$L_1 = 10mH$

$R_3 = 10\Omega$

$156\ Sin(377t + 30°)\ V$

$R_2 = 20\Omega$

Solution

Solution strategy: Convert the given inductance value of $L_1 = 10$ mH into its equivalent inductive reactance X_L. Convert the given capacitance value $C_1 = 10\mu F$ into its equivalent capacitive reactance X_C. Combine the load elements in each of the two parallel branch circuits into a single branch circuit impedance, and then combine the resulting parallel branch circuits into a single impedance, using parallel load combination formula. The impedance thus derived would be combined, through series combination approach, with resistor $R_1 = 10\Omega$ to arrive at the combined equivalent impedance representing all load elements driven by the 156 V_p AC source.

The 10mH inductor and the 10Ω resistive branch circuit, series combination:

$$Z_{R_3L} = R_3 + Z_L$$
$$= R_3 + jX_L$$
$$= R_3 + j\omega.L$$
$$= R_3 + j(2).(\pi).(f).(L)$$
$$= 10 + j.(2).(3.14).(60).(10 \times 10^{-3})$$
$$= 10 + j3.77 = 10.7\angle20.7°\Omega$$

The 10μF capacitor and the 20Ω resistive branch circuit, series combination:

$$Z_{R_2C} = R_2 + Z_C$$
$$= R_2 - jX_C$$
$$= R_2 - j\frac{1}{\omega C}$$
$$= 20 - j\frac{1}{(2).(3.14).(60).(10 \times 10^{-6})}$$
$$= 20 - j265.4 \quad = 266\angle-85.7°\Omega$$

The formula for parallel combination of the RC and RL parallel circuits:

$$\frac{1}{Z_{EqRLC}} = \frac{1}{Z_{R_2C}} + \frac{1}{Z_{R_3L}}$$

$$\text{Or, } Z_{EqRLC} = \frac{(266\angle-85.7).(10.7\angle20.7)}{266\angle-85.7 + 10.7\angle20.7} = \frac{2848\angle-65°}{263.3\angle-83.5°} = 10.8\angle18.5°$$

The final combination consisting of $R_1 = 10\Omega$ and the newly derived impedance $Z_{EqRLC} = 10.9\angle 18.5\Omega$:

$$Z_{Eq} = R_1 + Z_{EqRLC}$$
$$= 10 + 10.8\angle 18.5 = 10 + 10.26 + j3.43 = 20.26 + j3.43$$
$$= 20.55\angle 9.6\Omega$$

4.   A single phase 1 kVA resistive load, designed to operate at 240 VAC, has to be powered by a 480 VAC source. A transformer is applied as shown in the diagram below. Answer the following questions associated with this scenario:

  (a) Would the transformer be connected in a "step up" configuration or a "step down" configuration?
  (b) When installing the transformer, what turns ratio should it be connected for?
  (c) What would be the secondary current, I_s, when the load is operating at full capacity?
  (d) What would be the primary current at full load?

Solution

  The scenario described in this problem can be illustrated as follows:

(a) As discussed in this chapter, when voltage must be reduced to accommodate the load, the transformer must be connected in a step down configuration. Therefore, the answer is: **step down configuration**.

(b) **Turns ratio:** According to Eq. 3.3, the relationship between the turns ratio, the primary voltage and the secondary voltage is defined, mathematically, as:

$$V_s = \frac{V_p}{a} = \left(\frac{N_s}{N_p}\right).V_p$$

Eq. 3.12

Or,

$$a = \frac{V_p}{V_s}$$

Therefore,

$$\text{Turns Ratio} = a = \frac{480}{240} = 2:1$$

(c) **Secondary current:** As explained the single phase transformer section of this chapter, and mathematically expressed in form of Eq. 3.15:

$$\text{Single Phase AC Apparent Power} = |S_{1-\Phi}| = V_s.I_s = V_p.I_p$$

Rearranging of this equation yields:

$$I_s = \frac{|S_{1-\Phi}|}{V_s}$$

Or,

$$\text{Secondary Current} = I_s = \frac{|S_{1-\Phi}|}{V_s} = \frac{1000 \text{ VA}}{240 \text{V}} = 4.2 \text{ A}$$

(d) **Primary current at full load:** According to ideal transformer equation Eq. 3.4,

$$I_s = a.I_p = \left(\frac{N_p}{N_s}\right).I_p$$

Or,

$$I_p = \frac{I_s}{a}$$

Since the turns ratio was determined to be 2:1 in part (b), and the secondary current I_s was determined to be 4.2 A in part (c),

$$I_p = \frac{I_s}{a} = \frac{4.2}{2} = 2.1 \text{ A}$$

5. Calculate the equivalent impedance as seen from the vantage point of the AC source V_{ac} in the circuit shown below. The transformer in the circuit is assumed to be ideal. The values of the primary and secondary circuit elements are: $X_{lp} = 1\Omega$, $R_p = 4\Omega$, $R_s = 10\Omega$, $X_{Ls} = 5\Omega$, $X_{Cs} = 10\Omega$, $N_p = 100$, $N_s = 200$.

Solution

As in Example 3.2, for simplicity, the primary and secondary inductors are given in form of their reactances "X_L" instead of inductance values. In this problem, a capacitor is introduced on the secondary side. This capacitor is represented in form its X_{Cs}. As explained earlier in this chapter, in the process of reducing the AC circuits to an equivalent impedance form, the individual inductive and capacitive reactances are combined in form of their respective impedance.

Therefore, on the primary side:
$$Z_{Lp} = jX_{Lp} = j1\Omega$$
Hence, $Z_P = R_p + Z_{Lp} = 4 + j1\Omega$

On the secondary side:
$$Z_{Ls} = jX_{Ls} = j5\Omega, \text{ and}$$
$Z_{Cs} = -jX_{Cs} = -j10\Omega$; the reason for the negative sign, as explained in chapter 3, lies in the fact that $Z_{Cs} = (1/j)X_{Cs}$ and that $1/j = -j$

Hence, the total secondary impedance $\quad = Z_s = R_s + Z_{Ls} + Z_{Cs}$
$$= 10 + j5 - j10\Omega$$

a = Turns ratio = $N_p/N_s = 100/200 = 1/2$

$$Z'_s = a^2.Z_s$$

$Z'_s = (1/2)^2. (10 + j5 - j10\Omega) = 2.5 + j1.25 - j2.5\Omega$

Then, combination of the primary impedance and the total reflected secondary impedance would result in the equivalent impedance Z_{eq}:

$$Z_{eq} = Z_p + Z'_s = (4 + j1\Omega) + (2.5 + j1.25 - j2.5\Omega)$$
$$= 6.5 - j0.25\Omega$$

This final equivalent impedance Z_{eq} can be represented in rectangular complex form as:

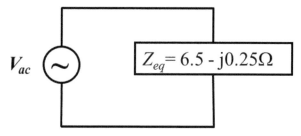

The equivalent impedance Z_{eq}, derived in the rectangular form above, can be stated in polar or phasor form as:

$$Z_{eq} = 6.5 - j0.25\Omega = 6.5\angle\text{-}2.2°$$

This conversion from rectangular to phasor form can be accomplished through a scientific calculator, with complex math feature or, as illustrated earlier in this chapter, through application of Pythagorean Theorem and trigonometry.

6.   The no load voltage at the main switch yard of a manufacturing facility is 13,400 V_{AC}. The voltage regulation of the main switch yard is 4%. What is the rated full load voltage that is most likely to be measured on the load side of the main switch yard.

Solution

$$V_{NL} = 13,400 \text{ V}$$

Voltage Regulation = **4% = 0.04**

$$V_{FL} = V_{Rated} = ?$$

Apply Eq. 3.19:

$$\text{Voltage Regulation in \%} = \left[\frac{V_{NL} - V_{FL}}{V_{FL}} \right] \times 100$$

$$\text{Voltage Regulation in \%} = 4 = \left[\frac{13,400 - V_{FL}}{V_{FL}} \right] \times 100$$

$$0.04 V_{FL} = 13,400 - V_{FL}$$

$$V_{FL} = \frac{13,400}{1.04} = 12,885 \text{ V}_{AC}$$

7.   Consider the power distribution system shown in the schematic below. Determine the following unknown parameters on the Y load side of the transformer given that the turns ratio is 2:1:

a) $|I_{L\text{-Sec}}|$ = Magnitude of load or secondary line current
b) $|I_{P\text{-Sec}}|$ = Magnitude of secondary phase current or load phase current
c) $|V_{P\text{-Pri}}|$ = Magnitude of Phase voltage on the source or primary side of the transformer
d) $|V_{L\text{-Sec}}|$ = Magnitude of Line voltage on the load or secondary side of the transformer
e) $|V_{P\text{-Sec}}|$ = Magnitude of Phase voltage on the load or secondary side of the transformer
f) $|V_{L\text{-N, Sec}}|$ = Magnitude of Line to neutral voltage on the load or secondary side of the transformer

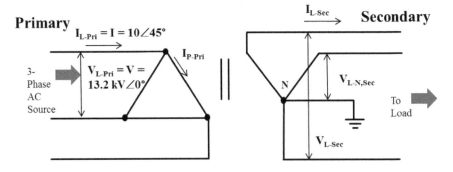

Solution

$$V_{P-Pri} = V_{L-Pri} \qquad\qquad \text{Eq. 3.28}$$

$$V_{L-Sec} = V_{P-Sec} \cdot \sqrt{3} \qquad\qquad \text{Eq. 3.29}$$

$$I_{P-Pri} = \frac{I_{L-Pri}}{\sqrt{3}} = \frac{I}{\sqrt{3}} \qquad\qquad \text{Eq. 3.30}$$

$$I_{P-Sec} = I_{L-Sec} \qquad\qquad \text{Eq. 3.31}$$

Where,
V_{P-Pri} = Primary phase voltage = **V**
V_{L-Pri} = Primary line voltage = $V_{Line-Delta}$ = **V**
V_{P-Sec} = Secondary phase voltage
V_{L-Sec} = Secondary line voltage = V_{Line-Y}

I_{L-Pri} = Primary line current = **I**
I_{P-Pri} = Primary phase current
I_{P-Sec} = Secondary phase current = I_{Line-Y}
I_{L-Sec} = Secondary line current = I_{Line-Y}

For a Δ-Y three phase transformers, as illustrated in Figure 3.26, the voltage and current transformations can be assessed using the following equations:

$$\frac{V_{Line-Y}}{V_{Line-Delta}} = \frac{\sqrt{3}}{a} \qquad\qquad \text{Eq. 3.32}$$

$$\text{or, } V_{Line-Y} = \frac{\sqrt{3}}{a} \cdot V_{Line-Delta} = \frac{\sqrt{3}}{a} \cdot V$$

$$\frac{I_{Line-Y}}{I_{Line-Delta}} = \frac{a}{\sqrt{3}}$$

$$\text{or, } I_{Line-Y} = \frac{a}{\sqrt{3}} \cdot I_{Line-Delta} = \frac{a}{\sqrt{3}} \cdot I \qquad\qquad \text{Eq. 3.33}$$

a)   According to Eq. 3.33:

$$I_{Line-Y} = \frac{a}{\sqrt{3}} \cdot I_{Line-Delta} = \frac{a}{\sqrt{3}} \cdot I$$

or, $\left| I_{L-Sec} \right| = \frac{a}{\sqrt{3}} \cdot |I| = \frac{2}{\sqrt{3}} \cdot |10\angle45°| = \frac{20}{\sqrt{3}} = 11.55 \ A$

b)   As characteristic of Y three phase AC circuits and in accordance with Eq. 3.31

$$I_{P-Sec} = I_{L-Sec}$$

or, $\left| I_{P-Sec} \right| = \left| I_{L-Sec} \right| = 11.55 \ A$

c)   As characteristic of Δ three phase AC circuits and in accordance with Eq. 3.28

$$V_{P-Pri} = V_{L-Pri}$$

and, $\left| V_{P-Pri} \right| = \left| V_{L-Pri} \right|$

$\therefore \ \left| V_{P-Pri} \right| = 13.2 \ kV$

d)   According to Eq. 3.32:

$$V_{Line-Y} = \frac{\sqrt{3}}{a} \cdot V_{Line-Delta} = \frac{\sqrt{3}}{a} \cdot V$$

or, $\left| V_{L-sec} \right| = \frac{\sqrt{3}}{a} \cdot |V| = \frac{\sqrt{3}}{2} \cdot |13.2\angle0°| = 11.432 \ kV = 11,432 \ V$

e)   As shown in Figure 3.26, on the secondary side of the transformer:

$$V_{P-Sec} = \frac{V_{Line-Delta}}{a} = \frac{V}{a}$$

and, $\left| V_{P-sec} \right| = \frac{|V|}{a} = \frac{|13.2\angle0°|}{2} = 6.6 \ kV = 6,600 \ V$

f)   As apparent from Figure 3.25 (a):

$$V_{\text{L-N, Sec}} = V_{\text{P-Sec}}$$

$$\therefore \quad |V_{\text{L-N, Sec}}| = |V_{\text{P-Sec}}| = 6.6 \text{ kV or } 6,600 \text{ V}$$

CHAPTER 4—SOLUTIONS

1.   Consider a hydroelectric reservoir where water is flowing through the turbine at the rate of 1100 ft³/sec. See the diagram below. The turbine exit point is 700 ft lower than the elevation of the reservoir water surface. The turbine efficiency is 90% and the total frictional head loss through the penstock shaft and turbine system is 52 ft.

   a) Calculate the power output of the turbine in MWs.
   b) If the efficiency of the Electric Power Generator is 92%, what would the electric power output be for this hydroelectric power generating system?

Solution

Given:

γ = specific weight = 62.4 lbf/ft^3

h_f = Frictional head loss in the penstock and elsewhere in the system, upstream of the turbine = 52 ft

Hz = Total head available due to the height of the water level in the reservoir = 700 ft

∴ Net head delivered by the water to the turbine = Head added

$= h_A = Hz - h_f = 648$ ft

$\overset{\bullet}{V}$ = Volumetric Flow Rate = 1100 ft^3/s

Turbine Efficiency 90 %

Generator Efficiency 92 %

a) Calculate the power output of the turbine in MWs.

$$P_p = WHP = \text{Hydraulic Horsepower in hp} = \frac{h_A \cdot \gamma \cdot \overset{\bullet}{V}}{550}$$

$$= \frac{(648)(62.4)(1100)}{550} = 80,870 \text{hp}$$

Eq. 4.39

P_{out} = Power Output of the Turbine

= (0.90 x 80,870 hp)

= 72,783 hp

$$P_{out} = \text{Power Output of the Turbine in MW} = \frac{72,783\text{hp} \times 0.746 \text{ kW/hp}}{1000 \text{ kW/MW}}$$

b) If the efficiency of the Electric Power Generator is 92%, what would the electric power output be for this hydroelectric power generating system?

$P_{\text{Gen. Syst.-out}}$ = Power Output of the Generator in MW

= (Power Output of the Turbine in MW).(η_{Turbine})

= (54 MW x 0.92)

= 50 MW

2. Which of the following two water heaters would cost the least to operate, on annual cost basis, under the given assumptions?

A.   Electric Water Heater:
Estimated annual energy required to heat the water: 9,000 kWh
Efficiency: **95%**
Cost Rate: **$0.10/kWh**

B.   Natural Gas Water Heater:
Estimated annual energy required to heat the water: Same as the Electric water heater
Efficiency: **98%**
Cost Rate: **$10.87/DT**

Solution
        Note the given 9000 kWh worth of electrical energy is the *energy actually absorbed by the water.* Since the electric water heater efficiency is 95%, the electrical energy *pulled from the utility* would be:

$$= \left(\frac{9000 \text{ kWh}}{0.95} \right) = 9{,}474 \text{ kWh}$$

And the annual operating cost for the electric water heater would be:
        =(9,474 kWh).($0.10/kWh) = $947.40

The *annual energy absorbed by the gas water heater*, assumed to be the same as the electric water heater = 9000 kWh. Since the gas water heater efficiency is 98%, the electrical energy pulled from the utility would be:

$$= \left(\frac{9000 \text{ kWh}}{0.98} \right) = 9{,}184 \text{ kWh}$$

Then, the annual energy consumption by the gas water heater, in DT or MMBtu would be = (9184 kWh).(3412Btu/kWh)(1DT/1,000,000Btu) = **31.33 DT.**

Since the natural gas cost rate is given as $10.87/DT, the annual operating cost for the gas water heater would be = (31.33 DT).($10.87/DT) = **$340.61.**

Answer: The gas water heater would cost substantially less to operate than the electric water heater.

3. A computer manufacturing company is testing a prototype for the amount

of heat it dissipates as wasted energy over a 10 hour period of operation. The computer is powered by a 24V DC power supply and is designed to draw 3A of current. Determine the total energy dissipated in Btu.

Solution

Apply Eq. 4.10:

Energy = V.I.t = (24V).(3A).(10hr) = (72W).(36,000 s)

$$= \left(72\frac{J}{s}\right)(36,000s) = 2,592,000 \text{ J} = 2,592\text{kJ}$$

Since 1.055 kJ = 1.0 Btu

$$\text{Energy Dissipated, in Btu} = (2,592\text{kJ})\left(\frac{1\text{Btu}}{1.055\text{kJ}}\right) = 2,457\text{Btu}$$

4. In response to a significant near miss incident and midair fire on a new commercial jet aircraft, a governmental agency is performing forensic analysis on the type of Lithium Ion aircraft battery suspected to be the root cause. Estimate the amount of *current* involved in the suspected fault on the basis of the following forensic data:

–   Total energy released in the catastrophic failure of the battery: **866kJ**
–   Estimated duration of fault: **2 seconds**
–   Rated voltage of the battery: $\mathbf{3.7V_{DC}}$

Solution

Apply Eq. 4.10:

Energy = V.I.t.

$$\text{or, } I = \frac{\text{Energy}}{V.t} = \frac{866,000\text{J}}{(3.7\text{V}).(2\text{s})} = \frac{866,000(\text{V.A.s})}{(3.7\text{V}).(2\text{s})} = 117,027\text{A}$$

5. A **156Sin377t** sinusoidal voltage is connected across a load consisting of a parallel combination of a **20 Ω** resistor and a 10Ω capacitive reactance.

(a) Determine the real power dissipated by the resistor.

(b) Determine the reactive power stored in a **10Ω** parallel capacitive reactance.

(c) Calculate the total apparent power delivered to this parallel **R** and **X** circuit by the AC voltage source.

Solution

The circuit diagram for this scenario would be as depicted below:

(a) We can apply Eq. 4.19 to determine the power dissipated or consumed in the 20Ω resistor. However, we must first derive the V_{RMS} from the given AC voltage of **156Sin377t.** This is due to the fact that coefficient of 156 stated in the give AC voltage function of **156Sin400t** is the peak or maximum voltage, V_m.

As discussed in Chapter 3 and stipulated by Eq. 3.3:

$$V_m = \sqrt{2}V_{RMS} \quad \text{or} \quad V_{RMS} = \frac{V_m}{\sqrt{2}}$$

$$\therefore V_{RMS} = \frac{156}{\sqrt{2}} = 110.3V$$

$$P = \frac{V_{RMS}^2}{R} = \frac{(110.3)^2}{20} = 608 \text{ Watts}$$

(b) Apply Eq. 4.20 to determine the reactive power sequestered in the 10Ω parallel capacitive reactance.

$$Q = \frac{V_{RMS}^2}{X} = \frac{(110.3)^2}{10} = \frac{12,166}{10} = 1217 \text{ VAR}$$

(c) Apply Eq. 4.12 to calculate the total apparent power **S** delivered to this parallel R and X circuit by the AC voltage source.

$$\bar{S} = P + jQ = 608 - j1217 = 1361 \angle -63.4°\Omega$$

Note that jQ reactive power entity is entered into the apparent power calculation above as "-jQ" because of the fact that capacitance in an AC circuit results in negative impedance contribution or "-jX." Therefore the reactive power Q due to a capacitor is applied as "-j1217," in overall S calculation.

We can also apply Eq. 4.22 to verify the magnitude of the total apparent power **S** delivered to this parallel R and X circuit:

$$|S| = \text{Magnitude of AC Apparent Power} = \sqrt{P^2 + Q^2}$$
$$= \sqrt{608^2 + 1217^2} = 1361 \text{ VA}$$

Ancillary: Reader is encouraged to verify the apparent power of 1361 VA by applying equation 4.21. **Hint:** The Z, in this case must be computed through parallel combination of R and Z_C as shown below:

$$|Z| = \left| \frac{R.Z_C}{R + Z_C} \right|$$

6. A **156Sin400t** sinusoidal voltage is connected across an unknown resistive load. If the power dissipated in the resistor is 1000 Watts, what is the resistance of the resistive load?

Solution

We can apply Eq. 4.19 to determine the value of the resistor using the given power dissipation value of 1000W. However, we must first derive the V_{RMS} from the given AC voltage of **156Sin400t**. This is due to the fact that coefficient of 156 stated in the give AC voltage function of 156Sin400t is the peak or maximum voltage, $\mathbf{V_m}$.

As discussed in Chapter 3 and stipulated by Eq. 3.3:

$$V_m = \sqrt{2}V_{RMS} \quad \text{or} \quad V_{RMS} = \frac{V_m}{\sqrt{2}}$$

$$\therefore V_{RMS} = \frac{156}{\sqrt{2}} = 110.3\text{V}$$

$$P = \frac{V_{RMS}^2}{R}$$

$$1000 = \frac{(110.3)^2}{R} \qquad \text{Eq. 4.19}$$
$$\therefore \ R = 12.17 \ \Omega$$

7. The AC circuit shown below depicts a three phase, one-line schematic of a hydroelectric power generating station. Assume that there is no voltage drop between the generator and the primary side of the transmission system transformer. The line current is indicated by an EMS system to be **10kA, RMS**. Calculate the following if the power factor is known to be **0.95**:

a)   Magnitude of the apparent power presented to the transmission lines.

b)   Magnitude of the real power presented to the transmission lines.

c)   The RMS line to neutral voltage at the source.

Solution

Magnitude of the apparent power presented to the transmission lines:
Note that the AC voltage function is NOT specified in RMS form; hence, by convention, it is "peak" or "maximum" voltage. Therefore, we need to derive the RMS voltage for subsequent computations. The line current is given in RMS form.

$$\left|V_{L\text{-}L, RMS}\right| = \left(\frac{V_{L\text{-}L, Peak}}{\sqrt{2}}\right) = \left(\frac{28,000V}{\sqrt{2}}\right) = 19,799V$$

According to Eq. 4.30:

$$\left|\overline{S}_{3\text{-}\phi}\right| = 3\left|\overline{S}_{1\text{-}\phi}\right| = 3\left(\frac{V_{L\text{-}L}.I_L}{\sqrt{3}}\right) = \sqrt{3}V_{L\text{-}L}.I_L, \text{ for three}$$

phase Y and Δ circuits.

Therefore,

$$\left|\bar{S}_{3-\phi}\right| = \sqrt{3}V_{L\text{-}L}.I_L = \sqrt{3}(19,799V).(10,000A) = 342,928,564VA$$
$$= 342,929 \text{ kVA} = 342.93 \text{ MVA}$$

Magnitude of the real power presented to the transmission lines can be determined by rearranging and using Eq. 4.23:

$$PF = \text{Power Factor} = \frac{|P|}{|S|}$$

or, $|P| = PF.|S| = (0.95).(342.929MVA) = 326MW$

(c) **Hint:** Use the $V_{L\text{-}L}$ computed in part (a). Line to neutral voltage, as introduced in Chapter 3, can be stated mathematically as:

$V_{L\text{-}N, Y} =$ Line to neutral voltage in a Y source or load

$$\text{configuration} = \frac{V_{L\text{-}L}}{\sqrt{3}} = \frac{19,799}{\sqrt{3}} = 11,431V = 11.43kV$$

8. A pump is to be installed on the ground floor of a commercial building to supply 200 ft³/sec of water up to an elevation of 100 ft. Determine the minimum size of the motor for this application. Assume that the efficiency of the pump is 80%. The weight density of water $\gamma = 62.4$ lbf/ft³

Solution
    Solution strategy in this case would be to use Eq. 4.39 to compute the WHP. Then, the amount of real power "**P**" delivered by the motor would be computed based on the given efficiency of the pump.

Given:
$h_A = 100$ ft

$\dot{V} =$ Volumetric Flow Rate $= 200$ cu – ft/sec

Pump Efficiency $= 80\ \%$
$\gamma = 62.4$ lbf/ft³

According to Eq. 4.39, Water Horsepower delivered by the pump to the water can be stated as:

WHP $= P_p =$ Fluid horse power delivered by the pump to the water

$$= \frac{(h_A)(\gamma)\left(\overset{\bullet}{V}\right)}{550} = \frac{(100 \text{ ft})(62.4 \text{lbf/ft}^3)(200 \text{ft}^3/\text{s})}{550 \text{ ft-lbf/s/hp}}$$

$$= 2269 \text{ hp}$$

Then, according to the Wire to Water (hydraulic pump) power flow diagram in Figure 4.3:

$$\text{BHP delivered by the motor to the} = \frac{\text{WHP}}{\eta_p} = \frac{2269}{0.8} \text{ hp} = 2836 \text{ hp}$$

Therefore, a commercially available motor size above 2,836 hp should be selected.

CHAPTER 5—SOLUTIONS

1. Determine the power factor of the circuit shown below, as seen by the AC source.

Solution

 This problem can be solved through multiple approaches. Two of those approaches are as follows:

i.   Calculate the equivalent or combined impedance of the circuit as seen by the source. Then take the cosine of the angle of that impedance.

ii.   Calculate the equivalent or combined impedance of the circuit as seen by the source. Apply the Ohm's law to compute the AC current. Then take the cosine of the angular difference between given AC voltage and the computed current.

We will utilize the first approach:

$$Z_{Eq} = 10\Omega + j20\Omega - j40\Omega = 10\Omega - j20\Omega = 22.36\angle\text{-}63.43°$$

$$\left\{\text{Note } |Z_{Eq}| = \sqrt{10^2 + 20^2} = 22.36 \ \Omega, \text{ and } \theta_Z = -\text{Tan}^{-1}\left(\frac{20}{10}\right) = -63.43\right\}$$

Therefore, PF = $\text{Cos}\theta_Z = \text{Cos}(\text{-}63.43°) = 0.45$ or 45%
Reader is encouraged to verify the result through approach ii.

2. Assume that the circuit depicted below represents one phase of a special power transmission line. Determine the power factor of the circuit shown below, as seen by the AC source.

Solution
    As with problem 1, this problem can be solved through multiple approaches. Two of those approaches are as follows:

i.   Calculate the equivalent or combined impedance of the circuit as seen by the source. Then take the cosine of the angle of that impedance.

ii.   Calculate the equivalent or combined impedance of the circuit as seen by the source. Apply the Ohm's law to compute the AC current. Then take the cosine of the angular difference between given AC voltage and the computed current.
    In this case will illustrate the application of approach (ii):

$Z_{Eq}= 10\Omega + j20\Omega - j20\Omega + 10\angle45°\Omega = 10\Omega + 7.07\Omega + j7.07\Omega$

$\quad = 17.07\Omega + j7.07\Omega = 18.48\Omega\angle22.5°\Omega$

$\left\{ \text{Note } |Z_{Eq}| = \sqrt{17^2 + 7.07^2} = 18.48\Omega, \text{ and } \theta_Z = \text{Tan}^{-1}\left(\dfrac{7.07}{17.07}\right) = 22.5° \right\}$

$I = \dfrac{V}{Z_{Eq}} = \dfrac{110kV\angle0°}{18.48\Omega\angle22.5°\Omega} = 5.95kA\angle\text{-}22.5°, \text{ or } \phi_1 = -22.5°$

Therefore, PF $= \text{Cos}(\theta_V - \phi_I) = \text{Cos}(0-(22.5°)) = \text{Cos}(22.5°) = 0.923$ or 92.3%

Reader is encouraged to verify the result through approach (i); which, actually, would require fewer steps.

3. If the power factor in problem 2 is less than 1.0, how much capacitance or inductance must be added in series to raise the power factor to unity?

Solution

This problem can be solved by simply focusing on the total or equivalent impedance $\mathbf{Z_{Eq}}$ and determining the amount of reactance needed to offset the reactance in the original impedance.

$Z_{Eq}= 10\Omega + j20\Omega - j20\Omega + 10\angle45°\Omega = 10\Omega + 7.07\Omega + j7.07\Omega$

$\quad = 17.07\Omega + j7.07\Omega = 18.48\Omega\angle22.5°\Omega$

Now the question is: how much and what type of reactance must be added such that Z_{Eq} reduces to its resistance portion, or 17.07Ω.

Or, $17.07\Omega + j7.07\Omega + jX\Omega = 17.07\Omega$

$\quad \therefore \quad jX = - j7.07\Omega$, the negative sign indicates that the reactance to be added must result in negative impedance; it must be capacitive.

And, $|X| = |X_C| = 7.07\Omega$

$X_C = 7.07\Omega$, and by definition, $X_C = \dfrac{1}{\omega C}$. Frequency, f, in the US, can be assumed to be 60 hz, and by definition, $\omega = 2\pi f$. Therefore,

$\quad \omega = 2\pi f = 2\,(3.14)(60hz) = 377$ radians/second

And, $C = \dfrac{1}{\omega X_C} = \dfrac{1}{(377).(7.07)} = 0.000375F$, or $375\ \mu F$

4. The output of a variable frequency drive, as shown in the circuit below, is 157Sinωt. The VFD output is currently set at 50Hz. This drive is connected to a resistive load, capacitive reactance, and an inductive reactance.

  (a) What should be the new frequency setting to attain a power factor of 1, or 100%.
  (b) What is the existing power factor, at 50Hz?
  (c) What would be the power factor if all circuit elements remain unchanged and the VFD frequency is lowered to 30Hz.

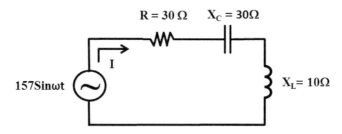

Solution

   (a) We must convert reactances X_C and X_B to corresponding capacitance, **C**, and inductance, **L**, values.

Since $X_L = 2\pi f L$, $L = \dfrac{X_L}{2\pi f}$, or, $L = \dfrac{10\Omega}{2\pi(50hz)} = 0.03185$ H

And,

$$X_C = \frac{1}{2\pi f C}, \quad C = \frac{1}{2\pi f X_C}, \text{ or, } C = \frac{1}{2\pi(50hz)(30\Omega)} = 106 \times 10^{-6}F$$

$$f = f_o = \frac{1}{2\pi\sqrt{LC}} = \frac{1}{2(3.14)\sqrt{(0.03185).(106\times10^{-6})}} \qquad \text{Eq. 5.8}$$

$$\therefore \quad f = f_o = 86hz$$

(b) The power factor can be calculated through the impedance angle, using equation 5.3. We must compute the circuit's total impedance first, **at 50 Hz**.

As derived in part (a), L = 0.03185H   and C = 0.000106F

$\therefore$ X_L = $2\pi fL$ = 2(3.14)(50hz)(0.03185H) = 10Ω {Given}

And, X_C = $\dfrac{1}{2\pi fC}$ = $\dfrac{1}{2(3.14)(50hz)(0.000106F)}$ = 30Ω {Given}

$X_t = X_Z$ =

$\therefore$ Z = 30Ω - j30Ω + j10Ω = 30Ω - j20Ω = 36$\angle$-33.7°Ω

And, Power Factor = Cos (θ_Z) = Cos (- 33.7°) = 0.83.3,

   or 83.3% leading.

(c) The power factor at 30Hz:

As derived in part (a), L = 0.03185H   and C = 0.000106F

$\therefore$ X_L = $2\pi fL$ = 2(3.14)(30hz)(0.03185H) = 6Ω

And, X_C = $\dfrac{1}{2\pi fC}$ = $\dfrac{1}{2(3.14)(30hz)(0.000106F)}$ = 50Ω

$\therefore$ Z = 30Ω - j50Ω +j6Ω= 30Ω - j44Ω = 53.3$\angle$-55.7°Ω

And, Power Factor = Cos (θ_Z) = Cos (- 55.7°) = 0.5631,

   or 56.31% leading.

5. The HMI (Human Machine Interface) monitor of an Automated HVAC system, monitoring an air washer supply fan motor is indicating a reactive power, Q_1, of 60 kVARs. This system is located in the United Kingdom, where the AC frequency is 50 Hz. Determine the amount of capacitance that must be added to improve the power factor of the motor branch circuit such that the reactive power is reduced to Q_2 of 20 kVARs. The branch circuit is operating at 240 V_{RMS}.

Solution

   Apply Eq. 5.4:

$$C = \frac{(Q_1 - Q_2)}{2\pi fV^2} = \frac{(60kVAR - 20kVAR)}{2\pi(50hz)(240V)^2}$$

$$= \frac{(60,000VAR - 20,000VAR)}{2\pi(50hz)(240V)^2} = 0.002211F = 2.211mF$$

6. Power Factor Improvement and Cost Savings: In conjunction with the local utility company DSM program, a manufacturing plant is being offered $2 per kVA for improvement in power factor from 0.75 to 0.85. The plant is operating at its contract level of 30 MW. Determine estimated annual pre-tax revenue if the plant accepts the offer.

Solution

   Since the objective is to assess the cost savings on the basis of apparent power (S) reduction, we must determine the apparent power S_1 being drawn by the air compressor motor at the existing power factor of 0.75 (75%), and the apparent power S_2 at the desired power factor of 0.85 (85%). Rearrange and apply Eq. 5.1.

$$\text{Magnitude of Apparent Power } |S| = \frac{|P|}{PF}$$

$$\text{Magnitude of existing apparent power } |S_1| = \frac{|P|}{PF_1} = \frac{30,000kW}{0.75}$$
$$= 40,000kVA$$

$$\text{Magnitude of improved apparent power } |S_2| = \frac{|P|}{PF_2} = \frac{30,000kW}{0.85}$$
$$= 35,294kVA$$

Annual DSM revenue expected from proposed power factor improvement

$$= \left(40,000kVA - 35,294kVA\right).\left(\frac{\$2}{kVA\text{ - Month}}\right).\left(12\text{ Months/Year}\right)$$
$$= \$112,941$$

7. The output of a variable frequency drive, as shown in the circuit below, is $157Sin\omega t$. The VFD output is currently set at 60Hz. This drive is connected to a resistive load, capacitive reactance, inductive reactance and a "black box" load, Z_B, of $10\Omega\angle 45°$. What should be the new frequency setting to attain a power factor of 1, or 100%?

Solution

Solution strategy: As explained earlier in this section, the power factor of an AC circuit, consisting of inductive and capacitive reactance, peaks to the maximum value of unity, or 100%, at resonance frequency, f_0. However, in this case, because of the presence of the black box impedance of $Z_B = 10\Omega\angle45°$, we cannot apply Eq. 5.8, directly, to compute f_0. We must convert impedance into its rectangular form to derive the reactance component $X_{L,B}$, combine it with the other inductive reactance in the circuit, derive the L and C values and then apply Eq. 5.8 to calculate the resonance frequency f_0.

$Z_B = 10\angle45° = 7.07\Omega + j7.07\Omega$. Therefore, $X_{L,B} = 7.07\Omega$.

Since the other inductive reactance is contributed by $X_L = 10\Omega$,

$X_{L,total} = X_L + X_{L,B} = 10\Omega + 7.07\Omega = 17.07\Omega$

$\therefore$ Total "equivalent L" in the circuit wuld be:

$X_{L,total} = 2\pi f L_{total} = 17.07\Omega$,

$\therefore L_{total} = \dfrac{17.07\Omega}{2\pi f} = \dfrac{17.07\Omega}{2(3.14)(60hz)}$, or, $L = 0.0453H$

And, $X_C = \dfrac{1}{2\pi f C} = \dfrac{1}{2(3.14)(60hz)(C)} = 30\Omega$

$\therefore C = \dfrac{1}{2(3.14)(60hz)(30\Omega)} = 88.46 \times 10^{-6}F$

Then, according to Eq. 5.8,

$f_0 = \dfrac{1}{2\pi\sqrt{L.C}} = \dfrac{1}{2(3.14)\sqrt{(0.0453)(88.46 \times 10^{-6})}} = 79.54hz$

CHAPTER 6—SOLUTIONS

1. The BMS System at a truck assembly plant, that operates 365 days a year, is displaying following electrical power and energy consumption data:

   Billing Days in the Current Month: 30
   On Peak Energy Consumption: 2,880,000 kWh
   Off Peak Energy Consumption: 11,520,000 kWh
   Three highest 30 minute energy usages for the billing month are
   (i) 12,500 kWh,
   (ii) 12,300 kWh and
   (iii) 12,290 kWh.

Assuming this facility is on OPT, Time of Use, contract with 30 minutes demand interval, determine the following:
   (a) Average demand.
   (b) Peak demand.
   (c) The load factor for the current month.
   (d) Average annual demand

Solution

(a) Average demand can be calculated by applying Eq. 6.1 as follows:

$$= \frac{\text{Energy (kWh or MWh) consumed during the billing Month}}{\text{Total number of hours in the billing month}}$$

$$= \frac{\text{On Peak Energy} + \text{Off Peak Energy Consumption}}{\text{Total number of hours in the billing month}}$$

$$= \frac{2{,}880{,}000 \text{ kWh} + 11{,}520{,}000 \text{ kWh}}{(24 \text{ hrs/Day}).(30 \text{ Days/Month})}$$

$$= 20{,}000 \text{ kW or 20 MW}$$

(b) Peak Demand can be calculated by applying Eq. 6.2 to the 30 minute interval during which the highest energy consumption is recorded. In this problem, three high 30 minute kWh consumptions are stated. However, we are interested in the "highest" kWh recorded; and that is 12,500 kWh. Therefore, the peak demand would be:

$$\text{Peak Demand} = \frac{\text{Peak Energy in kWh}}{\text{Time in hours}} = \frac{\text{Peak Energy in kWh}}{\frac{1}{2}\text{hour}}$$

$$= \frac{12,500 \text{ kWh}}{0.5 \text{ h}} = 25,000 \text{ kW or 25 MW}$$

(c) Load factor can be calculated by applying Eq. 6.3 as follows:

$$\text{Load Factor} = \frac{\text{Average Demand for the Month, in kW or MW}}{\text{Peak Demand for the Billing Month, in kW or MW}}$$

Using the Average Demand, calculated in part (a) as 20 MW, and the Peak Demand calculated in part (b) as 25 MW:

$$\text{Load Factor} = \frac{20 \text{ MW}}{25 \text{ MW}} = 0.80 \text{ or } 80\%$$

2. A 200 kVA transformer has been tested by the manufacturer to safely and continuously sustain a load of 230 kVA. What service factor should the manufacturer include on the nameplate of this transformer?

Solution

$$\text{Service Factor} = \frac{\text{Safe or Continuous Load, in kW, kVA or hp}}{\text{Nameplate rating of equipment, in kW, KVA or hp}}$$

$$= \frac{\text{Safe Operating Load}}{\text{Full Load Rating of the Transformer}} = \frac{230 \text{ kVA}}{200 \text{ kVA}} = 1.15$$

3. A 5 hp single phase AC motor, rated at a service factor of 1.10, is being tested at maximum safe load, powered by 230 V_{AC} source. Determine the amount of current drawn by this motor, from the power source if the motor efficiency is 90% and the power factor is 0.85.

Solution

The 5 hp motor, with a service factor of 1.10, operating at its maximum safe load is, essentially, delivering energy or performing work at the rate of:

$$P = (5 \text{ hp}) \cdot (\text{Service Factor}) = (5 \text{ hp}) \cdot (1.10) = 5.5 \text{ hp}.$$

Also note that in most current computations the power, in hp, must be converted to power in watts or kWs.

$$\therefore P_{Motor\text{-}watts} = (5.5 \text{ hp}).(746 \text{ W/hp}) = 4103 \text{ W}$$

As introduced in earlier chapters:

Magnitude of apparent power $S = |S| = \dfrac{P_{Motor-Watts}}{PF} = |V||I|_{Motor}$

And magnitude of apparent power drawn from the source =

$$\dfrac{P_{Motor-Watts}}{(PF).(Eff.)} = |V||I|_{Source}$$

$$\therefore |I|_{Source} = \dfrac{P_{Motor-Watts}}{(PF).(Eff.).(|V|)} = \dfrac{4103\text{W}}{(0.85).(0.9).(230V_{RMS})} = 23.32 \text{ A}$$

4. Consecutive electrical power meter readings at a home in Hawaii are listed below. Determine the total electrical power bill for the month of this residence if the flat \$/kWh cost rate is¢ 21/kWh. The renewable energy rider is \$15 and the energy sales tax rate is 4%.
  Previous reading: 45000
  Current or present reading: 46000

Solution
  According to Eq. 6.6,

$$\text{Baseline Charge} = (46000 - 45000).(\$0.21 / kWh)$$
$$= (1000).(\$0.21/kW) = \$210$$
$$\text{Total Bill} = (\$210 + \$15).(1 + 4\%) = (\$225).(1 + 0.04)$$
$$= (\$225).(1.04) = \$234.00$$

Note that the 4% sales tax is applied to the subtotal comprising of the baseline cost plus the renewable rider.

5. If the peak demand in Case Study 6.2 is reduced by 10% through implementation of peak shaving measures, what would be the baseline cost for the demand portion of the bill?

Solution

According to spreadsheet 6.1, the original peak demand for the month is 26,000 kW.

$$\therefore \text{ New, reduced, demand} = (26,000 \text{ kW}).(1 - 0.01)$$
$$= (26,000 \text{ kW}).(0.9) = 23,400 \text{ kW}$$

Then, revise the demand portion of spreadsheet 6.1 for the reduced peak demand of 23,400 kW as shown below:

Description	Demand and Energy Parameters	Rates	Line Item Charge
On-Peak Billing Demand	**23,400 kW**		
On-Peak Billing Demand Charge			
For the First 2,000 kW	2,000 x	$7.64 per kW =	$15,280.00
For the Next 3,000 kW	3,000 x	$6.54 per kW =	$19,620.00
For Demand Over 5,000 kW	**(23,400 – 5,000) x**	**$5.43 per kW =**	**$99,912**
Total Demand Cost for the Billing Month			$15,280.00 +$19,620.00 +$99,912 = **$134,812**

CHAPTER 7—SOLUTIONS

1. A gas powered prime mover is rotating the rotor of a single phase alternator at a speed of 1200 rpm. The alternator consists of six pole construction. The effective diameter of the coil is 0.15m and the length of the coil loop is 0.24m. The coil consists of 20 turns. The magnetic flux density has been measured to be 1.2T. Calculate the power delivered by this generator across a resistive load of 10Ω.

Solution

The RMS, effective or DC voltage produced through an alternator or generator can be computed by applying Eq. 7.6:

$$V_{DC} = V_{RMS} = V_{eff} = \frac{V_P}{\sqrt{2}} = \frac{\pi npNAB}{(60).\left(\sqrt{2}\right)}$$

Given:

n = 1200 rpm
p = 6
N = 20
B = 1.2T
A = (Eff. diameter of the coil conductor) x (Eff. length of the coil)
  = (0.24m) x (0.15m) = 0.036m²

$$V_{DC} = V_{RMS} = \frac{\pi npNAB}{(60).\left(\sqrt{2}\right)}$$

$$= \frac{(3.14)(1200rpm)(6)(20)(0.036m^2)(1.2T)}{(60).\left(\sqrt{2}\right)} = 230.2V$$

$$P = \frac{V^2}{R} = \frac{(230.2)^2}{10} = 5,299 \text{ W} = 5.3kW$$

2. A four pole alternator/generator is producing electrical power at an electrical frequency of 50 Hz. (a) Determine the angular speed corresponding to the generated electrical frequency. (b) Determine the rotational (synchronous) speed of the armature/rotor. (c) Determine the angular velocity of the armature/rotor (rads/sec).

Solution

(a) Angular speed, ω, corresponding to the generated electrical frequency, f, can be calculated using Eq. 7.8:

$$\omega = 2\pi f = (2).(3.14).(50 \text{Hz}) = 314 \text{ rad/s}$$

(b) The rotational or synchronous speed of the armature/rotor is given by Eq. 7.7:

$$n_s = \frac{120f}{p} = \frac{(120)(50 \text{ Hz})}{4} = 1500 \text{ rpm}$$

(c) Angular velocity of the armature/rotor is simply the rotational speed, in rpm, converted into rad/s. Since there are 2π radians per revolution:

$$\omega_s = 1500 \text{ rev./min} = \left(\frac{1500 \text{ rev.}}{\min}\right).\left(\frac{2\pi \text{ rad./rev}}{(60\text{sec/min})}\right) = 157 \text{ rad/sec}$$

3. A four pole single phase AC generator consists of windings constituting 80 series paths and is driven by a diesel engine. The effective or mean length of the armature is 18 cm and the cross-sectional radius of the armature is 5 cm. The armature is rotating at 1800 rpm. Each armature pole is exposed to a magnetic flux of 1.0 T. The efficiency of this generator is 90% and it is rated 2 kW. Determine the following:
   (a) The maximum voltage generated.
   (b) The RMS voltage generated.
   (c) The horsepower rating of the generator
   (d) The horse power output of the prime mover.

Solution

The maximum voltage, V_m, generated by this alternator is given by Eq. 7.7.

$$V_m = V_P = \left(\frac{\pi}{2}\right).\left(\frac{n}{30}\right).pNAB = \frac{\pi n p NAB}{60} \qquad \text{Eq. 7.7}$$

Given:
   n = 1800 rpm
   p = 4

N = Number of series paths = 80

B = 1.0 T

A = (Eff. diameter of the coil conductor) x (Eff. length of the coil)

= (2 x 5cm) x (18cm) = (0.1m) x (0.18m) = 0.018m²

(a) $V_m = \dfrac{\pi n p N A B}{60} = \dfrac{(3.14).(1800rpm)(4)(80)(0.018m^2)(1.0T)}{60}$

= 543V

(b) The RMS voltage could be calculated by using Eq. 7.6, or simply by dividing V_m, from part (a), by the square root of 2 as follows:

$$V_{RMS} = V_{eff} = \frac{V_m}{\sqrt{2}} = \frac{543}{\sqrt{2}} = 384 \text{ V}$$

(c) The horsepower rating of this generator is the power *output* rating specified in hp, premised on the stated *output capacity* of 2.0 kW. Therefore, application of the 0.746 kW/hp conversion factor yields:

$$P_{hp} = \frac{P_{kW}}{0.746kW/hp} = \frac{2.0kW}{0.746kW/hp} = 2.68hp$$

(d) The horsepower rating of the prime mover—or the propane fired engine—would need to offset the inefficiency of the AC generator. Therefore, based on the given 90% efficiency rating of the generator:

$$P_{hp\text{ - Prime Mover}} = \frac{P_{hp\text{ - Gen.}}}{Eff._{Gen}} = \frac{2.68 \text{ hp}}{0.90} = 2.98 \text{ hp}$$

4. A three phase, four pole, AC induction motor is rated 170 hp, is operating at full load, 60 Hz, 460 V_{rms}, efficiency of 90%, power factor of 80%, and a slip of 4%. Determine the (a) motor shaft speed, in rpm, (b) torque developed, in ft-lbf. (c) line current drawn by the motor and (d) the amount of reactive power, Q, sequestered in the motor under the described operating conditions.

Solution

Given:

$P_{L,3\text{-}\phi}$ = Real power or rate of work performed by the motor = 170 hp
= (170 hp).(746 W/hp) = 126,820 W

p = 4 poles

V_L = 460 V_{RMS}

Pf = 80% or 0.80

Eff. = 90% or 0.90

n_s = Synchronous speed, in rpm = ?

Slip, s = 4%

f = Frequency of operation = 60 Hz

(a) **Shaft or motor speed:** Rearrange and apply Eq. 7.10:

$$\text{Slip} = s = \left(\frac{n_s - n}{n_s} \right)$$

And, by rearrangement: $n = n_s (1 - s)$

Next, we must determine the synchronous speed of the motor by applying Eq. 7.9:

$$n_s = \text{Synchronous speed} = \frac{120f}{p} = \frac{(120).(60)}{4} = 1800 \text{ rpm}$$

$$\therefore n = (1800 \text{ rpm})(1 - 0.04) = 1728 \text{ rpm}$$

(b) **Torque developed, in ft-lbf:** There are multiple methods at our disposal for determining the torque developed. Formulas associated with two common methods are represented by Eq. 7.12, Eq. 7.13 and Eq. 7.14. Since the power is available in hp and the rotational speed in rpm, apply Eq. 7.12:

$$\text{Torque}_{(ft-lbf)} = \frac{5250 \times P_{(horsepower)}}{n_{(rpm)}} = \frac{5250 \times 170 \text{ hp}}{1728 \text{ rpm}} = 516 \text{ ft-lbf}$$

(c) **Line current drawn by the motor:** Full load line current is the current the motor draws from the power source or utility in order to sustain the full load. Application of Eq. 7.22 takes this into account in the computation of 3-phase motor line current, after the motor hp rating (or actual load) is converted into watts:

The 3-ϕ phase line curent, $|I_L| = \dfrac{|S_{L,3\text{-}\phi}|}{\sqrt{3}\,(|V_L|)} = \dfrac{P_{L,3\text{-}\phi} \text{ (in Watts)}}{\sqrt{3}\,(|V_L|)(Pf).(Eff.)}$

Three phase (total) real power was converted into watts under "Given" as
$P_{L,3-\phi} = 126{,}820$ W

Therefore,

The 3-ϕ phase line curent, $|I_L| = \dfrac{126{,}820 \text{ Watts}}{\sqrt{3}\left(460 \text{ V}_{RMS}\right)(0.90).(0.80)} = 221$ A

(d) **Reactive power, Q**: There are multiple approaches available to us for determination of reactive power Q for this motor application. We will utilize the power triangle or apparent power component vector method:

$$S = P + jQ, \text{ and } S^2 = P^2 + Q^2,$$
$$\text{or, } Q^2 = S^2 - P^2$$
$$\text{and } Q = \sqrt{S^2 - P^2}$$

Real power "P" was computed earlier as 126,820 W, and apparent power S can be assessed using Eq. 7.20:

$$\left|S_{L,3-\phi}\right| = \frac{P_{3-\phi} \text{ (in Watts)}}{(Pf).(Eff.)} = \frac{126{,}820\text{W}}{(0.90).(0.80)} = 176{,}139 \text{ VA}$$

Therefore,

$$Q = \sqrt{S^2 - P^2} = \sqrt{176{,}139^2 - 126{,}820^2} = \sqrt{14{,}941{,}595{,}779}$$
$$= 122{,}236\text{VAR}$$

5. A three phase, four pole, AC induction motor is tested to deliver 200 hp, at 900 rpm. Determine the frequency at which this motor should be operated for the stated shaft speed. Assume the slip to be negligible.

Solution
Given:
  p = 4 poles
  n_s = Synchronous speed, in rpm = 900 rpm = Shaft speed, since slip is zero
  f = Frequency of operation = ?

Rearrange and apply Eq. 7.9

$$n_s = \text{Synchronous speed} = \frac{120f}{p},$$

$$\text{Or, } f = \frac{(n_s).(p)}{120} = \frac{(900).(4)}{120} = 30 \text{ Hz}$$

6. A three phase induction motor delivers 600 kW at a power factor of 80%. In lieu of installing power factor correction capacitors a synchronous motor is being considered as a power fact correction measure. Determine the apparent power size of the synchronous motor—in kVA—that should be installed to carry a load of 300 hp and raise the (combined) power factor to 93%. The source voltage is 230 V_{rms}.

Solution

Given:

P_I = Real power delivered by the 3-ϕ induction motor = 600 kW
P_S = Real power contributed by the synchronous motors = 300 hp = (300 hp) x (0.746 kW/hp) = 223.8 kW
Pf_i = Initial Power Factor = 80% = 0.80
Pf_f = Final Power Factor = 93% = 0.93
V_s = Source voltage = 230 Vrms. However, the voltage information is not needed to solve this problem.

This problem is similar to Example 7.6. So, the strategy for solving it is the same as the one employed for Example 7.6. Overall, we need to determine the reactive power, Q_S, contributed by the synchronous motor. Then, by applying the Pythagorean Theorem to Q_S and P_S, we can derive the apparent power (kVA) of the synchronous motor.

$$\text{Since } S_T^{\,2} = P_T^{\,2} + Q_T^{\,2}, \quad Q_T = \sqrt{S_T^{\,2} - P_T^{\,2}}$$

The combined real power of the induction motor and the synchronous motor:

$$P_T = 600 \text{ kW} + 223.8 \text{ kW} = 824 \text{ kW}$$

$$\text{Since } S_T \text{Cos}\theta = P_T, \quad S_T = \frac{P_T}{\text{Cos}\theta},$$

And since θ_T, theh final power factor angle = $\text{Cos}^{-1}(0.93) = 21.58°$,

$$S_T = \frac{P_T}{Cos\theta_T} = \frac{824 \text{ kW}}{Cos(21.58°)} = 886 \text{ kVA}$$

Therefore,

$$Q_T = \sqrt{S_T{}^2 - P_T{}^2} = \sqrt{886^2 - 824^2} = \sqrt{106,058} = 326 \text{ kVAR}$$

Now, in order to determine Q_S, the reactive power contributed by the synchronous motor, we must subtract the original reactive power, Q_O, from the final, total, reactive power, Q_T. However, Q_O is unknown and can be determined through the power triangle as follows:

S_O = Original, induction motor apparent power

S_O

Q_O

$\theta_O°$ P_O

Q_O, reactive component of original induction motor apparent power S_O

P_O, real power component of original induction motor apparent power S_O

Since $Tan(\theta_O) = \dfrac{Q_O}{P_O}$, or, $Q_O = P_O Tan(\theta_O) = 600 \text{ kWTan}(\theta_O)$

And, $\theta_O = Cos^{-1}(Pf_O) = Cos^{-1}(0.80) = 36.87°$

Therefore,

$Q_O = 600 \text{ kWTan}(36.89°) = 600 \text{ kW}(0.75) = 450 \text{ kVAR}$

And, reactive power contributed by the synchronous motor would be:

$$Q_S = Q_O - Q_T = 450 \text{ kVAR} - 326 \text{ kVAR} = 124 \text{ kVAR}$$

Therefore,

$S_S = P_S + jQ_S$, and $S_S{}^2 = P_S{}^2 + Q_S{}^2$, or, $S_S = \sqrt{P_S{}^2 + Q_S{}^2}$

Or, $S_S = \sqrt{(224 \text{ kW})^2 + (124 \text{ kVAR})^2} = 256 \text{ kVA}$

CHAPTER 8—ANSWERS/SOLUTIONS

1. A substation in a manufacturing facility is being fed from a 13 kV transformer secondary. This switchgear in this substation would be categorized as:
    A. Medium voltage
    B. Low voltage
    C. High voltage
    D. Medium voltage
    E. None of the above

2. Power transmission lines would be categorized as:
    A. Medium voltage
    B. Low voltage
    C. High voltage
    D. None of the above

3. The breakers installed in residential breaker panels are:
    A. OCBs
    B. Thermal magnetic circuit breakers
    C. Low voltage thermal magnetic circuit breakers
    D. None of the above
    E. Both B and C

4. The vacuum circuit breakers tend to offer longer service spans between overhauls than do air circuit breakers.
    A. True
    B. False

5. The SF_6 type high voltage circuit breakers are not preferred due to environmental concerns.
    A. True
    B. False

6. MCCs are not designed to accommodate PLCs and VFDs.
    A. True
    B. False

7. The bus bars in MCCs are commonly constructed out of:
    A. Aluminum
    B. Silver plated copper
    C. Silver
    D. Iron

8. Pilot devices on MCCs:
    A. Control circuit breakers
    B. Indicate the status of MCC
    C. Indicate the status of motor/load
    D. Include "Start" and "Stop" controls
    E. Both (C) and (D).

9. Power control cubicles in MCCs are fixed and cannot be removed while the main fusible disconnect switch of the MCC is ON.
    A. True
    B. False

10. A control transformer, in a given MCC compartment:
    A. Steps down the voltage for control circuit operation.
    B. Provides power for MCC cabinet lighting.
    C. Is seldom needed.
    D. Serves as an isolation transformer.
    E. Both (C) and (D).

CHAPTER 9—SOLUTIONS

1. A given circuit is meant to carry a continuous lighting load of 16 A. In addition, four loads designed for permanent display stands are fastened in place and require 2A each when operating. What is the rating of the over current protective device (OCPD) on the branch circuit?

Solution

From Article 210.20(a) of the NEC ®, the over current protective device (OCPD) must be rated at 100% of the non-continuous load plus 125% of the continuous load.

∴ The Rating of OCPD = (1.00) (2 A + 2 A + 2 A + 2 A) + (1.25) (16 A)
= 28 A, minimum

A standard fixed-trip circuit breaker or a fuse rated at 30 A can be used (see Sec. 240.6).

2. A three-phase, four-wire feeder with a full-sized neutral carries 14 A continuous and 40 A non-continuous loads. The feeder uses an over current device with a terminal or conductor rating of 60°C. What is the minimum copper conductor size? Assume no derating applies. Use Tables 9.1 and 9.2.

Solution

Feeder conductor size, before derating, is based on 100% of the non-continuous load and 125% of the continuous load [Art. 215.2(a)].

∴ Amp load = (1.00)(40 A) + (1.25)(14 A)
= 40 A + 17.5 A
= 57.5 A ≈ 58 A

The total of 58 A is used since an ampacity of 0.5 A or greater is rounded up [Sec. 220.2(b)]. Using NEC ® Table 310.15—as presented in form of Tables 9.1 and 9.2 in this text—TW, UF, **AWG 4 from the 60 °C column should be selected.** The ampacity of AWG 4 is 70 amps.

Note: AWG 6 (the size below AWG 4), with ampacity of 55 amps, would be undersized.

3. Electrical specifications for a brewery company call for a fusible disconnect switch enclosure that must be able withstand occasional splashing of water during periodic wash downs required by the local health codes. This design will be applied in breweries in the US as well as Europe. The water flow from the 1-in wash down nozzles is expected to less than 60 GPM from a distance of 11 ft for less than 4 min. (a) Determine the NEMA rating of enclosure for the US installations. (b) Determine the IP rating of enclosure for the European installations.

Solution

(a) Examination of the NEMA—IP rating table in this chapter shows that NEMA 4 enclosure is rated for:

Table 9.1: An older version of ampacity table. Included for general format and general content reference only. Contd. Courtesy, NEC, NFPA.

Table 310.15(B)(2)(a) Adjustment Factors for More Than Three Current-Carrying Conductors in a Raceway or Cable

Size AWG or kcmil	Temperature Rating of Conductor						Size AWG or kcmil
	60°C (140°F)	75°C (167°F)	90°C (194°F)	60°C (140°F)	75°C (167°F)	90°C (194°F)	
	Types TW, UF	Types RHW, THHW, THW, THWN, XHHW, USE, ZW	Types TBS, SA, SIS, FEP, FEPB, MI, RHH, RHW-2, THHN, THHW, THW-2, THWN-2, USE-2, XHH, XHHW, XHHW-2, ZW-2	Types TW, UF	Types RHW, THHW, THW, THWN, XHHW, USE	Types TBS, SA, SIS, THHN, THHW, THW-2, THWN-2, RHH, RHW-2, USE-2, XHH, XHHW, XHHW-2, ZW-2	
	COPPER			ALUMINUM OR COPPER-CLAD ALUMINUM			
18	–	–	14	–	–	–	–
16	–	–	18	–	–	–	–
14*	20	20	25	–	–	–	–
12*	25	25	30	20	20	25	12*
10*	30	35	40	25	30	35	10*
8	40	50	55	30	40	45	8
6	55	65	75	40	50	60	6
4	70	85	95	55	65	75	4
3	85	100	110	65	75	85	3
2	95	115	130	75	90	100	2
1	110	130	150	85	100	115	1
1/0	125	150	170	100	120	135	1/0
2/0	145	175	195	115	135	150	2/0
3/0	165	200	225	130	155	175	3/0
4/0	195	230	260	150	180	205	4/0
250	215	255	290	170	205	230	250
300	240	285	320	190	230	255	300
350	260	310	350	210	250	280	350

Table 9.2: An older version of ampacity table. Included for general format and general content reference only. Contd. Courtesy, NEC, NFPA.

Table 310.15(B)(2)(a) Adjustment Factors for More Than Three Current-Carrying Conductors in a Raceway or Cable

Size AWG or kcmil	Temperature Rating of Conductor						Size AWG or kcmil
	COPPER			ALUMINUM OR COPPER-CLAD ALUMINUM			
	60°C (140°F)	75°C (167°F)	90°C (194°F)	60°C (140°F)	75°C (167°F)	90°C (194°F)	
	Types TW, UF	Types RHW, THHW, THW, THWN, XHHW, USE, ZW	Types TBS, SA, SIS, FEP, FEPB, MI, RHH, RHW-2, THHN, THHW, THW-2, THWN-2, USE-2, XHH, XHHW, XHHW-2, ZW-2	Types TW, UF	Types RHW, THHW, THW, THWN, XHHW, USE	Types TBS, SA, SIS, THHN, THHW, THW-2, THWN-2, RHH, RHW-2, USE-2, XHH, XHHW, XHHW-2, ZW-2	

CORRECTION FACTORS

Ambient Temp. (°C)	For ambient temperatures other than 30°C (86°F), multiply the allowable ampacities shown above by the appropriate factor shown below.						Ambient Temp. (°F)
21-25	1.08	1.05	1.04	1.08	1.05	1.04	70-77
26-30	1	1	1	1	1	1	78-86
31-35	0.91	0.94	0.96	0.91	0.94	0.96	87-95
36-40	0.82	0.88	0.91	0.82	0.88	0.91	96-104
41-45	0.71	0.82	0.87	0.71	0.82	0.87	105-113
46-50	0.58	0.75	0.82	0.58	0.75	0.82	114-122
51-55	0.41	0.67	0.76	0.41	0.67	0.76	123-131
56-60	–	0.58	0.71	–	0.58	0.71	132-140
61-70	–	0.33	0.58	–	0.33	0.58	141-158
71-80	–	–	0.41	–	–	0.41	159-176

* Small Conductors. Unless specifically permitted in 240.4(E) through (G), the overcurrent protection shall not exceed 15 amperes for 14 AWG, 20 amperes for 12 AWG, and

and copper-clad aluminum after any correction factors for ambient temperature and number of conductors have been applied.

Watertight (weatherproof). Must exclude at least 65 GPM of water from 1-in. nozzle delivered from a distance not less than 10 ft for 5 min. Used outdoors on ship docks, in dairies, and in breweries.

Therefore, a NEMA 4 enclosure would exceed the "1-in wash down nozzle, 60 GPM from a distance of 11 ft for less than 4 min" worst case exposure. Hence, a **NEMA 4 enclosure** should be specified for the US installation.

(b) Since the European installation would be exposed to the same worst case conditions, US NEMA 4's European counterpart, **IP 66 should be specified**.

Note: If the health code, in the US, were to require that chlorine solution be used for the wash down, since chlorine is corrosive, **NEMA 4X** should be specified.

4. Over current Protection & Conductor Ampacity:
Applicable Code/Codes: Articles 210.19 (A) (1), 210.20 (A) and 310.15. Table 310.15(B)(16) as presented in Tables 9.1 and 9.2 of this text (Note: This is not a current table and is only reproduced in this text for exercise and illustration purposes).

The branch circuit in the exhibit below consists of three continuous loads. Over current protection in the branch circuit is provided through a 20 amp circuit breaker. (a) Determine the size of copper conductor based on the ampacities given in Tables 9.1 and 9.2 assuming conductor temperature is at 75°C or less. Assume that 75°C operation and selection is allowed. (b) Verify the size/specifications of the circuit breaker. Assume no derating applies. (c) If the ambient temperature were to rise to 50°C, how would the conductor size be impacted?

Solution

(a) In accordance with article 210.19 (A) (1): "Branch circuit conductors shall have an ampacity not less than maximum load to be served......(and) shall have an allowable ampacity not less than the non-continuous load plus 125% of the continuous load.

∴ The conductor ampacity for the given branch circuit
= 1.25 x Continuous Load + 1.00 x Non-Continuous Load
= 1.25 x (16A) + 1.00 x (0) = 20 amps.

According to Table 9.1 (in this text) for 75°C operation, with Types RHW, THHW, THW, THWN, XHHW, USE, ZW insulation, AWG 12 conductor carries an allowable ampacity of 25 amps for conductors that are insulated, rated for 0-2000 volt operation, in situations with no more than three (3) current carrying conductors in raceway, cable earth (directly buried); under ambient temperature (not exceeding) 30°C (85°F); with no required/applicable derating. With all conditions remaining the same, an AWG 14 could carry 20 amps. However, the asterisk annotation on both AWG 14 and AWG 12, through the footnotes on Table 9.2, stipulates that if AWG 14 is selected it must be protected at no more than 15 amps. Selection of 15 amp protection for a 16 continuous could result in nuisance tripping or fuse clearing. Therefore, AWG 12, with the stated ampacity of 25 amps, at 75°C operation should be selected. Also, selection of AWG 12 would maintain the existing 20 amp breaker in compliance with the code.

(b) In accordance with article 210.20 (A):

"Branch-circuit conductors and equipment shall be protected by overcurrent protective devices that have a rating or setting that complies with 210.20(A) through (D).

(A) Continuous and Non-continuous Loads. Where a branch circuit supplies continuous loads or any combination of continuous and non-continuous loads, the rating of the overcurrent device shall not be less than the non-continuous load plus 125 percent of the continuous load." NEC® 2011.

The over current protection should be rated = 1.25 x Continuous Load
= 1.25 x (6 + 6 + 4) = 20 amps.

∴ The 20 amp circuit breaker as an over current protection device is adequate for the given branch circuit provided AWG 12 conductor is used.

(c) Ambient Temperature Rise and Conductor size:
According to Article 310, Table 310.15(B)(16), as presented under Tables 9.1 and 9.2 of this text, when ambient deviates from 40°C, the derating/up-rating multipliers, listed at the bottom of Tables 310.15(B)(2)(b)—Table 9.2 in this text—must be applied to adjust the ampacity of the conductor.

As recognized above, the ampacity of an AWG #12, under 75°C operation, is 25 amps. From Table 9.2, the multiplier for 50°C ambient, under 75°C terminal rating, is 0.75.

∴ The adjusted or derated ampacity of AWG #12 conductor, in this case, would be:

= 0.75 x 25 amps = 18,75 amps

Since the derated ampacity of AWG #12, for this application, falls below the 20 amp capacity mandated by 210.19 (A) (1), AWG #12 would no longer be adequate. Therefore, AWG #10, which is the next size above AWG#12, must be considered. According to Table 9.1, for 75°C operation, with Types RHW, THHW, THW, THWN, XHHW, USE, ZW insulation, AWG 10 conductor carries an allowable ampacity of 35 amps. Then, if the 50°C adjustment rating of 0.75 is applied to 35 amp ampacity of an AWG 10, the derated ampacity would be:

= 0.71 x 35 amps = 26.25 amps

Since the 26.25 amp derated ampacity of AWG #10 exceeds the 20 amp requirement, it would meet the code. AWG #10 should be selected for this application at 50°C ambient temperature.

5. A US appliance manufacturer is planning to market a new appliance in Mexico. The most appropriate safety certification for this appliance would be:

    E.   UL
    F.   ULC
    G.   ETL

H. **NOM: NOM Mark** (Norma Official Mexicana) NOM is a mark of product safety approval for virtually any type of product exported into Mexico

6. Assume that the alarm switch in the control circuit depicted below is opened after being closed for a prolonged period of time. Which of the following conditions would best describe the status of the annunciating lights and the horn when the switch is opened?

    E.   Alarm horn will turn off
    F.   Red light will turn off
    G.   Green light will turn on
    H.   **All of the above**

CHAPTER 10—SOLUTIONS

1.   Answer the following questions pertaining to the branch circuit shown in the schematic diagram below:

    (a) What is the maximum voltage the power distribution system for this branch circuit rated for?

    (b) How many wires and phases is the power distribution system for this motor branch circuit rated for?

(c) What would be the proper rating for the branch circuit disconnect switch?

(d) What should the solid-state overload device be set for at commissioning of this branch circuit?

480 V Bus, 3 φ, 4W, 600 A, 60 Hz,

Solution

(a) The maximum voltage rating of the power distribution system for this branch circuit is included in the specification stated at the very top of the schematic diagram, as 480 V, within the caption:

"480 V Bus, 3 φ, 4W, 600 A, 60 Hz."

Therefore, the answer is: **480 V.**

(b)   The number of wires and phases in this branch circuit can be assessed from the specification stated at the very top of the schematic diagram:

BUSS® SYSTEM 300 MOTOR PROTECTION GUIDE
"NO-DAMAGE" "TYPE 2" SHORT-CIRCUIT PROTECTION
OVERLOAD OR BACK-UP OVERLOAD PROTECTION

115VAC (120V), 1 ph. (LPN-RK-SP or LPJ-SP)

HP	Motor Full Load Amps	Overload 1.15 S.F. Or Greater Or Rated Not Over 40°C	Overload All Other Motors	Back-up 1.15 S.F. Or Greater Or Rated Not Over 40°C	Back-up All Other Motors	Switch Or Fuseholder Size	Minimum NEMA Starter Size	Minimum Copper Conductor Size‡	Minimum Conduit Size‡
1/6	4.4	5	5	5.6	5.6	30	00	14	1/2
1/4	5.8	7	6.25	7.5	7	30	00	14	1/2
1/3	7.2	9	8	9	9	30	0	14	1/2
1/2	9.8	12	10	15	12	30	0	14	1/2
3/4	13.8	15	15	17.5	17.5	30	0	14	1/2
1	16	20	17.5	20	20	30	0	14	1/2
1½	20	25	20	25	25	30	1	12	1/2
2	24	30	25	30	30	30	1	10	1/2

200V (208V), 3 ph. (LPN-RK-SP or LPJ-SP)

HP	Motor Full Load Amps	Overload 1.15 S.F. Or Greater Or Rated Not Over 40°C	Overload All Other Motors	Back-up 1.15 S.F. Or Greater Or Rated Not Over 40°C	Back-up All Other Motors	Switch Or Fuseholder Size	Minimum NEMA Starter Size	Minimum Copper Conductor Size‡	Minimum Conduit Size‡
1/2	2.3	2.6	2.5	3	2.8	30	00	14	1/2
3/4	3.22	4	3.5	4.5	4	30	00	14	1/2
1	4.14	5	4.5	5.6	5	30	00	14	1/2
1½	5.98	7	6.25	7.5	7	30	00	14	1/2
2	7.82	9	8	10	9	30	0	14	1/2
3	11	12	12	15	15	30	0	14	1/2
5	17.5	20	20	25	25	30	1	12	1/2
7½	25.3	30*	25*	35	30*	60	1	8	1/2
10	32.2	40	35	45	40	60	2	6	3/4
15	48.3	60	50	70**	60	60	3	4	1
20	62.1	75	70	80	75	100	3	3	1
25	78.2	90	80	100	90	100	3	1	1¼
30	92	100	100*	125	110	200	4	1/0	1¼
40	120	150	125	150	150	200	4	1/0	1¼
50	150	175	150	200	175	200	5	3/0	1½
75	177	200*	200*	225	225	400	5	4/0	2
100	221	250	250	300	300	400	5	300	2
125	285	350	300	400	350	400	6	500	3
150	359	400*	400*	450	450	600	6	2-4/0	2-2
150	414	500	450	600	500	600	6	2-300	2-2

230V (240V), 1 ph. (LPS-RK-SP or LPJ-SP)

HP	Motor Full Load Amps	Overload 1.15 S.F. Or Greater Or Rated Not Over 40°C	Overload All Other Motors	Back-up 1.15 S.F. Or Greater Or Rated Not Over 40°C	Back-up All Other Motors	Switch Or Fuseholder Size	Minimum NEMA Starter Size	Minimum Copper Conductor Size‡	Minimum Conduit Size‡
1/6	2.2	2.5	2.5	2.8	2.8	30	00	14	1/2
1/4	2.9	3.5	3.2	4	3.5	30	00	14	1/2
1/3	3.6	4.5	4	4.5	4.5	30	00	14	1/2
1/2	4.9	5.6	5.6	6.25	6	30	00	14	1/2
3/4	6.9	8	7.5	9	8	30	00	14	1/2
1	8	10	9	10	10	30	00	14	1/2
1½	10	12	10	15	12	30	0	14	1/2
2	12	15	12	15	15	30	0	14	1/2
3	17	20	17.5	25	20	30	1	12	1/2
5	28	35	30*	35	35	60	2	8	1/2
7½	40	50	45	50	50	60	2	6	3/4
10	50	60	50	70**	60	60	3	4	3/4

460V (480V), 3 ph. (LPS-RK-SP or LPJ-SP)

HP	Motor Full Load Amps	Overload 1.15 S.F. Or Greater Or Rated Not Over 40°C	Overload All Other Motors	Back-up 1.15 S.F. Or Greater Or Rated Not Over 40°C	Back-up All Other Motors	Switch Or Fuseholder Size	Minimum NEMA Starter Size	Minimum Copper Conductor Size‡	Minimum Conduit Size‡
1/2	1	1.25	1¼	1.25	1.25	30	00	14	1/2
3/4	1.4	1.6	1.6	1.8	1.8	30	00	14	1/2
1	1.8	2.25	2	2.25	2.25	30	00	14	1/2
1½	2.6	3.2	2.6	3.5	3	30	00	14	1/2
2	3.4	4	3.5	4.5	4	30	00	14	1/2
3	4.8	5.6	5	6	5.6	30	0	14	1/2
5	7.6	9	8	10	9	30	0	14	1/2
7½	11	12	12	15	15	30	0	14	1/2
10	14	17.5	15	17.5	17.5	30	1	14	1/2
15	21	25	25	30	25	30	1	12	1/2
20	27	30*	30*	35	35	60	2	8	1/2
25	34	40	35	45	40	60	2	6	3/4
30	40	50	45	50	50	60	3	6	3/4
40	52	60*	60*	70	60*	100	3	4	1
50	65	80	70	90	75	100	3	3	1
60	77	90	80	100	90	100	4	1	1¼
75	96	110	110	125	125	200	4	1/0	1¼
100	124	150	125	175	150	200	4	2/0	1½
125	156	175	175	200	200	200	5	3/0	1½
150	180	225	200*	225	225	400	5	4/0	2
200	240	300	250	300	300	400	5	350	2½

* Fuse reducers required. ** 100A switch required.
† THWN connected to 60°C terminations up to #1. AWG to 75°C terminations for 1/0 and larger. Consult equipment manufacturer for listed termination temperature rating. Higher equipment termination temperature ratings may allow smaller conductor and conduit size.
‡ Based on 3 conductors for 3 ph. circuits and 2 conductors for 1 ph. circuits.

MPG

"480 V Bus, **3 φ, 4W**, 600 A, 60 Hz."

Therefore, the answer is: **3 φ, 4W.**

(c)   The proper rating for the branch circuit disconnect switch can be determined by using the NEC ® or through the use of published tables such as the Buss ® table introduced earlier in this chapter. As stated in the schematic diagram—in the circled motor symbol—the motor's full load rating is **10 hp**. As specified at the top of schematic diagram, the motor is being powered by a **480 V, 3-phase** source (480 V Bus, 3 ϕ, 4W, 600A, 60 Hz). Therefore, according to the Buss ® table, and as highlighted (circled) below—under the "460 V(480V), 3-ph, section—the fusible disconnect switch size would be **3-ϕ or 3-pole, 30A, 480V.** See the 4th column from the right.

(d)   Using the Buss ® table above, for **3—phase, 10 hp** motor, operating at a **480 V,** the overload device should be set at **17.0 A.**

Note: If NEC ® tables were used here, with the full load current rating of 14A (See the circled segment of the Buss ® table below) the 115% multiplier by NEC ® would result in an overload setting of 16A. It is important to bear in mind that the NEC ® requires that full load amps stated on the nameplate of the motor be used for computations associated with the code. In this text, the Buss ® table is being used for simplicity and illustration purposes.

2.   Determine the sizes/specifications of the following components in the branch circuit shown below using the Bus ® Table and information included in the questions:

    (e)   Conductor size
    (f)   Conduit size
    (g)   Overload setting based on the 115% NEC ® stipulation.
    (h)   Disconnect switch size, safety and fusible.

Solution

(a)   **Conductor size:** Since the load in the given branch circuit is a 20-hp motor, at 460/480 V, we will focus on the lower right section of the Bus ® table. See the circled 20-hp row on the Bus ® table below. The rightmost columns in the Bus ® table represent NEC ® based switch, starter, conductor and conduit sizes. The second column from the right, for the 20-hp motor load, shows that **AWG # 8 conductor should be specified.**

(b)   **Conduit size:** According to the rightmost column in the Bus ® table, a ½" conduit is permitted by the code for housing AWG # 8 conductors associated with the 20-hp motor.

(c)   **Overload setting based on the 115% NEC ® stipulation**: First we must identify the FLA, Full Load Amps, for the 20-hp motor, using the Bus ® table. The FLA is listed in the column adjacent to the HP column. So, according to the

480 V Bus, 3 φ, 4W, 600 A, 60 Hz,

D5

480V

120V

20

Bus ® table, FLA for a 3-phase, 20-hp motor, operating at 460/480 V, would be 27 A. Then, as explained in this chapter, if the code requires that equipment be protected against overload at 115%, the overload should be set at:

$$1.15 \times (\text{Motor Full Load Amps}) = 1.15 \times 27\,\text{A} = \textbf{31 A}$$

(d)   **Disconnect switch size, safety and fusible:** In order to specify the sizes of the branch circuit fusible disconnect switch and the safety switch in the field, we will focus on the fourth column from the right in the Bus ® table. In this column we see that **60 A switch is required for a 20-hp motor.** This specification would apply to the branch circuit fusible disconnect switch and the safety switch in the field.

BUSS® SYSTEM 300
MOTOR PROTECTION GUIDE
"NO-DAMAGE" "TYPE 2" SHORT-CIRCUIT PROTECTION
OVERLOAD OR BACK-UP OVERLOAD PROTECTION

115VAC (120V), 1 ph. (LPN-RK-SP or LPJ-SP) / 200V (208V), 3 ph. (LPN-RK-SP or LPJ-SP)

HP	Motor Full Load Amps	Overload 1.15 S.F. Or Greater Or Rated Not Over 40°C	Overload All Other Motors	Back-up 1.15 S.F. Or Greater Or Rated Not Over 40°C	Back-up All Other Motors	Switch Or Fuseholder Size	Minimum NEMA Starter Size	Minimum Copper Conductor Size†	Minimum Conduit Size‡
1/6	4.4	5	5	5.6	5.6	30	00	14	1/2
1/4	5.8	7	6.25	7.5	7	30	00	14	1/2
1/3	7.2	9	8	9	9	30	00	14	1/2
1/2	9.8	12	10	15	12	30	0	14	1/2
3/4	13.8	15	15	17.5	17.5	30	0	14	1/2
1	16	20	17.5	20	20	30	0	14	1/2
1½	20	25	20	25	25	30	1	12	1/2
2	24	30	25	30	30	30	1	10	1/2
1/2	2.3	2.6	2.5	3	2.8	30	00	14	1/2
3/4	3.22	4	3.5	4.5	4	30	00	14	1/2
1	4.14	5	4.5	5.6	5	30	00	14	1/2
1½	5.98	7	6.25	7.5	7	30	00	14	1/2
2	7.82	9	8	10	9	30	0	14	1/2
3	11	12	12	15	15	30	0	14	1/2
5	17.5	20	20	25	25	30	1	12	1/2
7½	25.3	30*	25*	35	30*	60	1	8	1/2
10	32.2	40	35	45	40	60	2	6	3/4
15	48.3	60	50	70**	60	60	3	4	1
20	62.1	75	70	80	75	100	3	3	1
25	78.2	90	80	100	90	100	3	1	1¼
30	92	100	100*	125	110	200	4	1/0	1¼
40	120	150	125	150	150	200	4	1/0	1¼
50	150	175	150	200	175	200	5	3/0	1½
60	177	200*	200*	225	225	400	5	4/0	2
75	221	250	250	300	300	400	5	300	2
100	285	350	300	400	350	400	6	500	3
125	359	400*	400*	450	450	600	6	2-4/0	2-2
150	414	500	450	600	500	600	6	2-300	2-2

230V (240V), 1 ph. (LPS-RK-SP or LPJ-SP) / 460V (480V), 3 ph. (LPS-RK-SP or LPJ-SP)

HP	Motor Full Load Amps	Overload 1.15 S.F. Or Greater Or Rated Not Over 40°C	Overload All Other Motors	Back-up 1.15 S.F. Or Greater Or Rated Not Over 40°C	Back-up All Other Motors	Switch Or Fuseholder Size	Minimum NEMA Starter Size	Minimum Copper Conductor Size†	Minimum Conduit Size‡
1/6	2.2	2.5	2.5	2.8	2.8	30	00	14	1/2
1/4	2.9	3.5	3.2	4	3.5	30	00	14	1/2
1/3	3.6	4.5	4	4.5	4.5	30	00	14	1/2
1/2	4.9	5.6	5.6	6.25	6	30	00	14	1/2
3/4	6.9	8	7.5	9	8	30	00	14	1/2
1	8	10	9	10	10	30	00	14	1/2
1½	10	12	10	15	12	30	0	14	1/2
2	12	15	12	15	15	30	0	14	1/2
3	17	20	17.5	25	20	30	1	12	1/2
5	28	35	30*	35	35	60	2	8	1/2
7½	40	50	45	50	50	60	2	6	3/4
10	50	60	50	70**	60	60	3	4	3/4
1/2	1	1.25	1¼	1.25	1.25	30	00	14	1/2
3/4	1.4	1.6	1.6	1.8	1.8	30	00	14	1/2
1	1.8	2.25	2	2.25	2.25	30	00	14	1/2
1½	2.6	3.2	2.6	3.5	3	30	00	14	1/2
2	3.4	4	3.5	4.5	4	30	00	14	1/2
3	4.8	5.6	5	6	5.6	30	0	14	1/2
5	7.6	9	8	10	9	30	0	14	1/2
7½	11	12	12	15	15	30	1	14	1/2
10	14	17.5	15	17.5	17.5	30	1	14	1/2
15	21	25	20	30	25	30	2	10	1/2
20	27	30*	30*	35	35	60	2	8	1/2
25	34	40	35	45	40	60			3/4
30	40	50	45	50	50	60	3	6	3/4
40	52	60*	60*	70	60*	100	3	4	1
50	65	80	70	90	75	100	3	3	1
60	77	90	80	100	90	100	4	1	1¼
75	96	110	110	125	125	200	4	1/0	1¼
100	124	150	125	175	150	200	4	2/0	1½
125	156	175	175	200	200	200	5	3/0	1½
150	180	225	200*	225	225	400	5	4/0	2
200	240	300	250	300	300	400	5	350	2½

* Fuse reducers required. ** 100A switch required.
† THWN connected to 60°C terminations up to #1. AWG to 75°C terminations for 1/0 and larger. Consult equipment manufacturer for listed termination temperature rating. Higher equipment termination temperature ratings may allow smaller conductor and conduit size.
‡ Based on 3 conductors for 3 ph. circuits and 2 conductors for 1 ph. circuits.   MPG

3.   Consider the wiring diagram for the 75-hp motor shown in Figure 10.6 and answer the following questions based on the control logic explained in this chapter:

(a)   What would be the likely outcome if the START switch is depressed while the motor is operating?

Answer: If the START switch is depressed while the motor is operating, the START switch contacts will provide a *redundant* 120 V supply to the starter coil and the motor ON light. Therefore, the motor will continue to operate, with no change.

(b)   What would be the outcome if START and STOP switches are depressed simultaneously?

Answer: If the START and STOP switches are depressed simultaneously, the motor will stop. This is due to the fact that the 120 V supply to the motor starter coil is *interrupted* whenever the normally closed contacts of the STOP switch are opened.

(c)   What would be the likely outcome if the main disconnect switch for the MCC is opened?

Answer: If the main disconnect switch for the MCC is opened, the 480 V supply to the control transformer primary is removed. This de-energizes the secondary, turns off the 120 V supply for motor control circuit and turns off the motor branch circuit.

(d)   Can the motor be stopped if the motor starter latching contact "welds" shut due to overheating?

Answer: When the spring loaded STOP switch is depressed, the 120 V supply to the motor control circuit is removed and the motor shuts off. However, because the latching contact is sealed shut, the moment the STOP switch is released, the *motor control coil would re-energize and restart the motor.* This is due to the fact that with the coil latching contacts sealed closed, the START switch is *not needed* to start the motor.

4.   Consider the logic associated with Timer P105.TD in Rung # 11 of the PLC relay ladder logic program shown in Figure 10.10 and answer the following questions:

a) What would be status of the timer **Done** bit: "--| |--" 105.TD_ CASCADE_ START. DN, if the XIC System E-Stop bit turns "**False,**" 1 second after Timer P105.TD in Rung # 11 is turned on?

Answer: The "--| |--" 105.TD_ CASCADE_ START. DN bit is OFF when the timer begins the count. This bit remains OFF until the timer has timed out and Accumulated time becomes equal to the Preset time of 2500 milliseconds, or 2.5 seconds. The XIC System E-Stop bit is representing the state of the System Emergency Stop switch. When this switch is depressed, the E-Stop bit turns "**False**" and the Timer P105.TD_CASCADE_START rung (Rung #11) is "broken," and the timer stops. Since, at 1 second, or 1000 milliseconds, the accumulated time is not equal to the present time of 2500 milliseconds, the 105. TD_ CASCADE_ START. DN, **Done** bit will remain **OFF**.

b) Would the "P105 Conveyor _CMD" bit—commanding P105 Conveyor to turn ON—be **True** if and when the system E-Stop switch is engaged/pressed?

Answer: The answer is no. This is because the XIC System E-Stop bit, representing the state of the System Emergency Stop switch, will turn "**False**" and stop the timer before it times out and turns ON the "P105.TD_ CASCADE_ START. DN" bit. From examination of Rung #11, we can see that if the P105. TD_CASCADE_START. DN **Done** bit is **not True**, P105 Conveyor _CMD" coil will *not* energize and P105 Conveyor will remain **OFF**.

CHAPTER 11—SOLUTIONS

1. HP, or hourly pricing, program is a standard feature in all OPT, or Time of Use, schedules.
A. True
B. False

2. The energy charge rate structure with Electrical OPT schedule is:
A. Flat, year round
B. Tiered
C. Exponential
D. Is a function of time and season.

3. A large industrial electricity consumer set the peak demand in July's billing month at 40 megawatt. The demand rate structure is same as that included in

Duke Energy ® OPT-I, Time of Use, rate schedule, as shown in Table 11.2.
Determine the demand cost for the month.

A. $367,000
B. $505,000
C. $407,000
D. $476,579

Solution

As explained earlier in this chapter, the demand charges under OPT-I rate
schedule are tiered. The tiered levels of demand charges are shown in Duke
Energy's OPT-I rate schedule introduced in this chapter under Table 11.2.
As apparent from the Table 11.2 excerpt below, July rates, under the OPT-I rate
schedule falls under the summer months. Therefore, the shaded section of the
excerpted table will be used for the tiered demand calculation.

II. Demand Charge	Summer Months	Winter Months
A. On-Peak Demand Charge	June 1 – September 30	October 1 – May 31
For the first 2000 kW of Billing Demand per month, per kW	$14.0767	$8.2981
For the next 3000 kW of Billing Demand per month, per kW	$12.8972	$7.1075
For all over 5000 kW of Billing Demand per month, per kW	$11.7067	$5.9064

Charge for the first tier, or 2000 kW = (2000 kW).($14.0767/kW)
$\quad$ = $28,153
Charge for the second tier, or 3000 kW = (3000 kW).($12.8972/kW)
$\quad$ = $38,691
Charge for the remaining demand = (40,000 kW - 5000 kW).($11.7067/kW)
$\quad$ = $409,734
Total demand charge for the month = $28,153 + $38,691 + $409,734
$\quad$ = **$476,579**

4. Calculate the *energy charge* for the month of July considered in Problem 3
if all of the energy is consumed during On-Peak hours, for 10 hours per day.
Assume that there are 30 days in the billing month and that the load factor is
1, or 100%.

A. $902,000
B. $416,808
C. $2,064,187
D. None of the above

Solution

As explained earlier in this chapter, the energy charges under OPT-I rate schedule are classified into two categories: (a) On-Peak energy charges per month, per kWh, and (b) Off-Peak energy charges per month, per kWh. The energy cost rates for On-Peak and Off-Peak usage are shown in the shaded fields of Table 11.2 excerpt below:

III. Energy Charges	All Months
A. All On-Peak Energy Per Month, Per kWh	**5.7847¢**
B. All Off-Peak Energy, Per Month, Per kWh	**3.4734¢**

Since the total energy consumption is not given, we can derive it based on the parameters identified in the problem statement.

Since the load factor is unity, or "1," as discussed earlier in this text:

$$\text{Load Factor} = \frac{\text{Average Demand}}{\text{Peak Demand}}$$

Since Load Factor, in this case = 1,

$$1 = \frac{\text{Average Demand}}{\text{Peak Demand}}, \text{ or Average Demand} = \text{Peak Demand } = 40,000 \text{ kW}$$

The energy consumed during the billing month
= (number of hours of operation).(Average Demand)
= (30 days/billing month x 10 hours of operation/day).(Average Demand)
= (300 hours of operation/month).(40,000 kW)
= 12,000,000 kWh per month

Total energy cost for the month:
= (12,000,000 kWh).(3.4734¢/kWh).(100¢/$)
      = **$416,808**

5. Which of the following statements describes the role of EPC and ESCO most accurately?

A   The terms EPC and ESCO are synonymous

B. EPC is method for implementing energy projects and ESCOs are entities that offer this alternative.

C. EPC is required by Department of Energy, ESCOs are not.

D. None of the above

CHAPTER 12—SOLUTIONS

1. Consider the lighting scenario described in Example 12.1. Determine the impact on the luminous flux at the 1 m circular target if the distance increased to 10 m. See the diagram below.

Solution

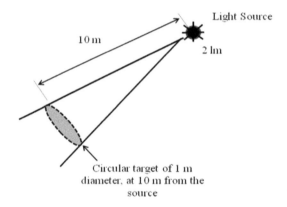

Luminous flux emitted by the source and received by the target at a distance of 10 m is calculated using Eqs. 12.1, 12.2 and 12.3.

The ω, **I** and luminous flux Φ values calculated above, with the 1 m target at a 10 m distance, can be compared with the ω, **I** and luminous flow Φ values

$$\Phi = 4.\pi.I \qquad\qquad \text{Eq. 12.1}$$

$$I = \frac{\Phi}{\omega} \qquad\qquad \text{Eq. 12.2}$$

$$\omega = \frac{A}{r^2} \qquad\qquad \text{Eq. 12.3}$$

By applying Eq. 12.3, with A = 0.785 m², and r = 10 m:

$$\omega = \frac{A}{r^2} = \frac{0.785 \text{ m}^2}{(10 \text{ m})^2} = 0.00785 \text{ sr}$$

By applying Eq. 12.2, with ω = 0.00785 sr, and the original source flux of 2 lm:

$$I = \frac{\Phi}{\omega} = \frac{2 \text{ lm}}{0.00785 \text{ sr}} = 255 \text{ lm/sr}$$

By applying Eq. 12.1, with I = 255 lm/sr as calculated before:

$$\Phi = 4.\pi.I = 4.\pi \text{ sr}.(255 \text{ lm/sr}) = 3200 \text{ lm}$$

calculated for the target distance of 5 m as shown in the table below.

At r = 10 m		At r = 5 m	Ex. 12.1
0.00785	sr	0.0314	sr
254.78	lm/sr	63.69	lm/sr
3200	lm	800	lm

It is apparent from the table above that, with all other parameters held constant, as the **distance of the 1 m target is doubled**, from 5 m to 10 m, the **luminous flux Φ quadruples**.

2. A 180W low pressure sodium vapor lamp is being used to illuminate an exterior space. Determine the luminous flux that would be emitted by this lamp.

Solution
Given:

   P = 180 W

   From Table 12.2, efficacy of a low pressure vapor lamp is 180 lm/W

   $$\eta = \text{Luminous Efficacy} = K = 180 = \frac{\Phi}{180}$$

   $$\Phi = \left(180 \, \frac{\text{lm}}{\text{W}}\right).(180 \text{ W}) = 32,400 \text{ lm}$$

   ∴ Luminous Flux, Φ = 32,400 lm

3.　Consider the scenario depicted in Example 12.2 and assume that the manufacturer's specifications show tested illuminance of the lamp, at 3.0 ft, to be 1000 lx. Determine the amount of illuminance, E, at the floor elevation, directly below the lamp.

Solution

　　Apply the single source special case interpretation of the inverse square law in form of Eq. 12.7:

$$E_1 r_1^2 = E_2 r_2^2$$

In this case,
　　$E_1 = 1000$ lx
　　$r_1 = 3.0$ ft
　　$r_2 = 2.33$ ft $+ 15$ ft $= 17.33$ ft
　　$E_2 = ?$

Then, by rearranging Eq. 12.7:

$$E_2 = \frac{E_1 r_1^2}{r_2^2} = \frac{(1000 \text{ lx}).(3\text{ft})^2}{(2.33+15)^2} = 30 \text{ lx}$$

4　　Consider the situation stated in Example 12.7. Proposed layout of the lighting system for the work space is shown below. The luminous intensity, I, for lamp **Y** is 700 cd. The luminous intensity for lamps **X** and **Z** is 600 cd. Determine the following:

a) Illuminance E_{Y-B}, at point B, due to light source Y.

b) Total Illuminance, E_B, at point B, due to light sources X, Y and Z.

Solution

a) Illuminance E_{Y-B}, at point B, due to light source Y.

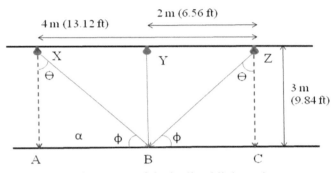

Geometry of the inclined light path

a) Apply the cosine-cubed law. Since point B lies directly below source Y, angle $\theta = 0$. Apply Eq. 12.9:

$$\text{Illuminance} = E = \frac{I}{h^2}\text{Cos}^3\theta$$

$$E_{Y-B} = \frac{I}{h^2}\text{Cos}^3\theta = \frac{700\ \text{cd}}{(3m)^2}\text{Cos}^3(0)$$

Eq. 12.9

$$= \frac{700\ \text{cd}}{(3m)^2}(1) = 77.78\ \text{cd/m}^2 \text{ or } 77.8\ \text{lx}$$

b) Total Illuminance, E_B, at point B, due the light sources X, Y and Z can be expressed as a sum of illuminance contributions by the three sources, at point B. In other words:

$$E_B = E_{X-B} + E_{Y-B} + E_{Z-B}$$

$E_{Y-B} = 77.8$ **lx**, as calculated in part (a)

E_{X-B} would be calculated by applying the cosine-cubed law. However, this would require knowledge of value of angle θ portended by the light path X-B with respect to the orthogonal line X-A as shown in the diagram.

Applying Pythagorean theorem to the triangle **XAB**:

$$\theta = \mathrm{Tan}^{-1}\left(\frac{BA}{XA}\right) = \mathrm{Tan}^{-1}\left(\frac{2}{3}\right) = 33.7°$$

Then, by applying the cosine-cubed law, or Eq. 12.9:

$$E_{X-B} = \frac{I}{h^2}\mathrm{Cos}^3\theta = \frac{600\ \mathrm{cd}}{3^2}\mathrm{Cos}^3(33.7°)$$

$$= \frac{600\ \mathrm{cd}}{(3\mathrm{m})^2}(0.576) = 38.4\ \mathrm{cd/m}^2\ \text{or}\ 38.4\ \mathrm{lx}$$

Due to symmetry, by inspection:

$$E_{Z-B} = E_{X-B} = 38.4\ \mathrm{lx}$$

Therefore,

$$\begin{aligned}
E_B &= E_{X-B} + E_{Y-B} + E_{Z-B}\\
&= 38.4\ \mathrm{lx} + 77.8\ \mathrm{lx} + 38.4\ \mathrm{lx}\\
&= 154.7\ \mathrm{lx}
\end{aligned}$$

CHAPTER 13—SOLUTIONS

1. What type of battery would be most suitable for the following applications:
    a. UPS, Uninterruptable Power Supply System
    b. DC power for substation breaker operation
    c. Remote wind turbine
    d. Remote solar PV farm

Answer: All four applications stated in this question would require large amounts of energy (Wh). Also, weight of the battery would not be a factor. While, a few years ago, with the cost of Li-Ion batteries being substantially higher than the Lead-Acid batteries, the choice would have been, decisively, in favor of the Lead-Acid batteries, today, as evident through Table 13.1, Li-Ion or Li-Po batteries would be the best choice, from almost all points of view.

2. In Example 13.1, if the volume of space reserved for the battery is 3 liters (3,000 cubic centimeters) of less, and the mass constraint is 10 kg, based

on energy density considerations, which of the given battery designs would be suitable?

Answer: As mentioned in the energy density discussion in this chapter, and as shown in Figures 13.6 and Figure 13.7, Li-Po has the best energy density, at 2.3 L (2,300 cc) per kWh. In addition, as stated in the solution for Example 13.1, the mass of 1 kWh Li-Po battery would not exceed 6 kg, which is well below the limit of 10 kg. Therefore, Li-Po would be the most suitable choice and would occupy the least volume.

Note: Based on charts in Figures 13.6, Figure 13.7, and other information provided in this chapter, a Lead-Acid battery system would occupy 11.8 L (11,800 cc) of space and a LiPO4 battery would have a mass of 4.6 kg. The Lead-Acid battery would also be rejected due to the three dimensional motion of the spacecraft.

3. It has been disclosed that the battery being selected for the mission to Mars, as outlined in problem 2, must be capable of serving as an emergency source of power for 1 kW of ignition power without overheating. However, the mass of the battery must not exceed 3.5 kg. Based on the specific power data provided in this chapter, which of the three batteries would meet the specifications?

Answer: As mentioned in the specific power discussion in this chapter, and as shown in Figures 13.6, Li-Ion battery would offer the best specific power (W/kg) among all battery choices discussed. Note that while capacitors might offer higher specific power than Li-Ion batteries, they would not meet the specific energy criteria.

4. A LiFePO4 cell battery is discharged at three times the nominal discharge rate. Will the energy density of the battery:
    (a) Remain 220 Wh/L?
    (b) Be greater than 220 Wh/L?
    (c) Be less than 220 Wh/L?
    (d) None of the above?

Answer: Based on the discussion on energy density in this chapter, energy density of a battery will vary with abnormal demand, due to that fact that energy delivered by a battery, in Watt-hours, can vary based on the rate of discharge. Therefore, the energy density of a LiFePO4 cell battery will decline when discharged at an excessive rate. Hence, the best answer is (c).

5. A Li-Po, or Lithium polymer cell battery, is discharged at half the nominal discharge rate. Will the specific energy of the battery:
    (a) Remain 180 Wh/kg?
    (b) Be greater than 180 Wh/kg?
    (c) Be less than 180 Wh/kg?
    (d) None of the above?

Answer: As implied in the discussion on specific energy in this chapter, specific energy of a battery will remain at or around the rated value when battery is discharged at, or less than, the nominal discharge rate. Therefore, the specific energy of a Li-Po battery will remain at, approximately, 180 kWh/kg. Hence, the best answer is (a).

6. A double-layered capacitor is being considered as an energy storage device to in an electrical system. The amount of initial energy needed by the system is 10 Wh. Based on the specific energy and specific power chart in Figure 13.6, the approximate weight of the double-layered capacitor would be:
    (a) 1kg
    (b) 10 kg
    (c) 100 kg
    (d) 1000 kg

Answer: Based on the chart depicted in Figure 13.6, the specific energy of a double layered capacitor is approximately 1Wh/kg. Therefore, a double layered capacitor capable of storing 10 Wh would have a mass of 10 kg. Correct answer: (b)

7. If, in problem 6, the maximum weight allowed for the charge storage system is 1kg, based on the specific energy and specific power chart in Figure 13.6, which charge storage system would be the most suitable?
    (a) Lead acid battery
    (b) Double-layered capacitor
    (c) Ultra-capacitor
    (d) None of the above

Answer: Based on the chart depicted in Figure 13.6, the specific energy of a Lead-Acid battery is 10 Wh/kg. Therefore, with the 10 Wh energy storage requirement, and the1kg limit on mass, Lead-Acid battery would be the best choice among the alternatives given in this problem. Correct answer: (a)

Appendix B

COMMON UNITS AND UNIT CONVERSION FACTORS

Power

In the SI or Metric unit system, DC power or "real" **power** is traditionally measured in watts and:

kW = 1,000 Watts
MW = 1,000,000 Watts = 10^6 W
GW = 1,000,000,000 Watts = 10^9 W
TW = 10^{12} W

Where k = 1000, M = 1000,000, G = 1 billion, and T = 1 trillion.

Some of the more common **power conversion factors** that are used to convert between SI System and US system of units are listed below:

1.055 kJ/s = 1.055 kW = 1 Btus
1–hp = One hp = 746 Watts
        = 746 J/s
        = 746 N-m/s
        = 0.746 kW
        = 550 ft-lbf/sec

Energy

In the SI or Metric unit system, DC energy or **"real" energy** is traditionally measured in Wh, kWh, MWh, GWh, TWh (10^{12} Wh).

kWh = 1,000 Watt-hours
MWh = 1,000,000 Watt-hour = 10^6 Wh
GWh = 1,000,000,000 Watt-hours = 10^9 Wh
TWh = 10^{12} Wh

Some mainstream conversion factors that can be used to convert electrical energy units within the SI realm or between the SI and US realms are referenced below:

1000 kW x 1h = 1 MWh
1 Btu = 1055 J = 1.055 kJ
1 Btu = 778 ft-lbf

Energy, Work and Heat Conversion Factors:

Energy, Work or Heat		
Btu	1.05435	kJ
Btu	0.251996	kcal
Calories (cal)	4.184	Joules (J)
ft-lbf	1.355818	J
ft-lbf	0.138255	kgf-m
hp-hr	2.6845	MJ
KWH	3.6	MJ
m-kgf	9.80665	J
N-m	1	J

Power Conversion Factors:

Power		
Btu/hr	0.292875	Watt (W)
ft-lbf/s	1.355818	W
Horsepower (hp)	745.6999	W
Horsepower	550.*	ft-lbf/s

Temperature Conversion Factors/Formulas:

Temperature		
Fahrenheit	(°F-32) /1.8	Celsius
Fahrenheit	°F+459.67	Rankine
Celsius	°C+273.16	Kelvin
Rankine	R/1.8	Kelvin

Common Electrical Units, their components and nomenclature:

Force	Newton	N	$kg\ m\ s^{-2}$
Energy	joule	J	$kg\ m^2\ s^{-2}$
Power	watt	W	$kg\ m^2\ s^{-3}$
Frequency	hertz	Hz	s^{-1}
Charge	coulomb	C	$A\ s$
Capacitance	farad	F	$C^2\ s^2\ kg^{-1}\ m^{-2}$
Magnetic Induction	tesla	T	$kg\ A^{-1}\ s^{-2}$

Common Unit Prefixes:

1.00E-12	pico	p
1.00E-09	nano	n
1.00E-06	micro	μ
1.00E-03	milli	m
1.00E+03	kilo	k
1.00E+06	mega	M
1.00E+09	giga	G
1.00E+12	tera	T

Wire Size Conversions:

A circular mil can be defined as a unit of area, equal to the area of a circle with a diameter of one mil (one thousandth of an inch), depicted as:

1 circular mil is approximately equal to:

•   0.7854 square mils (1 square mil is about 1.273 circular mils)

•   7.854×10-7 square inches (1 square inch is about 1.273 million circular mils)

- 5.067×10-10 m2
- 506.7 μm²
- 1000 circular mils = 1 MCM or 1 kcmil, and is (approximately) equal to:
- 0.5067 mm², so 2 kcmil ≈ 1 mm²

AWG to Circular Mil Conversion
The formula to calculate the circular mil for any given AWG (American Wire Gage) size is as follows:
A_n represents the circular mil area for the AWG size *n*.

$$A_n = \left(5 \times 92^{\frac{36-n}{39}}\right)^2$$

For example, a AWG number 12 gauge wire would use **n = 12**; and the calculated result would be 6529.946789 circular mils

Circular Mil to mm² and Dia (mm or in) Conversion:

kcmil or, MCM	mm²	Diameter in.	Diameter mm
250	126.7	0.5	12.7
300	152	0.548	13.91
350	177.3	0.592	15.03
400	202.7	0.632	16.06
500	253.4	0.707	17.96
600	304	0.775	19.67
700	354.7	0.837	21.25
750	380	0.866	22
800	405.4	0.894	22.72
900	456	0.949	24.1
1000	506.7	1	25.4
1250	633.4	1.118	28.4
1500	760.1	1.225	31.11
1750	886.7	1.323	33.6
2000	1013.4	1.414	35.92

Appendix C

GREEK SYMBOLS COMMONLY USED IN ELECTRICAL ENGINEERING

Greek Alphabet			
Aα	Alpha	Nν	Nu
Bβ	Beta	Ξξ	Xi
Γγ	Gamma	Oo	Omicron
Δδ	Delta	Ππ	Pi
Eε	Epsilon	Pρ	Rho
Zζ	Zeta	Σσς	Sigma
Hη	Eta	Tτ	Tau
Θθ	Theta	Yυ	Upsilon
Iι	Iota	Φφ	Phi
Kκ	Kappa	Xχ	Chi
Λλ	Lambda	Ψψ	Psi
Mμ	Mu	Ωω	Omega

Index

Symbols

3-φ phase line current 223

13 kV Loop Switch 259

Δ-Y three phase transformers 126

τ denotes torque 138

φ—neutral, 112

A

AAC (area access control) 294

AC 87, 88

AC alternator 214, 215

AC alternator construction and output waveform 215

AC asynchronous motors 231

AC circuit with a lagging power factor 171

accumulated or elapsed time 338

AC current 3, 4

AC frequency 98, 191

AC generator 209, 215

AC induction motor 217
   construction 218

AC inverters 230

AC motor speed control 230

AC power 143

AC sequence from ABC to ACB 250

AC sinusoidal voltage 216

AC source 37

AC synchronous motor 231

AC transformer circuit 107

actual discharge current 418

actual power demand 196

actual speed 219

actual time to discharge the battery 418

actual vs. forecasted power demand (kW) 197

actuator lever 260

actuator mechanism 260

addition of capacitance or capacitive reactance 183

addition of inverses 19

adjustable speed motor 231

adjusted or derated ampacity 276

adjustment factors 274

affinity laws 251

aggregate weighting factor 372, 373, 374

AGM (absorbed glass mat) 395

air circuit breakers 261

alarm horn 301, 332

alarm light 301, 332

alarm switch 300

algorithm 330

alkaline batteries 389

alkaline-manganese 389

Allen Bradley Control Logix Programmable Controller 333

alligator clip 11

alternating current (AC) 87

alternating current and impedance 94

alternating current, or AC 88

aluminum conductors 269

aluminum-electrolytic capacitors 402

ambient temperature rise and conductor size 276

amp-hour 389

ampacity 267, 272
   correction factors 272

analog 253
AND gates 79
angle 144
angle computation formula 174
angle of AC current 95,169
angle of AC voltage 169
angle of apparent power 145
angular difference 170
angular frequency 190
anions 397
annual energy consumption 140
anode 12
ANSI (American National Standards
    Institute) 286, 290
apparent power 1, 82, 146, 175, 180,
    195
    drawn by single phase motor 223
    drawn by three phase motor 223
    in kVA 188
appliance load—dwelling unit(s) 278
application of KVL 67
arc divider or extinguisher 261
arc fault 279
arc flash 279
arc flash energy 280
    intensity 280
arc flash hazard analysis 270
arc flash hazard calculation 291
arc flash hazard label 270
arc flash hazards 188
arc flash incident 42, 44
arc flash PPE 283
arc quenching system 262
area, in the work plane 383
area of cross-section 210
armature control 230
Art. 100—Definitions 267
Art. 110.6—Conductor Sizes 269
Art. 110.16—Arc-Flash Hazard

Warning 269
Art. 110.26—Clearances and Work-
    ing Space Requirements 270
Art. 210.9—Circuits Derived from
    Autotransformers 270
Art. 210—Branch Circuits 270
artificial light sources 381
ASD (adjustable speed drives) 230
assessment of power factor 179
augmented matrix 67
automated HVAC systems 354
automated manufacturing cell 302
automated power factor correcting
    capacitor system 185, 186
automotive battery 89
autotransformer 114, 270
average demand 199
average power 198
average value 93
    of AC 90
AWG 269

B
back plane 254
base 77
base clipper 73
baseline billing determinants 203
baseline charge 201
base reflectance 384
base to emitter junction 77
basic characteristics of inductive
    circuits 31
basic facilities charge 203, 345
batch MCR 333
battery 387
battery discharge rate and nominal
    capacity 412
battery equivalent circuit 415
battery voltage regulation 416

bill calculation 203
billing adjustments rider 350
billing demand for the month 203
bimetallic strip 260
biomass and geothermal sources 358
biomass energy 358
biomass heat 358
biomass power 358
bipolar (FET) junction, npn transistor 77, 78
blowing of the fuse 118
BMCS, building management control systems 355
BMS, building management systems 196
bolted pressure switch 244
bonded (bonding) 267
bottoming cycle cogeneration system 355
brake horsepower 141, 154, 156, 161
    delivered 162
branch circuit 267, 310, 311
branch currents 56, 57
breakers in two phase systems 112
breaker trip threshold 118
break horsepower 3, 4
break horsepower-hour 3
brightness 366
brilliant flash 282
Btu (British thermal unit) 137, 139
buck boost transformers 114
buck-boost voltage regulating transformers 116, 117
building envelope improvements 355
Bulletin 800T 249
bus-bar 11
Bus® Table, used for electrical design verification 313

C
C20 battery 410
cable pull safety switches 294
calculation for large electrical power bills 201
calibration screw 260
Calories/cm^2 280
candela 364, 365
capacitance 12, 13, 15, 16, 17, 19, 168
    required to improve the power factor 184
capacitive reactance 1, 23, 181, 189
    overdose 186
capacitor 12, 13, 14, 19, 80
    charging phase 18
capacitor-run 228
capacitors in parallel 21
capacitors in series 18, 19, 20
cascade start timer 337
catastrophic failure of fuses 245
cathode 12, 63
CAT I 239
CAT II 239
CAT III 239
CAT III, 150 V 238
CAT IV 239
CAT IV, 1000 V 238
cations 397
cavity ratio 384
CE certification 287
centrifugal switch 229
ceramic resistor 80
charged inductor 25
charged particles 2
charge storage capacity 14
charging and discharging rate 28
charging characteristics 389
charging electrochemical reaction 399

charging phase 399
CHP, combined heat and power 355
circuit breaker 258, 259
    and fuse types 271
circular mils 269
clamping circuit 73
clamp-on ammeter 1, 41, 44
Class B insulation windings 232
classification based on the winding
    insulation type 231
clipped 11
clockwise magnetic flux 104
coefficient matrix 68
coefficient of utilization 383, 384
cogeneration 355
coil windings 209
coils 25, 26
coincident AC voltage and current
    and perfect power factor 167
collector 76, 77
    to base junction 77
combination of the two impedances
    110
combined cycle cogeneration system
    355
commercial 342
commercial/governmental classifica-
    tion 350
commercial law 353
common commutator 209
common rail-mounted thermal-mag-
    netic miniature circuit breaker 261
common single phase AC transform-
    er applications 111
commutator 214
comparison between SMCs, VFDs
    and soft starts 253
complex AC 50
complex circuits 49

complex numbers and vectors 121,
    144
complex, vector, representation of
    line RMS current 151
complex, vector, representation of
    line to line or phase to phase RMS
    151
complex, vector representation of
    single phase AC power 145
complex, vector, representation of
    single phase AC voltage 145
complex, vector, representation of
    three phase AC power 150
compliance with OSHA 280
computation of large commercial
    electrical power bills 201
concentration of H_2SO_4 398
concussion (noise) 282
conductor 6, 7, 25, 272
    sizes 269
conduit and conduit size 312, 317
conjugate of vector 151
constant current stage 423
constant speed motors 231
constant voltage stage 423
construction management and super-
    vision 352
construction of a solenoid 26
contactor 25, 249
contacts 260
continuity 11
    tests 11
continuous 253
    load 267, 271
contracts 349
control contactor 187
control devices in the field 331
control diagram/drawing 327
control relay 300

control sequence 327
control system 255, 355
control transformer 187, 254, 325
control wires 249
conventional hydroelectric dams 358
conversion of AC voltage into DC
    voltage 91
copper 7
correction factors 274
cosine-cubed law 379
cosine law 376, 378
cosine of the angle of the impedance
    179
Coulomb 4, 92
counter and timer based control logic
    332
CPU, central processing unit 327
    microprocessor 328
Cramer's Rule 64, 67, 69
C-rating 389
    of batteries 401, 410
cross-wired 325
CSA (CSA Group) 286
CT I/O block 120
CU, coefficient of utilization 355
current 3, 4
current carrying conductor size 312
current comparison, AC versus DC
    90
current division 60
current function follows voltage
    function 168
current lagging behind voltage 90,
    97, 168, 175
current lagging voltage in phase
    angle domain 172
current lagging voltage in time do-
    main 172
current leading the voltage 181

in phase angle domain 177
in time domain 177
current protection and minimum
    conductor size 277
current response 17
    of a diode 71
    of a Zener diode 73
current transformer (CT) 44, 118,
    119, 120, 256
    applied to a single phase AC 121
    applied to a three phase 121
    operating principle 119
current vs. time plot 89

D
daylight based design considerations
    381
daylight factor method 382
DC 88
DC and AC currents 3, 39, 89
DC circuit analysis 49
DC clamp-on ammeters 46
DC electrical power 142
DC generator or dynamo 209, 210,
    213
DC motor 213, 214
    speed control 230
DC output 93
DC power 141
DC source 37
DC voltage 2, 3, 212
DC voltage output of a DC generator
    or dynamo 211
DC voltage source 39

deep cycle battery 394
dekatherms 139
Delta-Delta, 3-phase transformer
    configuration 123

Delta-Delta, voltage and current transformation 124
Delta-star, 3-phase voltage and current transformation 126
Delta-wye transformer wiring diagram 125
Delta (Δ)-Delta (Δ) combinations 122
demand 195
  computation 196
  elasticity 356, 357
  and the effect of demand response 357
  forecast 196
demand (in kW) 198
demand response 356
  measures 357
  or load management technologies 356
demand-side management 203
depth of discharge (DOD) 390
design engineering 351
determinant 68
determination of billing demand 347
DeviceNet® 293
diagnostics 294, 301
diagrammatic logic code 333
dielectric medium 13, 14
differential pressure 156, 158
digital clamp-on ammeter 45
digital multi-meter 43
diode 49, 69, 70, 79
diode-resistor circuit 74
direct current (DC) 87, 88
direct digital control (DDC) 120
  systems 355
direction of flow of AC power 104
direct-on-line starting 229
discharge 17

discharging 389
  characteristics of batteries 408
discrete 253
discrete 115 VAC output module 328
discrete signals 328
distribution segment 235
distribution service 350
distributions of daylight 381
DNet module 293
doping 69
dot convention for transformers 104
double-layered capacitors 402
DSM (demand-side management) 184, 352, 356
dynamo 209, 210

E
economy demand 347
  charge 205
effective (AC) current 4
effective (AC) voltage 2
effective resistance 91
effective value of AC voltage 3
efficacies of various lighting systems 363
efficacy values for common light sources 369
efficiency 152
  in percent 153
  of the luminaire 383
  ηTurbin 153
efficient and effective application of three phase AC 150
egg-line nanoparticles 425
electrical 195
  and controls drawings 307
  and electronic safety devices 291
  circuits 39
  conduit 317

drawings 307
energy 2, 357
  stored in a capacitor 14
electrical frequency 190, 210, 212
electrical metallic tubing: type EMT
  272
electrical power 139, 156
  baseline parameters 203
  distribution systems 312
  drawn by the motor 158
  generated 162
  generator 154
  rate schedules 341
electrical resistance 5
electrical safety certifications 286
electrical shielding 251
electrical shock 283
electrical short 11, 325
electric arc 24
electric horsepower 162
electric motor horsepower required
  140
electric power rates 343
electric utility companies 349
electric utility rate schedules 342
electric vehicles (EVs) 214, 388
electric water heater 140
electrochemical engine 2
electrodeposition 3, 90
electrodes 12
electrolytes 14, 389
electromagnetic field 217
electromagnetic interference 113,
  251
electromagnetism 26, 39
electromechanical equipment 239
electromechanically controlled
  device 249
electromechanical realm 153

electromechanical relay 332
electromechanical safety circuits 297
electromechanical systems 220
electromotive 1
  force 2, 39, 40
electronic controls 230
electronic devices 49
electronic safety devices 235
electronic switch 78
electrons 2
electroplating 91
ELI the ICE man 168
emergency standby generators 348
emergency stop relays 301
EMF (voltage) 215, 229
EMI 113
emitter 77
  junction 77
EMS based electrical power bill 201
EMS, energy management systems
  354
EMS, energy monitoring systems 196
energized conductors 117
energy 138, 142
  and charge storage systems 402
  audit 351
  charges 205, 345
  conservation 341
  conservation projects 354
  cost rate 342
  cost savings opportunities 341
  density 389
  efficiency enhancement
    technologies 356
  flow and power flow
    principles 158
  loss reduction 188
  performance contracting 351, 353
  procurement 353

productivity 168
energy star rating 355
energy storage capacity 26
energy storage device 12
energy usage (in kWh) 198
enthalpy 158, 159
EPC 341, 351, 352
EPC and ESCO alternatives 351
equivalent 19
    capacitance 20, 22
    capacitor 19, 21
    impedance 99, 101, 109, 111, 171
    inductance 34, 35
    resistance 12, 52
    resistor 9, 10
ESCO 341, 351, 352, 353
    opportunities 351
    projects 353
E-Stops operator 292
E-stop switch 303
Ethernet connection 335
Ethernet LAN 256
ETL and UL Listings 286
ETLCM 286
examination of branch circuit 312, 324
Examine if ON 336
Example of a Classification "0" label 285
Example of NFPA 70 E Class 4 PPE 284
examples of a safety certification labels 286
exclusive OR gates 79
explosive volumetric increase 279
exponential effect 8
exponential form 94, 98
externally reflected component 382
external work 220

extra facilities charge 203
extra high voltage (EHV) systems 237
extra low voltage 236

F
fail safe 328
fan laws 251
farad 13
Faraday, Michael 24, 387
Faradays 91, 92
feeder 268
Ferraz Shawmut class J and RK 245
ferromagnetic cylindrical core 25
ferromagnetic substances 41
ferroresonant, magnetic saturation based, transformers 116
field control 230
field current 213
final power factor 226
Finance and Accounting for Energy Engineers 168, 201, 203
financial justification 353
financial return and payback associated with power factor correction projects 187
fixed charges 201
flexibility in load schedules 350
flip flops 79
floating 129
float voltage stage 424
flooded (FLA) 395
florescent lamps 354
flow of energy 153
flow of power and energy 156
flux, in lumens 368
flux to current and current to flux conversion 113
flywheel 24

foot-candles 371
forward biased 70, 77
four diode bridge 75
frequency manipulation 182
   approach for correction of power
      factor 189
frequency of the AC power source 37
friction 5
frictional head loss 6, 153
ft-c 365
ft-lbf 137
fuel cells 403
full load 219
   amp 312, 314
   condition 115
   line current 224
   voltage 115
full wave rectifier 73, 75, 76
fuse 79
fusible disconnect switch 310, 312,
   319
FVNR, full voltage non-reversing
   motor starter 248
FVR, full voltage reversing starter
   and variety 229, 250

G

gassing voltage 420
gate 72
gel batteries 395
general duty 231
general service schedule 345
general wiring 271
generation segment 235
generator efficiency ηGenerator 153,
   156
generators 209
geometric hypotenuse 171
geometric representation of complex

AC entities 94
geometry of the inclined light path
   380
geothermal energy 358
geothermal heat 357, 358
geothermal power 358
graphical analysis of lagging power
   factor 173
graphical analysis of leading power
   factor 178
graphical contrast between AC voltage and current waveforms 97
graphical representation of impedance Z in vector form 99
ground 268
ground conductor size 312
grounded/neutral conductors 271
grounded three phase AC service 312
grounding 268
   and bonding 271
ground light contributions 382
GuardEdge™ Safety Edges 296

H

half hour intervals 196
half wave rectifier 73, 74
Hall Effect Principle 46
hazards of arc flash 281
heat sink 144
heat source 144
high efficiency chillers 355
higher efficiency 199
higher wattage ceramic resistor 80,
   82
high-pressure four-point connection
   245
high UV emission 282
high-voltage circuit breakers 262
high voltage systems 237

HMI (human machine interface) 184, 355
holes 1, 39, 391
Hopkinson's Law 40
horizontal and vertical components 96
horizontal ground bus 242
    bars in an MCC 243
horn-acknowledge and silence button 332
hot conductors 117
hourly pricing option/schedule 348
HP 342
    charges 205
    hourly pricing, billing determinants 205
human response factors 382
HVAC, heating, air conditioning and ventilation 355
hybrid capacitor combination 22
hybrid, parallel-series, network 50
hydraulic horsepower or fluid horsepower 155
hydraulic or fluid system 158
hydraulic pump systems 158
hydroelectric power 358
    generating system 161

I
I²R losses 317
ICI, International Commission on Illumination 363
ideal 104
    battery 408
    transformer 103, 104
    transformer equivalent circuit 108
    transformer model 105, 108
IEC 267
IEC enclosure IP44 241

IEC enclosure IP65 241
IEC, International Electrotechnical Commission 287, 290
IEEE: Institute of Electrical and Electronic Engineers 291
IESNA 363
IESNA illumination guidelines 372
IESNA illumination level weighting guidelines 371
illuminance 365, 366, 370, 373, 376, 377, 379, 380
illuminance at the source 378
illuminance cosine-cubed law 378
illuminance distributions 382
illuminance from the unobstructed sky 382
illuminance in the work plane 383
illuminance inverse square law 377
illuminated spherical area 368
illumination 363, 366
    contribution 377
Illumination Engineering Society of North America 363
illumination level 370
    guidelines for important tasks 375
    weighting guidelines 373
illumination standards 363
imbalanced load condition 118
impedance 1, 38, 229
impedance analysis 98
impedance contribution 23, 171
    by the capacitance 39
    by the inductor 37
incident energy 280
incoming lug compartment 243
    on a typical MCC 244
incoming power circuit breaker in a typical MCC 245
incoming power fusible disconnect

switch in a typical MCC 244
inductance 24, 26, 27, 30, 168, 189
   or inductive reactance to
      correct for leading power
      factor 188
inductive device 24
inductive reactance 1, 37, 38, 171
inductive-resistive circuit 28
inductor 24
industrial 342
industrial byproduct technologies
   356
industrial electrical power bill—EMS
   system based 204
industrial/public authority classification 350
inefficient lighting systems 354
ingress protection 287
initial enthalpy 160
initial power factor 226
input control diagram 328
input control drawing 329
input module 328
input/output (I/O) blocks 120
inputs and outputs 333
inrush current 253
inrush limiting contactors 187
in sight from, within sight from,
   within sight 268
instrumentation 235
insulators 258
integrated circuit (IC) 79
interlock contacts 325
interlocking or duplexing feature 112
interlock on MCC cubicle 248
internally reflected illuminance component 382
internal or external inputs 333
internal resistance 415

International Electrotechnical Commission Standard 267
international protection 287
International Society of Automation
   (ISA) 291
International Standards 267
interruptible 342
interruptible contract demand 349
interruptible power service schedule,
   IS 349
inverse square law 379
   illustration of 377
inverter driven AC motors 214
invisible (infra-red) light beams 293
I/O blocks 254
ionized air 24
IP56 233
IPMVP, International Performance
   Measurement and Verification
   Protocol 352
IP rating 287, 288
irradiance 366

J
J (Joules) 14, 137
jumper leads 11

K
kcmil 269
key interlock solenoid switches 295
kinetic energy 24, 138, 220
Kirchhoff's Current Law 49, 54
Kirchhoff's Voltage Law 49, 51
KVL method 62

L
labeled 268
lagging 168
   power factor 90, 170, 175

power factor from impedance
      perspective 170
power factor graph 169
power factor graph as a function
      of phase angle 169
Lamberts Law 376
      illustration of 376
large industrial electric bill calcula-
      tion 206
latching contact 327
lead-acid (SLI) battery 388
leading 168
      currents 226
      power factor 177
      power factor from impedance
            perspective 175
lead sulfate 398
LED, light emitting diode 80, 81, 354
life span of batteries 410
lighting design 371
      consideration 381
lighting efficacy 355
      comparison 371
lighting load 312
lighting principles 363
lighting system design 363
lighting terms, concepts and stan-
      dards 364
light loss factor 383, 384
Li-Ion Battery 391
Li+ ions 391
linear force 141
line current drawn by the motor 224
line of sight 319
lithium cobalt battery 394
Lithium Ion 391
Lithium Manganese Oxide Battery
      393
Lithium Phosphate Battery 392

Lithium Polymer Battery 393
load factor 1, 195, 198
load factor of unity 198
load impedance 108
lockable disconnect switch 320
locked rotor 230
logical functions 333
loop switches 257
loose voltage regulation 116
lossless 104
LOTO feature in the incoming power
      disconnect switch 247
low efficiency 199
lowest marginal cost 357
low load factor 199
low voltage applications 42
low voltage circuit breakers 260
low voltage connectors 236
low voltage differential signaling 236
low voltage in power supplies for
      fluorescent lamps 236
low voltage overhead power lines 236
low voltage systems 236
low wattage resistor 80
lumanious efficacy 371
lumen method 382
lumens 364
luminaire 377
luminous efficacy or efficiency 368,
      369
luminous efficiency factor 369
luminous flux 364, 365, 366, 368,
      370
luminous intensity 364, 365, 366,
      367, 368, 380
lux 365, 377
LVDS 236
      voltage level 236

M

magnetic circuits 39
magnetic energy 41
magnetic field 24, 25, 40, 209
    intensity 211
magnetic flux 25, 26, 40, 119
    density 119, 210
magnetic reluctance 41
magnetized rotor 215
magnetomotive force 40
magnitude 144
    of AC current 95
    of apparent power 146
    of the line to line, or phase to
        phase RMS 151
    of the real power 149
    of the single phase AC power 149
    of the total apparent power 148
    of three phase AC power 150
main circuit breaker 244
main disconnecting means 243
main fusible disconnect switch 243
main switch yard 258, 259
    and medium voltage switchgear
        256
maintained illuminance 383
maintenance of lead acid secondary
    batteries 420
manufacturing processes 345
master control relay 333
maximum allowable operating tem-
    perature 231
maximum power generated 161
maximum torque 213
maximum value 98
    of the AC 90
MCC (motor control center) 239, 308
MCC bus bars 310
    and bus connections 241

network 242
MCC cabinet 310
MCC compatible power monitoring
    equipment 256
MCC enclosure NEMA specifica-
    tions 240
MCC power distribution system 310
MCR (master control relay) 336
mean length of the core 27
measures typically included in ESCO
    or EPC endeavors 354
mechanical energy 2, 156
mechanical stress or strain 44
mechanical work 143
mechanically commutated electric
    motor 213
medium or high voltage 44
medium voltage switchgear—loop
    switch 257
medium voltage system 237
memory effect 392
mercury lamps 354
metal oxide varistor 82
metering or display of AC current
    120
micro-Farad 13
Micro-Logix series 254
microphone preamp voltage level
    237
mid-range illumination level 375
minimum area of cross-section 318
minimum conductor size 277
modems 256
modified constant voltage charging
    423
    process 424
monitoring and verification of energy
    savings 352
motor branch circuit 314

motor classifications 231
motor control center (MCC) 235, 239
motor dynamometer 221
motor efficiency 158, 223, 321
motor frame size 233
motor line current computation,
    single and three phase 221
motor nameplate information, large
    three phase induction motor 232
motors 209
motors add inductive reactance 186
motor slip 219, 231
motor soft starts 252
motor starter—A-B Bulletin 2100
    247
motor starter coil 327
motor starter contactor 327
motor starter details 248
motor starter pilot devices 250, 251
motor torque 229
    and power 220
MOV 82
multi-cell battery 408
multi-loop circuit analysis 61
multi-meter 1, 41
multiphase AC systems 44
multi-step constant current charging
    422
municipal 342

N
NAESCO, National Association of
    Energy Service Companies 352
NAND gates 79
n and p doping 76
nano-Farad 13
National Electric Code 265
n-doped 76
NEC® National Electrical Code 242,

272
ampacity table 273, 274
Article 240.6 (A) 324
Articles 266
handbook 265
introduction and outline 266
negative electrode 63
negative half of the AC 75
negatively charged ions 1
NEMA and IP electrical enclosure
    ratings 289
NEMA based efficiency of the motor
    233
NEMA enclosure types 240
NEMA® ratings 287, 288
NEMA "Size 1" starter 318
NEMA Type 1 241
NEMA Type 12 241
NEMA versus IP enclosure ratings
    288
net capacitance 24
net head added 156
    by the pump 158
neutral conductor 268
neutral leg of the three phase system
    118
NFPA® 188
    70 E 279
    70 E 265
Ni-Cad 391
Nickle Cadmium 391
Nickle Metal Hydride 405
N-m (Newton-meters) 137
nodal analysis technique 55
node 54, 55
no load condition 115
NOM Mark: (Norma Official Mexi-
    cana) 287
non-invasive 42

non-linear 16, 17, 28
  current response 30
  fashion 14
  voltage response 29
non-safety control systems 303
NOR gates 79
normally closed contact 335
normally open 328
  contacts 325
npn 76
NPV, net present value analysis 363
n-type semiconductor 69
number of poles 210, 212
  in the design 211
number of turns 27, 102, 211

O
ocean or tidal energy 358
OCPD 271
off-peak hours 199, 347
Ohm-meter 41
ohms 6
Ohm's Law 6, 39, 49, 50, 214
  application of 53
Oil circuit breaker 262
omnidirectional light source 365, 368
one-line drawing 308
one-line schematic 308, 310, 311, 316
  diagram 308
  for a power distribution system 309
one time constant 32
on-peak billing demand 347, 350
  charge 205
on-peak demand 199
on-peak hours 347
onsite generation 350
open circuit 11

opening 260
operating frequency 226
operations and maintenance 352
opportunities for electrical energy cost savings 348
OPT 201, 203, 360, 498
  time of the day 196
optional power industrial service 345
OR gates 79
oscilloscope 88
OSHA, Occupational Safety and Health Administration 280
outlets in dwelling unit(s) 278
output capacity 217
output control diagram 328, 331
output of a variable frequency drive 190
overcurrent 268
  conditions 245
  protection 271, 272
    and conductor ampacity 304
  protection devices 95, 271
  protection fuses 249
  protection fuse size 312, 319
  protection fuse specification 322
overload 268
  current setting 319
  protection device 327
  protection setting 312, 318

P
p and n doped materials 69
panelboards 272
parallel combination of inductors 34
parallel DC circuit 56
parallel plate capacitor 13
parallel resistor network 10
payback period 363
PCB 79, 81

peak clipper 73

peak current, IP 4

peak demand 196, 198, 199

   interval 196

peak-hold 42

peak hours 357

peak inverse voltage 71

peak shaving programs 196

peak voltage 2

performance of work 137

permeability 26, 27

   of free space or vacuum 27, 41

   of the core material 41

permittivity 13

personal protective equipment (PPE) 281

Peukert constant 418

Peukert's exponent 419

Peukert's Law 418

phasor representation 94

   of the source voltage 117

physical and thermal background of arc flash 281

physical size specifications of a typical MCC 240

pilot devices 249, 325

pilot light 327

plasma 24

PLC and I/O chassis 254

PLC based control circuits 297

PLC based control system 307, 331

PLC control program 336

PLC input module 327, 331

PLC output controls drawing 335

PLC output module 78

   based control drawing 330

PLC power supply 254

PLC program 336

PLC, programmable logic controller 120, 298, 327, 355

PLC programming software 332

PLC relay ladder logic programming example 334, 337

PLCs incorporated into an Allen-Bradley MCC 254

plug-in ground bus 243

p-n junction 69

pnp transistors 76

polar or phasor form 94, 98

polyphase induction motors, small 228

portable digital multi-meter 42

positive half of the AC cycle 75

positive impedance contribution 148

potential difference 12

potential energy 138, 220

power 137, 221

power bill computation 195

power conversion from steam to electrical form 153

power conversion from water to electrical form 155

power conversion in hydraulic pump systems 157

power dissipated 147

power distribution equipment 235

   measurement categories 238

power distribution systems 119

power factor 1, 146, 167, 168, 169, 174, 180, 198

power factor correcting capacitors 183, 187, 258, 259

power factor correcting capacitors, application, installation, operation and maintenance 188

power factor correction 182, 348

power factor goal for industrial, commercial and institutional facilities

187
power factor improvement and cost savings 184
power isolation transformer 113
power monitor 256
power output of the generator 159
power source 16
power transistor, npn (or NPN) 80
PPE and voltage rated tools and gloves 281
predominantly capacitive 167, 176
AC electrical system 176
predominantly inductive 167
circuit 183
preset time 338
pressure differential 2
pressure drop 6
pressure loss 6
pressures on the intake and exit ends of a turbine 161
pressure switch 331
preventive maintenance program 188
primary and secondary side of a transformer 102
primary and secondary windings 103
primary batteries 389, 390
primary stator 217
prime mover 217
printed circuit boards 78, 81
programmable logic controller (PLC) 120, 298, 327, 355
project financing 352
projectiles 282
p-type semiconductor 69
pull-apart terminal blocks 249
pump efficiency 158
pump horsepower 155, 157
purely resistive 167, 189

R
raceway 269
rack-out assemblies 251
rated capacity 418
rated discharge time 418
rate-of-change of slip 220
rate of generation of energy 137
rate schedules offered by Duke® Energy 344
rates under OPT-I schedule 346
rating of the over current protective device (OCPD) 304
RC circuit 15, 17
time constant 18
reactance 146
reactive power 1, 145, 147, 225
reactive power (Q, in VARs) 188
real batteries 408
real power 1, 145, 195
demanded by a single phase motor 223
demanded by a three phase motor 223
drawn by the motor 163
rear termination points on an Allen-Bradley Power Monitor 257
rechargeable alkaline batteries 389
rectangular form 94
rectification 70
rectifiers 70, 71
reduced market price 357
reduced-voltage reactor or auto-transformer starting 229
reduces transient voltage drops 253
redundant coils 301
reflectance 373, 374
reflectance of task background 374
reflective properties of the space 383
regular SLI battery 394

relative permeability 27
relative permittivity 13
relay 25, 187
    ladder logic 332, 333
reluctance 41
renewable electricity 359
    generation breakdown 359
renewable energy measures 357
renewable energy portfolio standard
    (REPS) adjustment 350
renewable energy technologies 358
renewable natural sources 357
residential breaker panel 112
residential electrical power bill 202
residential rate structures 342
residential service 343
resistance 5, 6, 7, 9, 16, 17, 30
resistive circuit 15
resistivity 6, 7, 8
resistors in parallel 9, 10
resistors in series 9, 21
resonance angular 190
resonance electrical frequency 190
resonance frequency 189, 191
response 28
return on investment 363
reverse biased 70
RIA—Robotics Industries Associa-
    tion 291
Rider 349
rigid metallic conduit 317
RL circuit 29, 31
RLC (Resister, Inductor and Capaci-
    tor) circuit 189
RMS 211, 212
    clamp-on ammeter 5
    current 4
    value 44
    voltage 2, 3

voltmeter 5
Rockwell® Area Access Control 295
Rockwell-Automation Soft Start 253
Rockwell® GaurdEdge System 298
Rockwell® GaurdMat Switch System
    297
Rockwell® Key interlock solenoid
    safety switch system 296
Rockwell® Safety Light Curtain 293
Rockwell® Safety PLC (Program-
    mable Logic Control) System 299
Rockwell® Safety Relays and Con-
    tactors 301
romagnetic circuit 40
root mean square value 2
rotating 214
rotational speed 23, 37, 210, 221
    of the armature 219
    of the dynamo 211
    of the generator 212
    of the rotor 219
rotor 218
    windings 217, 218
routers 256

S
safe operating load 200
SafeShield DeviceNet 294
safety cable pull switch 303
Safety Compliance Statement 287
safety control circuits 296
safety E-stop devices 292
safety guard 303
safety interlock 332
    contacts 325, 332
safety laser scanner 303
safety light curtain system 292
safety limit switch 303
safety PLC 293, 297, 303

based control circuits 298
safety relays 300
schedule OPT-I (NC) 345, 347
schedule or rider SG, on-site generation 348
schedule RS (NC) 343
SCRs 71, 72
sealing contacts 327
secondary batteries 389
secondary side 103
secondary turns, NS 107
secondary windings 218
self-actuating electronic switch 70
self-commutated motors 231
self-discharge 392
    rates 410
self-monitoring contact blocks 292
self-starting polyphase induction motors 229
semiconductor 69
series 9
series inductor combination 33
series inductor hybrid circuit 36
series RC circuit 15
series reactor 253
series RL circuit 27, 28, 30
series RL current 32
service contract 201
service factor 195, 200
SF6 circuit breakers 262
shaded pole 228, 229
shock wave 282
short 11, 31
short-circuit current rating 269
short circuit protection fuse 332
shrapnel injuries 282
shunt 60
silicone controlled rectifier 71
silver plated copper conductors 241

simulated arc-flash incident 281, 283
simultaneous equations 64
sine waveform 89
single phase AC alternator/generator 216
single phase AC current 145
single phase AC generator 216
single phase AC induction motor current 222
single phase AC power 143, 144, 145
single phase AC system 118
single phase alternator 212
single-phase capacitor-start 228
single phase configuration 150
single phase induction motors 119
single step constant current charging 421
sinusoidal AC waveform 74, 93
sinusoidal form 94, 215
sinusoidal or trigonometric form 96
sinusoidal representation 97
sinusoidal voltage 96
size of the OCPD 277
size of the service and feeder conductors 278
skin effect 230
sky component 382
SLC-500 254
slip 217
slow blow fuses 187
smart motor controller (SMC) 251, 252
SMC 50, Rockwell-Automation Smart Motor Controller 252
sodium vapor lamps 354
solar energy 358
solar heat energy 358
solar photovoltaic electrical energy 358

solenoid 261
  coil 78
  locking switches 295
  operated valve 25
  valve 78
solid angle 364
  in steradians 365
solid state motor overload device 318
solid-state soft starting systems 229
special purpose 231
  diodes 71
special riders 201
specification of RS service 343
specific energy 389
specific gravity of the electrolyte
  398
specific power 389
specific weight 156, 158
speed of a DC motor 213
speed/torque regulation characteristics 213
split-phase 228
squirrel cage rotor 218
  bars 230
stab assembly 245
  behind a typical MCC cubicle 246
standard (nameplate) load 200
standard temperature 8
star-Delta, 3-phase voltage and current transformation 129
star-delta starting 229
star-star, 3-phase transformer wiring
  diagram 130
star-star, voltage and current transformation 131
starter size 312, 319
starting current 268
starting methods for iduction motors
  228

start-up and commissioning 352
star (Y)-Delta (Δ) 122
star (Y)-Star (Y) 122
state of charge, or SOC 398
stator 218
stator serves as an armature 214
steam power generation system 154
steam to wire power and energy flow
  154
steam turbine efficiency 158
step down, residential power distribution transformer 112
  with load 112
step-down single phase transformer
  configuration 111
step down transformer 107, 258
  configuration 107
step down voltage transformation
  111
stepping down of voltage 88
stepping up 88
step-up or step-down single phase
  transformer application 111
step up transformer 107
steradian 364
stiff 115
storage 410
substitutional matrix 68
sulfur hexafluoride gas 262
sulfuric acid 397
superheated steam 144, 154
supply side 341
sustainable materials 360
  and associated operations 360
switchboards 272
symbols, units and conversion factors
  in illumination 366
symmetrical fault current 269
synchronous motors 226

synchronous speed 212, 217, 218, 219, 231
system interlocks 296
system power factor 182

T
target illumination levels 375
TEFC, totally enclosed fan cooled, motor 218
temperature 8, 410
    and battery discharge characteristics 412
    in ARC plasma 280, 282
    switch 331
terminal blocks 249
terminal strip 81
thermal coefficient of resistance 8
thermistor 229
three phase AC 117
    alternator/generator 216
    indusction motor current 222
    power 149
    rectification 119
    system 118
three phase configuration 150
three phase transformers 121
three phase configuration 150
    Δ-Δ transformer 122
    Δ-Y transformer 124
    Y-Δ transformer 127
    Y-Y transformer 128
three pronged approach to electrical energy cost reduction 341
threshold voltage 72
thyristor 72
time constant 17, 24, 28
time-delay fuse 319, 323
time of use industrial service 345
time of use rate schedule 196

timer based starting control logic 336
timer setting 336
topping cycle cogeneration system 355
toroid 40
toroidal core transformer 103
toroid type inductor 79
torque 137, 138, 220, 221
    developed in ft-lbf 224
total apparent power 147
total baseline charge 205
total bill 201
    for the month 206
total illuminance emitted by the light source 383
total impedance 182
total weighting factor 375
transduced 44
transformation discounts 350
transformation of power 154
transformer 88, 102
transient 16, 17, 28
    behavior 14, 28
    response 15
transistor 49, 76, 79
transmission lines 8
transmission segment 235
transmission service 350
    transformation discount 351
trapped key safety switch 303
treatment 410
trigonometric approach 95
trip 260
tripping of the breaker 118
true RMS ammeters 105
true RMS voltmeters 105
turbine efficiency η turbine 156
turbine entrance and exit enthalpies 160

turns ratio 105

U
UHV 237
UL® 286
ULC, Underwriters Laboratories of
    Canada 287
ultra-capacitors 402
ultra-high voltage systems 237
ultra-violet light 282
unequal phase impedances 118
unit for impedance 98
unit grounding provisions 246
unit ground stab 246
unit handle 246
unit interlock 247
units for illuminance 365
unit vector 146
unity power factor 189
U.S. electrical outlet voltage 89
U.S. electricity generation 359
utility company account managers
    349
utility rate schedule and contract tips
    349

V
vacuum circuit breakers 261
valve-regulated lead-acid (VRLA)
    395
vaporized copper 282
variable frequency drives (VFD) 113,
    220, 230, 235, 250, 252, 255, 335
vector addition of vectors 179
vector representation of phasor value
    of AC current 95
velocity, change in 141
vertical and horizontal ground bus
    bars in a typical MCC 242

vertical ground bus 242
VFDs or variable frequency drives
    113, 220, 230, 235, 250, 252, 255,
    335
virtual symbols 332
volta 387
voltage 2, 16
    applied to the armature 213
    categories 235
    comparison, AC versus DC 89
    difference 2
    division 58
    drop 29, 53, 115
        polarities 66
    in terms of time as well as the
        angle 171
    nominal 269
    regulating transformers 116
    regulation 114
        in percent 116
    regulator 79, 114
    response 16
    transformation 88
voltaic pile 387
Volt-Ohm-Meter, or VOM 42
volumetric flow rate 158, 162
    of water 156

W
wall reflectance 384
wasted heat 6
watchdog circuit 299
water horsepower 156, 157, 161, 162
water to wire 155, 156
water to wire power and energy flow
    155
Weber, Wilhelm Eduard 40
weight density 156
weighting factors 371, 372, 373

WHP 155, 156
wind energy 358
wind turbines 358
winding coils 209
wire to water 137
    flow of energy 158
    power flow 157
    transformation of power and
        energy 157
wiring diagram 311, 324
    for power distribution system 326
wiring numbers 325

work 138
wound rotor 218
wye-Delta, transformer wiring diagram 127

Z
Zener, Clarence 72
Zener diodes 71, 72, 73
zener knee voltage 73
zinc-carbon 389, 390
zinc-manganese 389
zonal cavity method 381